HELEN HALL LIBRARY
LEAGUE CITY, TEXAS

D1441038

INVENTING THE AMERICAN FIRE ENGINE

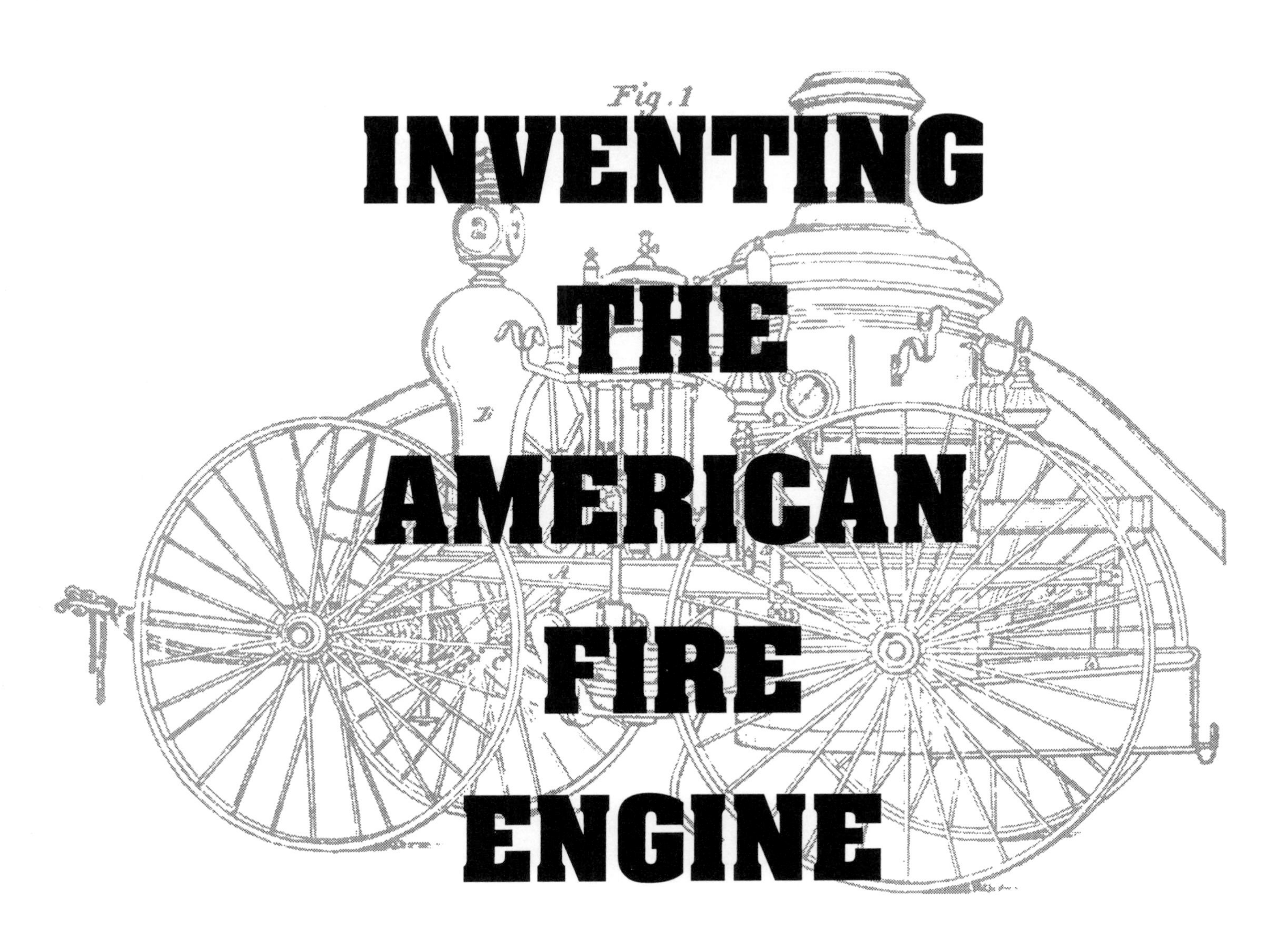

AN ILLUSTRATED HISTORY OF PATENTED IDEAS FOR FIRE PUMPERS

By M.W. Goodman, M.D.

Includes 384 photographs, illustrations and U.S. Patent drawings spanning three centuries.

Fire Buff House Publishers
New Albany, Indiana

Library of Congress Cataloging in Publication Data
Goodman, M.W., M.D.

Inventing The American Fire Engine – An Illustrated History of Patented Ideas For Fire Pumpers
Library of Congress Catalog Number: 94-070747
ISBN 0-925165-16-6

Fire Buff House Publishers, P.O. Box 711, New Albany, IN 47151-0711

Printed in the United States of America

Dust Jacket Design: Ron Grunder
Typography and Layout: Pam Campbell-Jones

ACKNOWLEDGEMENTS

Since childhood I have been fascinated with fire engines, and these wonderful vehicles and volunteer firefighting constitute one of the prominent interests of my life. My profession, one of genuine great love, has occupied most of my attention during the last thirty plus years, and it has enabled me to earn enough to underwrite all of my firematic interests. My three children and their spouses and my little grandchildren are collectively and individually my second great loves. My first love is and always will be Sheila, my wife. Along with everything else which she does so capably she has been a competent patent researcher in this effort, and a candid critic and a consummate partner.

My appreciation for numerous kindnesses goes to a number of student staffers and research librarians of the Morris Library of the University of Delaware. Sheila and I used that facility and its Patent Repository often while gathering data for this book. Others who helped me in locating information include Ms. Arlene Barnhart, Librarian of the Charles S. Morgan Technical Library of the National Fire Protection Association; Ms. Stacy Pomeroy Draper, Curator/Registrar of the Rensselaer County Historical Society of Troy, New York; Mr. Ernest N. Day of Dunellen, New Jersey, who I have known for many years and who is without doubt the fire apparatus dealer of longest tenure in the entire country; and my friend since we were kids together, Attorney Neal Abrams, who is a Judge with the Board of Patent Appeals of the United States Patent and Trademark Office as well as a veteran firefighter and paramedic.

My thanks also to Executive Publisher David McLaughlin and Editor Bill Stevenson of *Firefighters News* for permission to use copyrighted material previously published in my regular column, "Its Patented." Finally, I wish to recognize the efforts of Pam Campbell-Jones who worked out all of the graphic styles and page layouts, and to acknowledge the excellent and easy relationship I have had with publisher Fred Conway.

Mike Goodman, M.D.
Bethany Beach, Delaware
1994

TABLE OF CONTENTS

The history of fire engines and the tools used by firefighters have been addressed by many authors. This book, however, is the very first one in which those documents called "patents" have been employed and studied in an effort to understand the history and the evolution of fire apparatus. Hopefully this book will be the first of a set of four volumes in which virtually all of the hardware of firefighting and fire protection will be looked at in the same way, with patents as the principal reference sources.

In this volume the history of fire pumpers is presented by dividing them into categories according to their sources of power: pumps worked by men; pumps worked by horses or other animals; pumps worked by steam; pumps worked by electricity; chemical fire engines; and, lastly, internal combustion engines and modern fire pumpers.

Patents, of course, are documents issued by governments so as to grant to an inventor the exclusive right to make, to use, or to sell his invention. And so for any particular period of time the patent records for any particular discipline or hardware area, be it flash bulbs, flower pots, or fire engines, are the basic documentary account of the history of developments within that field.

Under our current United States laws and codes patents are granted for seventeen-year terms to anyone who has been able to convince the highly erudite examiners of patent applications that they have invented or discovered a new and useful art, machine, manufacture, or composition of matter. It is Article 1, Section 8 of the United States Constitution which gives to Congress the power, "To promote the Progress of Science and useful Arts, by securing for limited times to Authors and Inventors the exclusive Right to their respective Writing and Discoveries." Upon this authority the very first U.S. patent law was enacted in 1790. And it was in that same year that President George Washington remarked in his first message to the Congress that, "There is nothing which can better deserve your patronage than the promotion of science and literature. Knowledge is in every country the surest basis of public happiness." This concept that the law of the land can protect the right of ownership of an inventor is a quite recent phenomenon, especially when compared to the infinity of times past during which people have been inventing things.

The records of the United States Patent Office make it obvious that ingenuity and the desire to expand the boundaries of technology have, thankfully, although not exclusively, always been American characteristics. Consider the following records of patents granted:

U.S. PATENT NUMBER	YEAR OF ISSUE	TOTAL PATENTS FOR THAT YEAR
1	1836 (July 13)	103
1,000	1838	514
10,000	1853	884
100,000	1870	12,137
1,000,000	1911	32,856
2,000,000	1935	40,618
3,000,000	1961	48,368
4,000,000	1976	70,226
5,000,000	1991 (March 19)	—

Another way to think about these figures is this: for every single day since July 13, 1836 just about 90 inventions have been officially recognized by the granting of a patent!

Sometimes it is not quite clear as to who ought to be acknowledged as the inventor of some significant machine. This is the case, for example, with the steam fire engine. And in fact in his book, "The History of the Growth of the Steam Engine," author Robert H. Thurston commented that, "Great inventions are never, and great discoveries are seldom, the work of any one mind. Every great invention is really either an aggregation of minor inventions, or the final step of a progression. It is not a creation but a growth — as truly so as that of the trees in the forest. Hence, the same invention is frequently brought out in several countries, and by several individuals, simultaneously. Inventions only become successful when they are not only needed, but when mankind is so advanced in intelligence as to appreciate and to express the necessity for them, and to at once make use of them."

The United States patents which are the historical documentation for this book have been selected from a collection of about 1,000 sets of patent papers. Hopefully most of the really important relevant inventions have been included, but no claim is made that all such have been found. The author is solely responsible both for such errors by omission and for all mistakes, prayerfully small, infrequent, and insignificant, committed in this composition. Material and illustrations from old issues of Scientific American magazine have also been incorporated as information sources on patents and inventions, and the contemporary reader should know that inventors sometimes paid fees to the editors for descriptive articles and the front-page placement thereof! And finally these words regarding the illustrations used in the following pages: some are of less than optimal quality because of aging and deterioration of the originals; a source attribution for each illustration is given within the accompanying text; and all of the photographs placed together as "albums" are by the author.

HAND-POWERED FIRE PUMPS

INTRODUCTION

Sweating men with pungent body odors, their sleeves rolled up so that biceps muscles were visible during the summer months, their collars raised to shield neck and ears from the frost of the colder seasons, worked the pump levers of early American hand-powered fire engines. Respite and recuperation periods separated each physically exhausting time interval on the pump. Some fire companies responded with kegs of brandy buckled onto the shafts of their vehicles, but frequently it was the young boys of the community who distributed beer and spirits to the thirsty firefighters during their rest periods. Alcoholic malt beverages were consumed from ladles or were swilled directly from the beer buckets themselves. Thus lubricated the firemen found that succeeding trials with the pump levers were, somehow, less tiring than anticipated, although decidedly more rowdy.

Such a scenario does not unfairly represent fireground behavior during the paramount fireground task of those days, which was moving water. It was just this fatiguing job which dictated that the fire companies maintain large membership rosters, since large numbers of men, even into the hundreds, might be needed for the bucket brigades and for pumping duties.

Historically the prototype hand-worked piston pumps probably were of the small "siphona" type, dating to about the 4th century, B.C., and usually credited to a gent named Ctesibius. Then came the hand squirts, large syringe-like devices with barrels and plungers, able to hold only a couple of quarts of water. The next significant development was a cart mounted, quite large syringe pump, in which a hand-cranked screw was used to push the plunger. Hand-powered piston pumps reappeared during the 17th century, now in the form of force pumps placed within tubs or barrels, and relying upon buckets to supply the water. About 1670 Nicholas and Jan van der Heyde, or Heiden, invented, marketed, advertised, and distributed their own Dutch-made fire engines which were supplied, practically in "turn-key" condition, complete with intake and output hoses known as water snakes.

Meanwhile, here at home in the Colonies, just as in the Mother Country and in the rest of Europe, the cities burned. So, following one of Boston's great conflagrations, in 1653, a contract to build a fire engine was awarded to one Joseph Hencks. Unfortunately what he built was just another syringe pump, barely able to produce a predictably puny and intermittent stream. Thus it was that a London-made fire tub was imported, and this was the very first force pump to be used on the North American continent.

The Newsham Fire Engine

The effective date of Richard Newsham's fire engine patent was December 26, 1721. It has been said of Newsham, who hailed from London, that he had given "a nobler present to his country than if he had added provinces to Great Britain." Newsham-type hand-powered fire engines were used for many years, especially in England, until about 1865. The basic machine was fitted with a pair of single-acting pump cylinders which were of 4-1/2 inches diameter and 8-1/2 inches stroke. Water displacement was, therefore, about two quarts per downstroke. Thus, with a stroke rate of 60 times a minute the pump output was 30 gallons. A Newsham machine is shown in FIGURE 1.

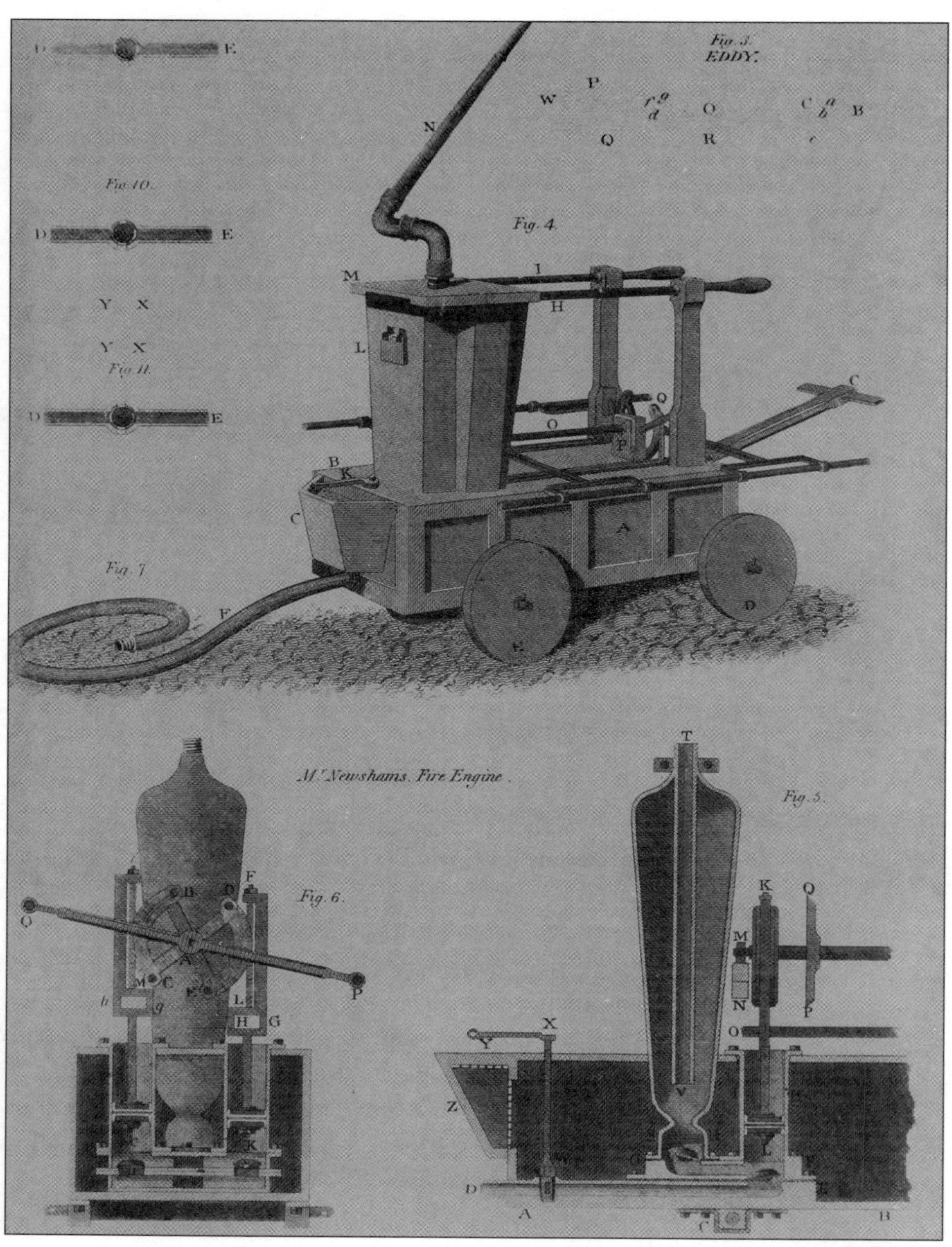

FIGURE 1

Early Builders of Hand-Powered Pumpers

The history of early native American hand fire engines can be found in such works as Kenneth H. Dunshee's "Enjine-Enjine, A Story of Fire Protection," and "An Illustrated History of Fire Apparatus with Emphasis on 19th Century American Pieces," by George A. Daly and John J. Robrecht. In fact these last-named authors compiled a list of more than 100 domestic builders of hand-powered fire engines, and they remarked that their information was, at best, only partial.

Among the prominent builders of fire engines during the 100 years before 1850, all of whom are noted within the pages of those two books, there were men who designed side-stroke and end-stroke pumps, double-deck side-stroke and double-deck end-stroke pumps, as well as rowing machine, coffee mill and cider mill style pumps. There was Thomas Lote who built "Old Brass Back," in 1743; Richard Mason who produced his first end-stroke pumper in 1768; Pat Lyon who began to make his "Philadelphia"-style pumpers in 1794; John Agnew, who perfected Lyon's style and built machines of this configuration between 1835 and 1860; William C. Hunneman, an apprentice of Paul Revere, who built "Boston"-style pumpers and sold them for export as well as here at home.

And there was James Smith who manufactured "New York"-style fire engines from 1813 to about 1856; Lysander Button who, beginning about 1835, built large "piano box" side-stroke pumps; Dudley Farnum and Franklin Ransom who, in 1847, introduced their "rowing machine" style fire pumper. Cowing and Company, Gleason and Bailey, Gould Manufacturing Company, Pine and Hartshorne, Remington Agricultural Company, Rumsey and Company, and Silsby Manufacturing Company are each examples of well-known builders of hand-powered fire pumps.

The oldest of these machines have been termed "gooseneck fire engines," because, in the style of Newsham, a curved eduction pipe, a sort of nozzle, protruded from the top of the large air chamber. Usually built above the rear wheels, the air chamber commonly measured four to six feet above the deck of the pumper.

For the most part all of these early builders of fire pumpers were just that: they were builders, and they were well-known and respected as builders, but they were not inventors. Searching the very oldest patent records, those for the years 1790-1836, reveals that during those 46 years only two patents were issued for fire engines (as well as but one for a fire alarm, one for a fire extinguisher, and one for a fire ladder). Contrast these arid years to the productivity of the period 1836-1873, an era marked by surging inventive activity: patents awarded for fire engines, 22; patents awarded for fire alarms, 57; patents awarded for fire ladders, 42; patents awarded for fire extinguishers, 124; and, patents awarded for fire escapes, 126. And to help in placing this activity into some perspective, consider that during these same years there were 1,250 patents granted for firearms and 1,500 for railroad applications.

Inventions of man-powered fire pumps continued to receive patent recognition until nearly the year 1915. And so, beginning with this brief background review, the contributions made by inventors during the nearly 75 years since 1845 can be scrutinized.

The First Patented Fire Pumpers

One of the very earliest United States patents for fire engines was numbered 4,316, dated December 20, 1845. The inventor was Ernest Marx, a resident of the city of New York, although not yet a citizen of this country. His patent drawing is shown as FIGURE 2. Although it is neither marked nor labeled well, it does illustrate those principal features which were espoused by the inventor: horizontally-aligned piston pumps operated by handles attached to the rear axle. In order to pump water it was, therefore, necessary to first lift the rear wheels up, off the ground. While they were in this position they were able to function as fly wheels for the pump.

The Farnum-Ransom fire engine, invented by this pair of New Yorkers, already mentioned, was assigned patent number 5,077 and was dated April 17, 1847 (FIGURE 3). Notice that the seats for the operators and the handles for each one to push and to pull was set at right angles to the long axis of the extended piston rod. The inventors remarked that, "By this arrangement of the pump and the benches or seats on the carriage the men can work the pump for a much greater length of time, and exert much more force than with any other known arrangement, as it is a well-known fact, ascertained by numerous experiments and long experience that when a man's strength is applied as in the act of rowing, the effect produced is nearly 150 percent more than in moving a pump lever or brake by the usual means."

The next illustration, FIGURE 4, shows 106 members of a fire company operating the apparatus patented on November

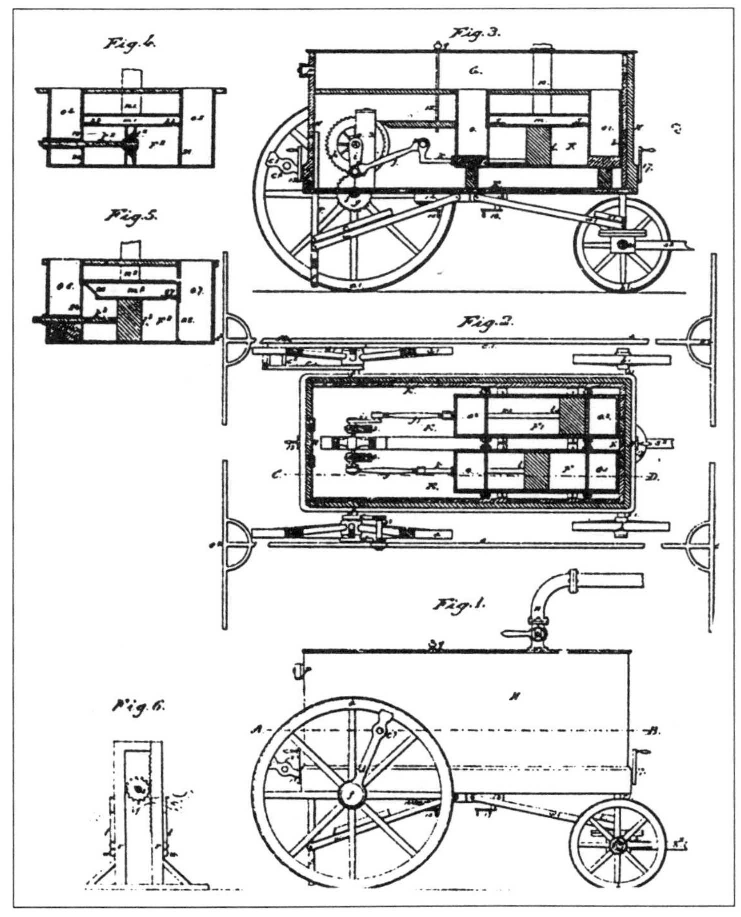

FIGURE 2

28, 1854 by Gottleib Backstein of Philadelphia, Pennsylvania. Curiously, the artist who produced the drawings for this patent, number 11,987, allowed for an imbalance in the pumping efforts, there being 51 men working the left-side handles and but 48 men on the right side. But notice also that there are two extra men on the right side, just beyond the far rows of pumpers. One can assume that these were the chief officers of the fire company, not because they were just standing around

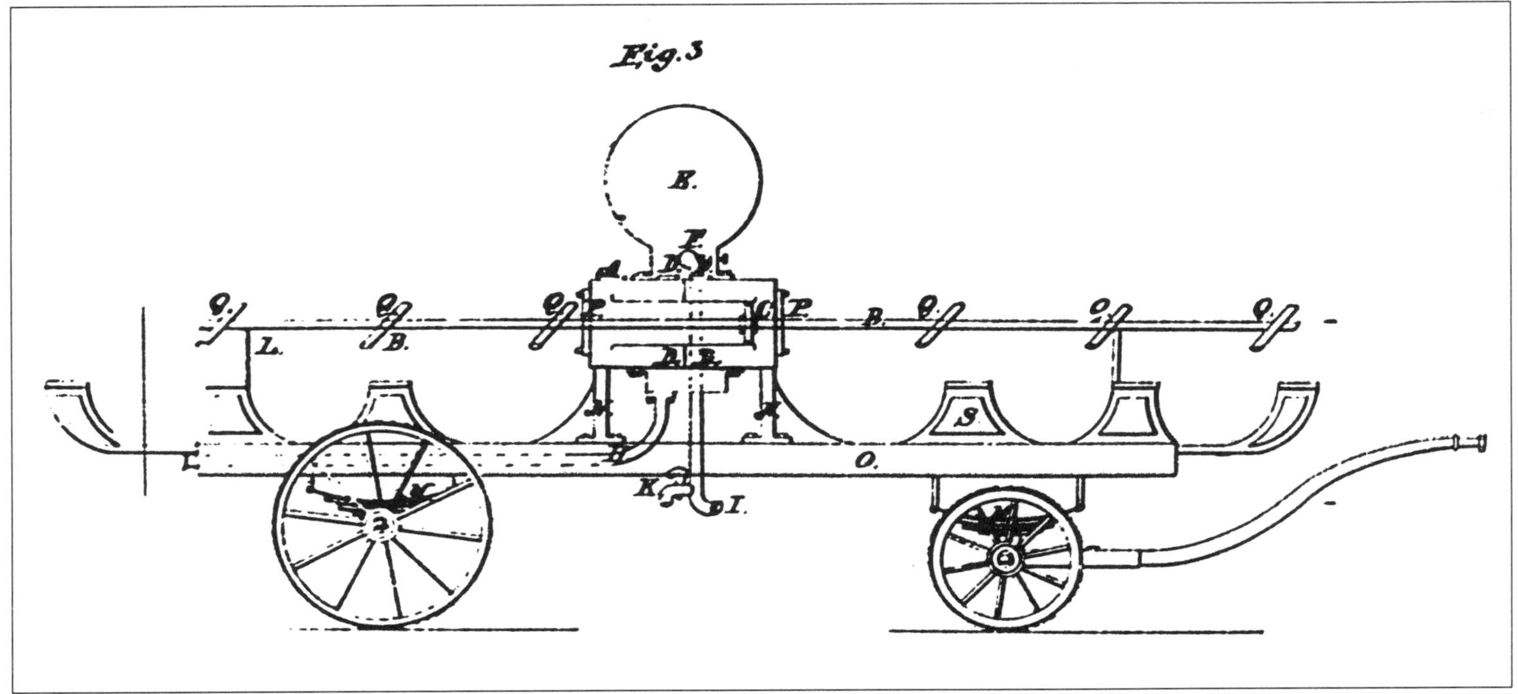

FIGURE 3

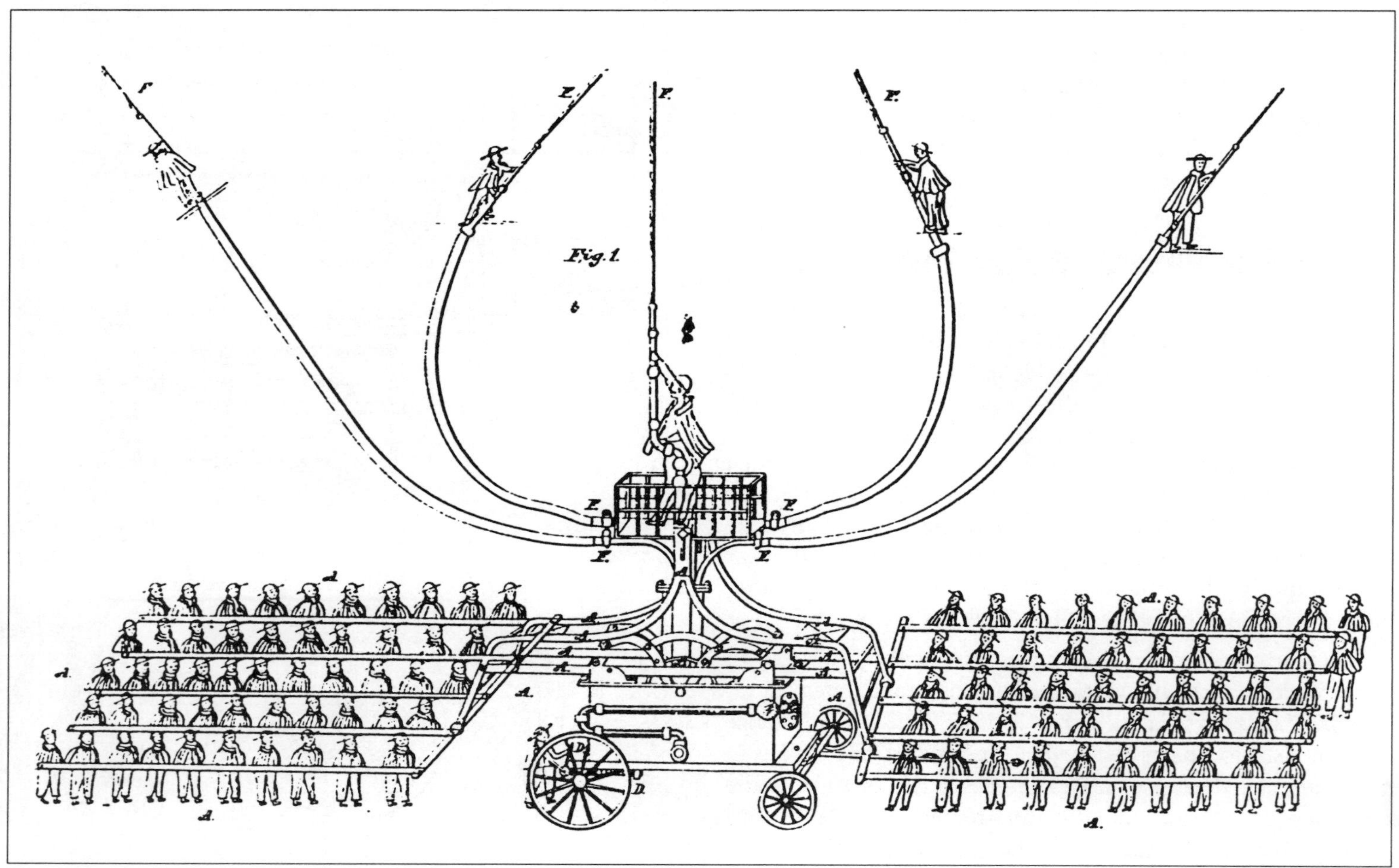

FIGURE 4

and looking bewildered, but rather because they were just standing around and looking on while everyone else in the sketch was hard at work. FIGURE 5 is another Bachstein patent drawing, showing a differing perspective.

Cowing and Company

The drawing for patent number 14,089 dated January 15, 1856 and credited jointly to the Cowings, John P., Philo, and George, all of Cowing and Company and all from the town of Seneca Falls, New York, is shown in FIGURE 6. Now, why would Cowing, or for that matter any manufacturer, situate a factory in such a tiny village, literally–in those days–in the middle of nowhere?

Located in the northwestern part of New York state, between the northern ends of Lake Cayoga and Lake Seneca, about five miles to the south of present-day Interstate Route 90 and 35 miles to the south of Lake Ontario, Seneca Falls was once the fire engine building capital of the country. Perhaps it was partially coincidental that so many fire engine manufacturers–Clapp, Cowing, Gleason and Bailey, Holly, LaFrance, Rumsey, and Silsby–had located in this small town. But more likely it was because of water power. Seneca Falls was for a time a preferred location for any kind of manufacturing activity because the Seneca River dropped about fifty feet, and this was more than adequate to turn the water wheels which powered the factories.

Cowing and Company was started there in 1840 by John Cowing and Henry Seymour, makers of pumps and one of the

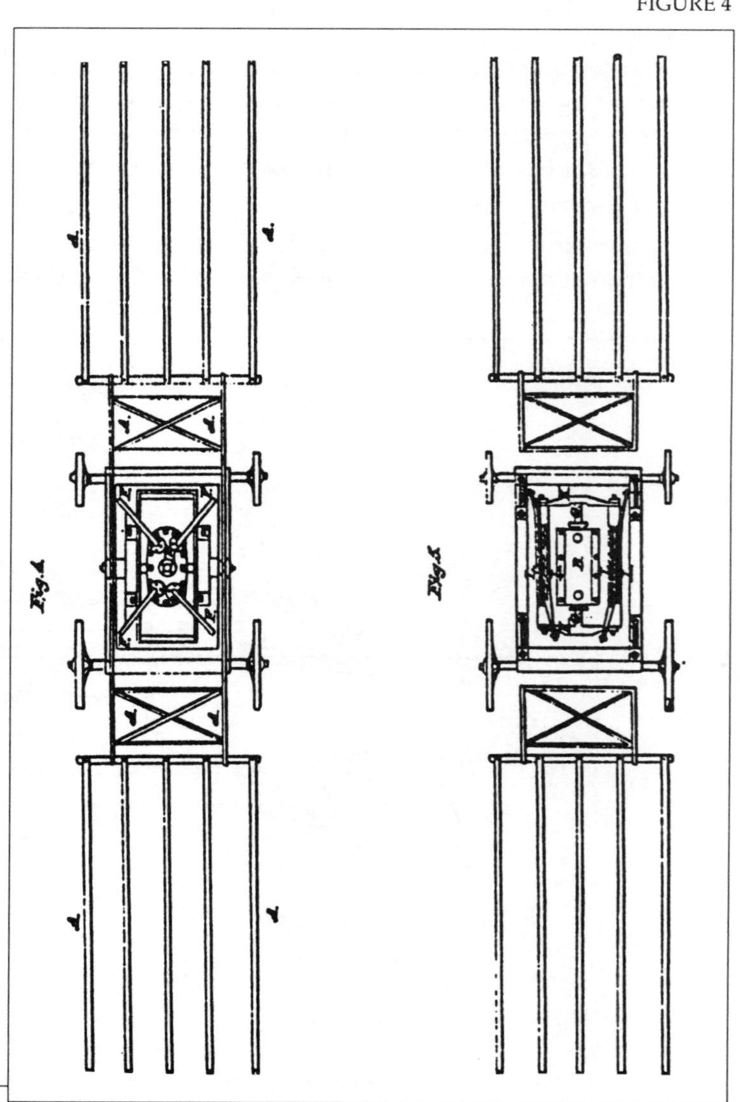

FIGURE 5

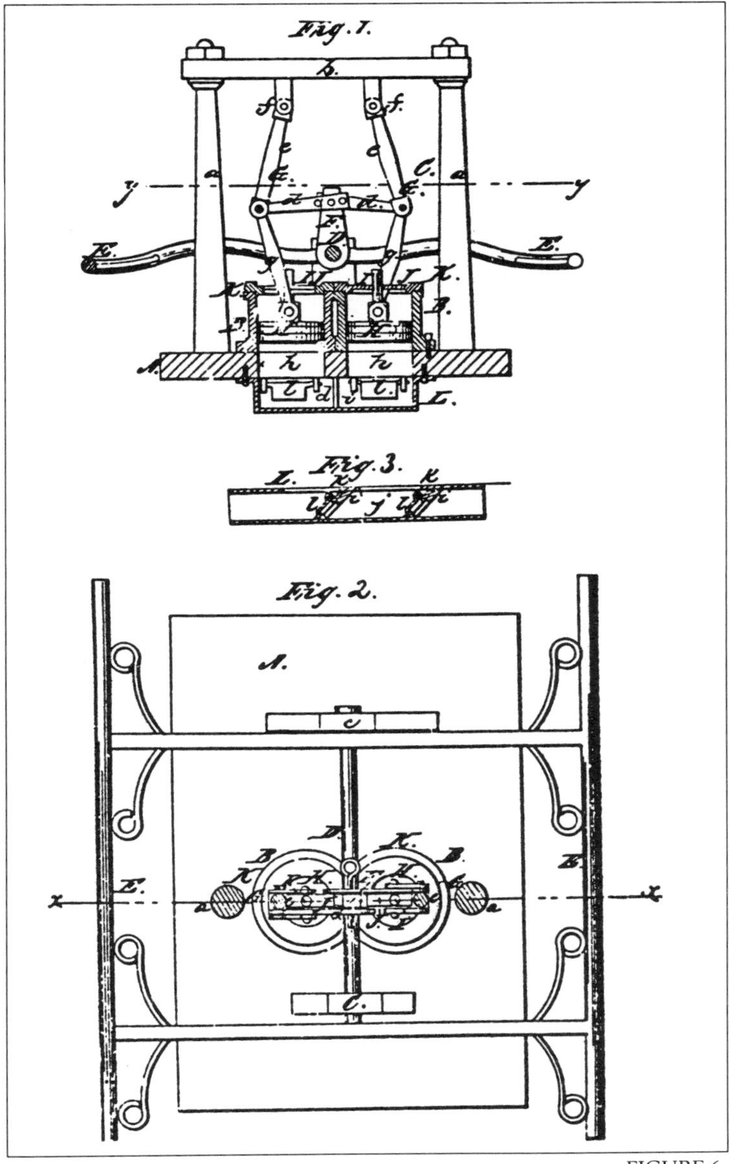

FIGURE 6

first manufacturers of hand-powered fire engines. The demise of this company occurred just 35 years later, in 1875. Notice in the drawing for this first Cowing patent that the upper sketch, represents a vertical plane along the line x-x of the lower figure, and that the lower figure is a horizontal section along the line y-y of the upper one. It is easily recognized that these drawings show a two-cylinder force pump. As the pump lever frames, designated by the letter E, were moved up and down by firemen, the pistons, H, were moved in the same directions within their cylinders by the toggles, G. As one piston was moved upwardly the other was moved downward. Water was drawn into the compartment under the cylinders during each upstroke and it was ejected with each downstroke.

There was, by the way, at least one additional Cowing fire engine patent. This was number 102,229 dated April 26, 1870, and granted to George Cowing. This was his design of a most peculiar chemical extinguisher attachment for fire engines. Remember that it was around this time that so many fire officials accepted as gospel the fact that the fire extinguishing properties of water could be greatly enhanced by the addition of such chemicals as ammonium sulfate or sodium bicarbonate. So Cowing described this device as consisting of "an air-tight case, chest, box, or other vessel suitable for containing the chemical substances provided with a tubular passage through it and near to the bottom, and adapted for connection of hoses." This hardware, shown in FIGURE 7, as well as the concept, was clearly unworkable, and perhaps one can point out as a kind of tribute to George Cowing that he never actually built this contraption. FIGURES 8 and 9, on the other hand, do illustrate fire engines which were built by Cowing and Company.

Frederic Kettler of Milwaukee, Wisconsin was the inventor of the rowing-machine type of hand pumper which is shown in FIGURE 10. This was patent number 24,876 dated July 26, 1859. Notice that Mr. Kettler provided enough seating accommodations for a crew of 24.

The "Improved Chemical Fire Annilhilator" invented by a pair of Bostonians, Rufus Lapham and George Clark, Jr., is illustrated in FIGURE 11 (patent 65,240 dated May 18, 1867). Obviously these inventors were believers in the symbiotic effect of carbon dioxide added to a stream of water. And apparently Messrs. Lapham and Clark also believed that they could actually immerse a force pump inside of a vat which contained sulfuric acid. No doubt this carelessly contrived contraption was never constructed.

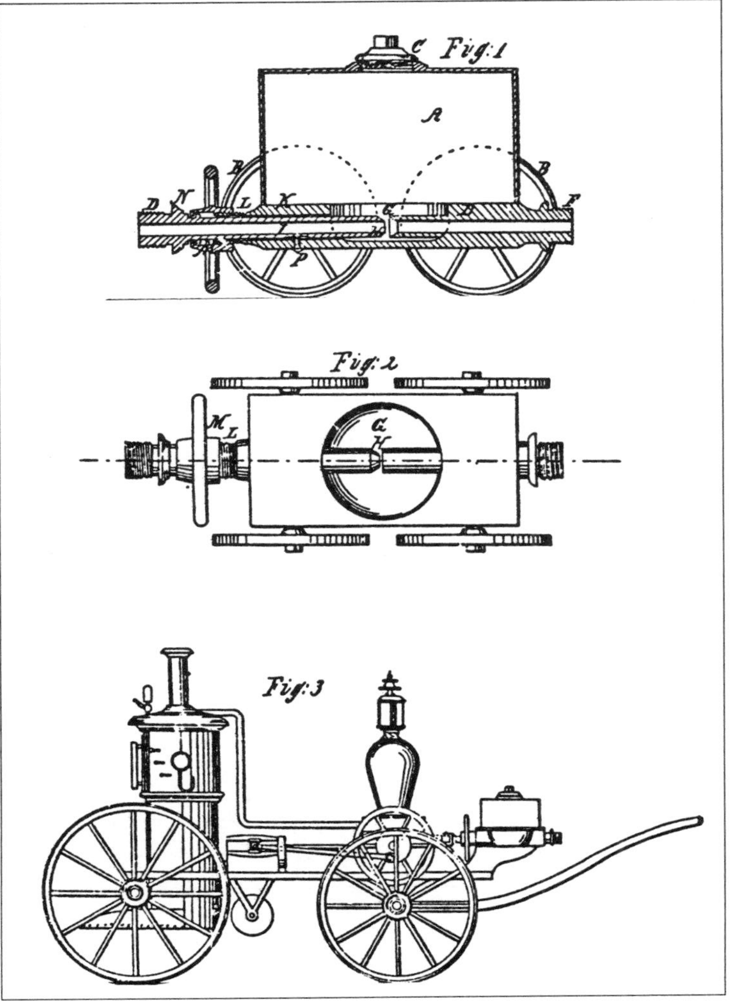

FIGURE 7

FIGURE 8

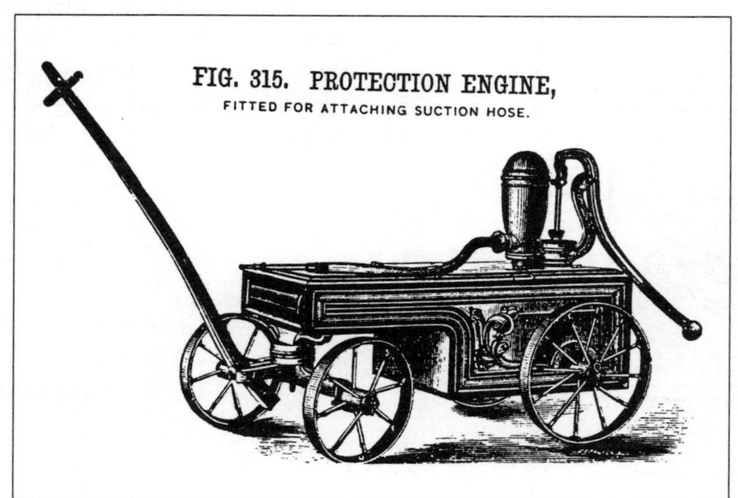

FIGURE 9

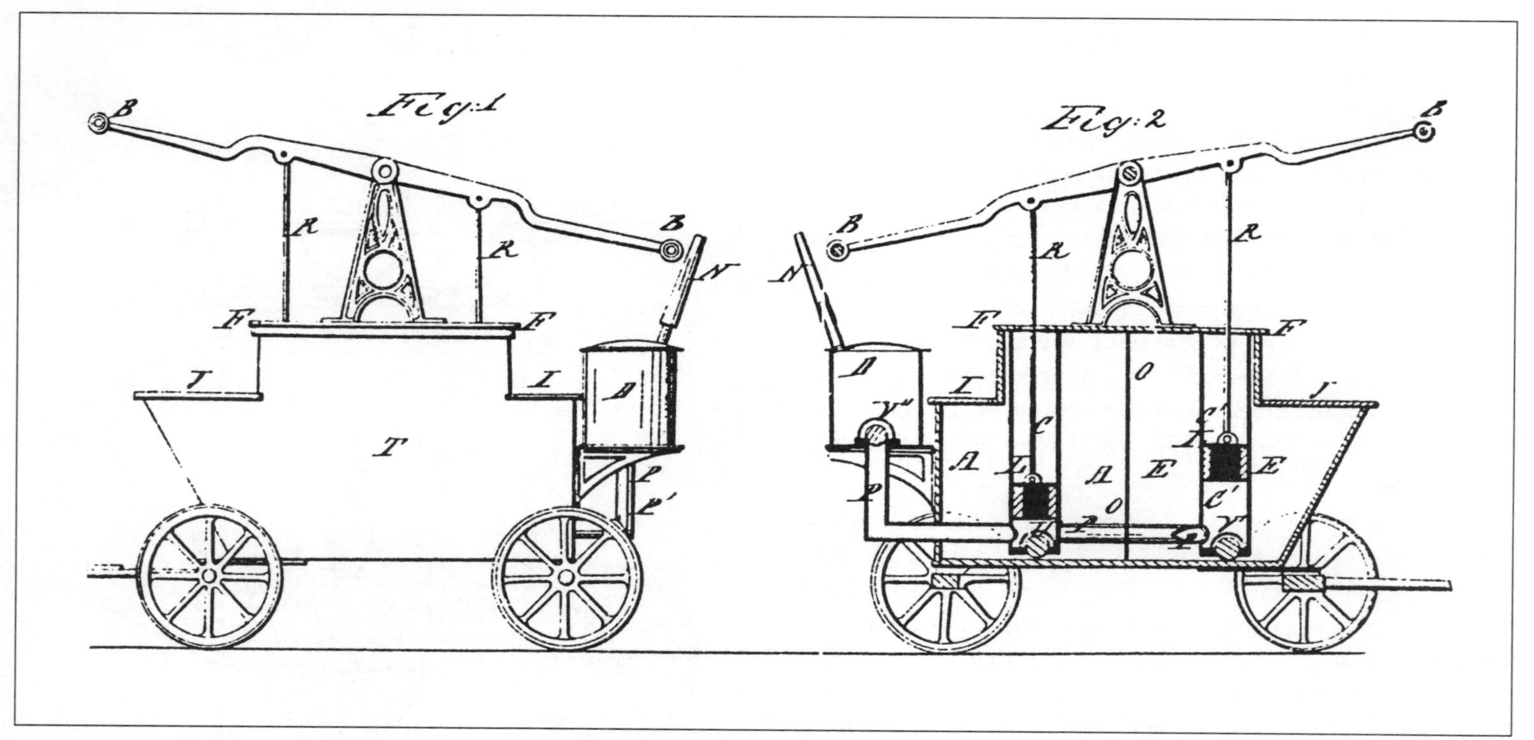

FIGURE 11

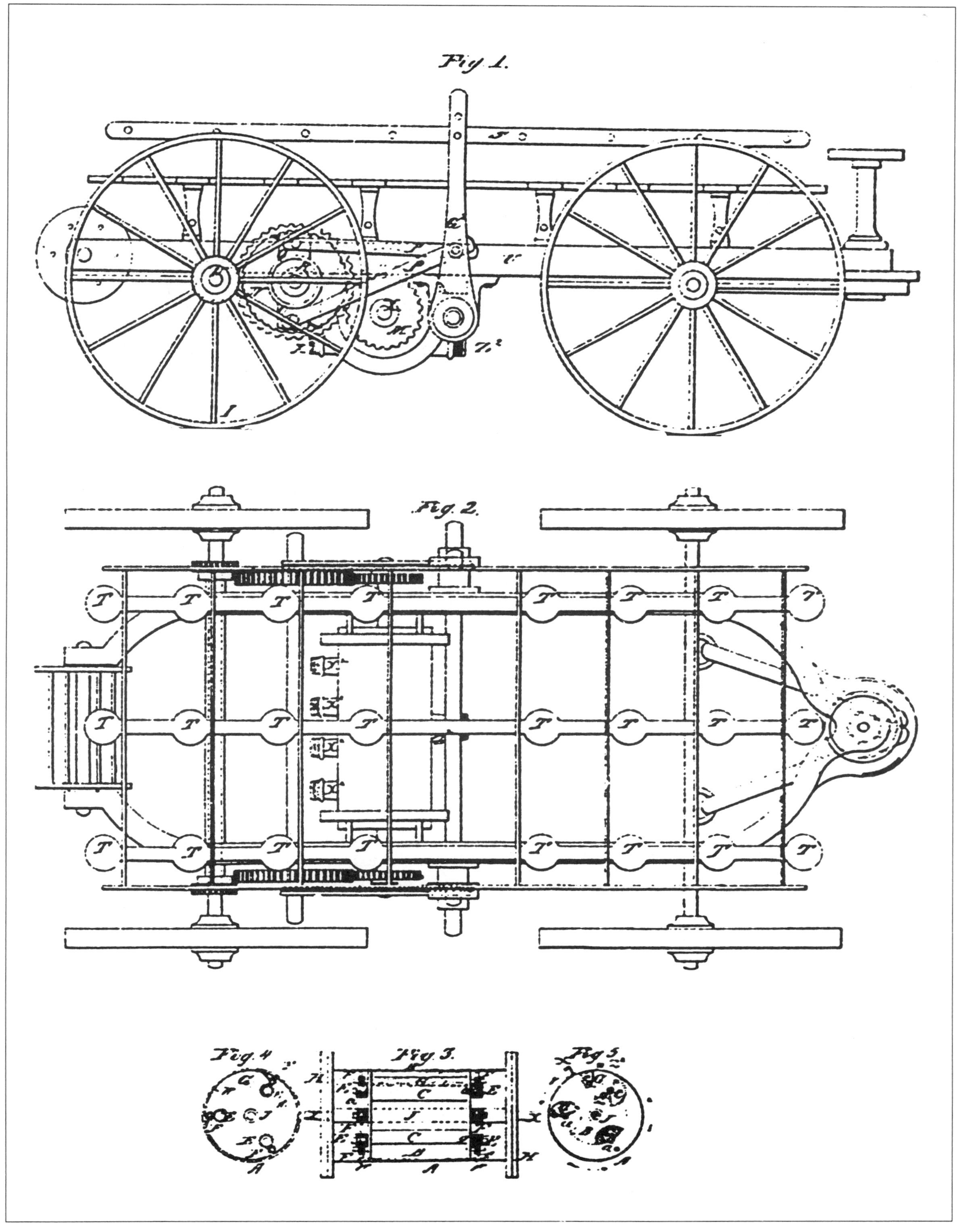

FIGURE 10

A Back-Pack Pump

Joseph W. Douglas' "knapsack-engine," patent number 93,974 dated August 24, 1869, is the subject of FIGURE 12. This Middletown, Connecticut inventor set forth the background for this patent in a brief narrative which included these lines:

"It is well known, and often stated in the newspapers, and by experienced firemen, that no inconsiderable portion of fires, when first discovered, might have been extinguished by a pitcher of water, applied in the right place."

"This invention covers a legitimate combination of individual devices not before used in the same way, to produce the same effect. It is a machine to all intents and purposes."

"It will not prevent a man from carrying it up a pair of stairs, or through the scuttle of a roof, or up a ladder and through a window."

"The reservoir holds about a cubic foot of liquid."

But a cubic foot of water weighs about 63 pounds. Add to that the weight of the apparatus itself, say 15 to 20 pounds, and one quickly realizes that not every fireman, no matter how motivated, could carry it up all these ladders and stairs. Nevertheless, it was a neat idea and a dandy machine.

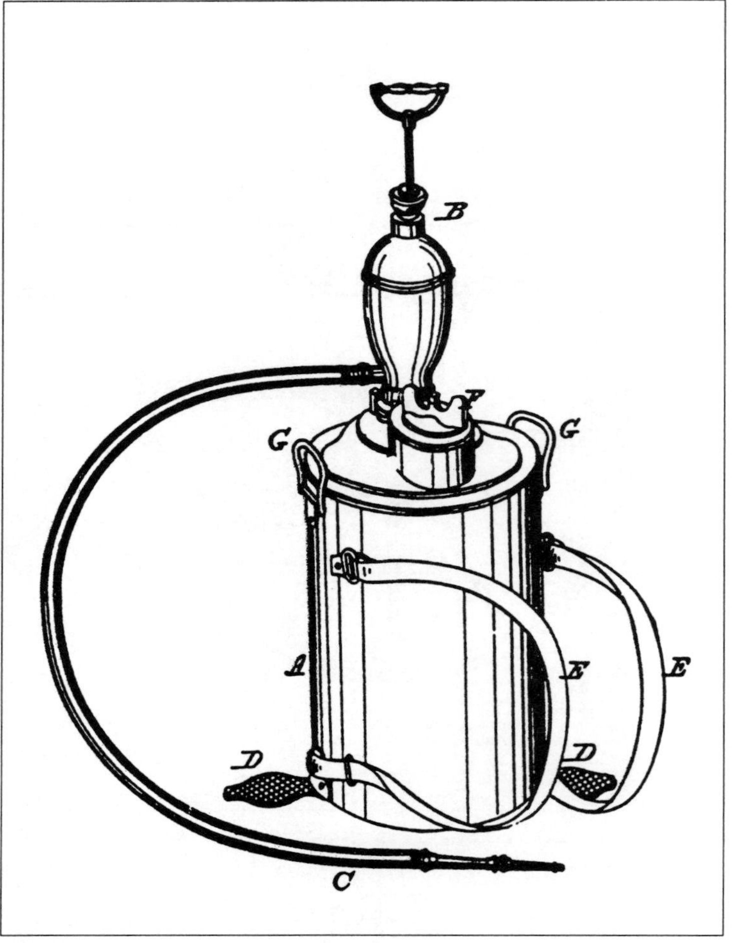

FIGURE 12

The Cleveland Brothers

The next invention, illustrated by FIGURE 13, was an important contribution, not so much perhaps for its hardware design but rather for its scientific basis. And it should be noted in passing that neither contemporary nor venerable inventors practically ever discoursed to any extent, within their patent presentations, as to the scientific premises for their ideas. But in their patent number 125,883 dated April 28, 1872, a New York city physician named Clement Cleveland, M.D., and his brother William, of Ithaca, New York, elucidated the theory of water use in firefighting, and they explained it just as clearly as State Fire School instructors do today.

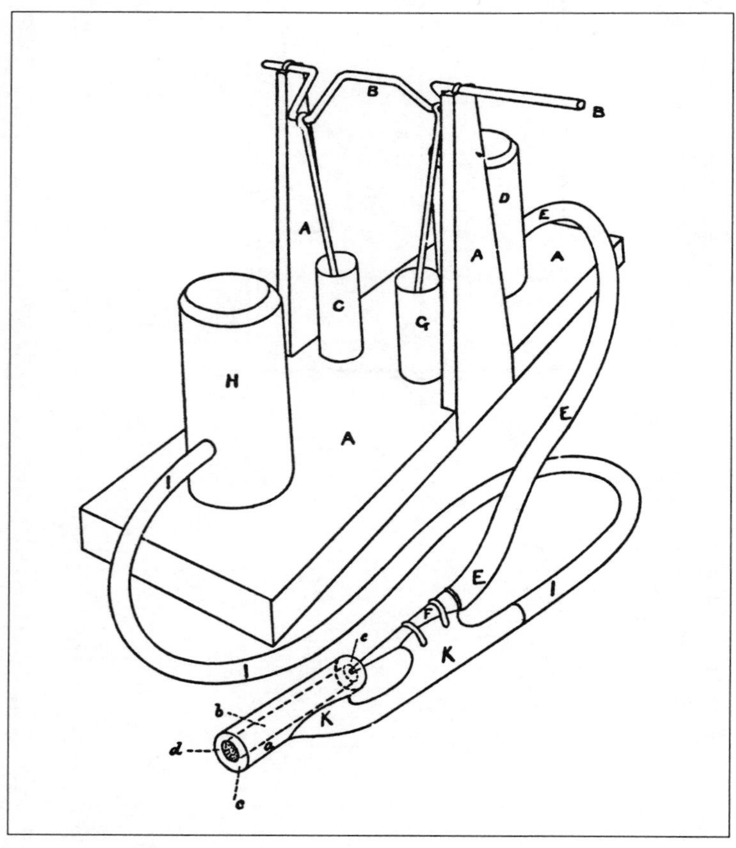

FIGURE 13

Said the Clevelands: "The object of our invention is the more ready and immediate extinction of fire. The method now in general use is that in which the water is discharged in a continuous stream from a nozzle attached directly to a hose. By this method the stream remains compact until it strikes the object upon which it is directed, thus presenting the least possible amount of surface to the fire. The physical effect of water upon fire is the reduction of the temperature of the combustibles–gases or whatever they may be–to below the point of combustion, when fire ceases. As water requires a very large quantity of heat for its vaporization it is one of the best, as well as being the most readily attainable medium for this purpose. The greater the amount of surface exposed by a given body of water to the source of the heat the greater will be the amount of vaporization, and consequently the greater will be the reduction of temperature at the source of heat. Steam has little effect in quenching fire, as it has already received its heat of vaporization; but, if water, still in the liquid state can be presented to the fire in a very-finely divided condition, as in a dense cloud of mist or spray, the amount of surface displayed will be the greatest possible, the amount of heat received the greatest possible, and thus will be produced the greatest possible reduction of temperature, quickly extinguishing the fire. Our method is to present the water to the fire in this finely-divided or atomized condition."

Then setting out their procedure for producing clouds of atomized water Clement and William C. Cleveland described their combination of a hand-powered force pump with an air chamber for ensuring uniform water flow, together with a hand-powered air pump.

Referring now to FIGURE 13, letter C designed the air pump and letter G the water pump. Each were connected by appropriate piping and valves to the air receiver, D, and the water receiver, H. Hose line E carried air to the blast pipe of the atomizing apparatus, F. Hose line I conducted water to the water outlet, K. This was composed of an outer cylinder, a, an inner cylinder, b, which was open at each end, and c, which denoted end plates enclosing a water space between the two cylinders. This water space communicated with the water outlet, K, and with the inner cylinder by way of numerous perforations, d, through which water was forced in fine streams. In this way water met the blast of air from the blast pipe, as it passed directly through the inner cylinder, thereby atomizing it and giving direction to its flow.

Foot Power To Pump Water

Next we turn to the legs and the feet of firemen, given that healthy and functional fire-fighters come so equipped. And since the inventors, sooner or later, took notice of these anatomical parts it was inevitable that foot-operated fire pumps would be developed. Examples of these devices are illustrated in FIGURES 14, 15, and 16. The first-depicted of these inventions, patent 133,278 dated November 19, 1872 was credited to James William Whitaker who lived in Kenosha, Wisconsin, the town which later became the home of the Peter Pirsch Company. The second patent, number 154,149, with a date of August 18, 1874, was designed by Arthur Paget of Loughborough, England. He assigned a one-half interest in his patent to New Yorker Hamilton E. Towie, his patent attorney, probably as payment for services. The third one of this trio of illustrations shows Paget's invention as presented in Scientific American, issue of February 22, 1873.

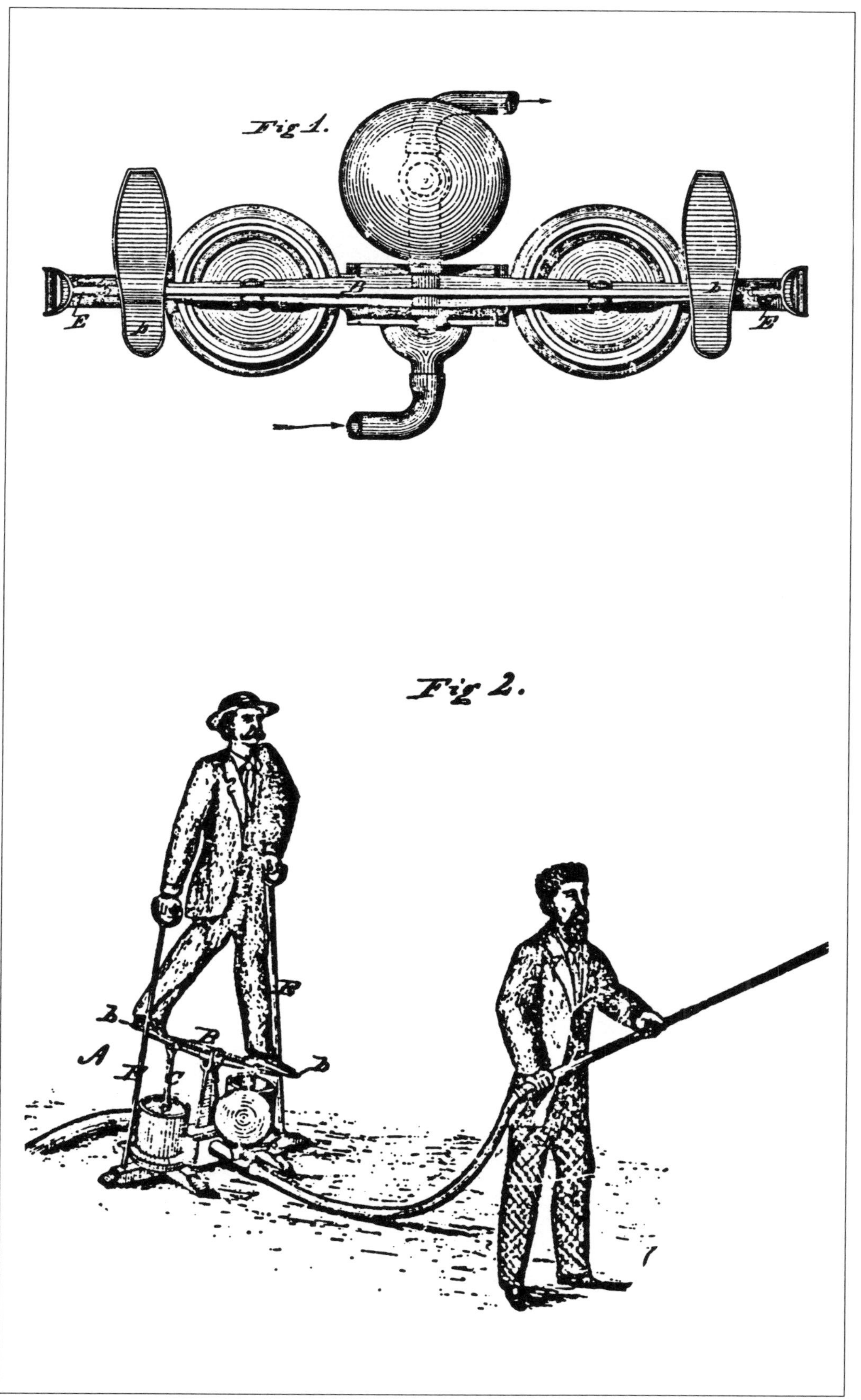

FIGURE 14

Looking back over these drawings, it ought to be pointed out that inventor Whitaker made no claim that his pump operator had to be bearded or mustached, nor that he had to be fully-dressed with tie and hat, as pictured. What he did claim as original ideas were the handles and the foot lever, designated respectively as E and B. And so he wrote, "As the operator throws his weight on either end of the lever, he at the same time pulls strongly on the corresponding rod, so as to

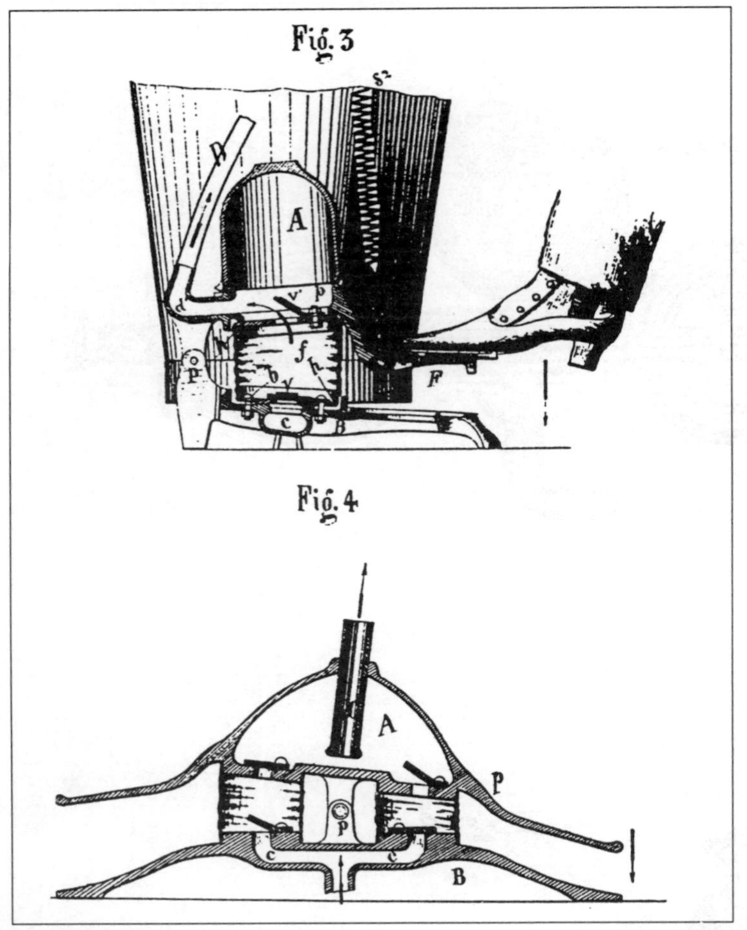

FIGURE 15

FIGURE 16

pull himself downward, and thus apply the muscular force of his arm in addition to his weight toward forcing the end of the lever down. When operating in this manner his legs are kept straight and his arms bent at each stroke; when his arms become fatigued he may hold them stiff and operate the lever by bending his legs. By thus changing the mode of operation from time-to-time, so as to throw the work upon arms and legs alternately, the operator is relieved from fatigue and enabled to work the pump for a long time" (and perhaps even longer if he loosens his tie, removes his hat, and shaves off his beard!).

The other two illustrations, which are of Mr. Paget's bellows pump, also show off the booted foot of a proper English gentleman.

Henry Neumeyer of Millerstown, Pennsylvania was another inventor who was aware of and made use of the feet of the pump operator. In FIGURE 17, the drawing for his patent 158,729 dated January 12, 1875, there was a foot-piece, designated V, which was intended to steady the pump as the operator turned the crank handle, R. This invention was a simple three-cylinder force pump, probably intended to be mounted and carried to fires upon a wagon or a cart.

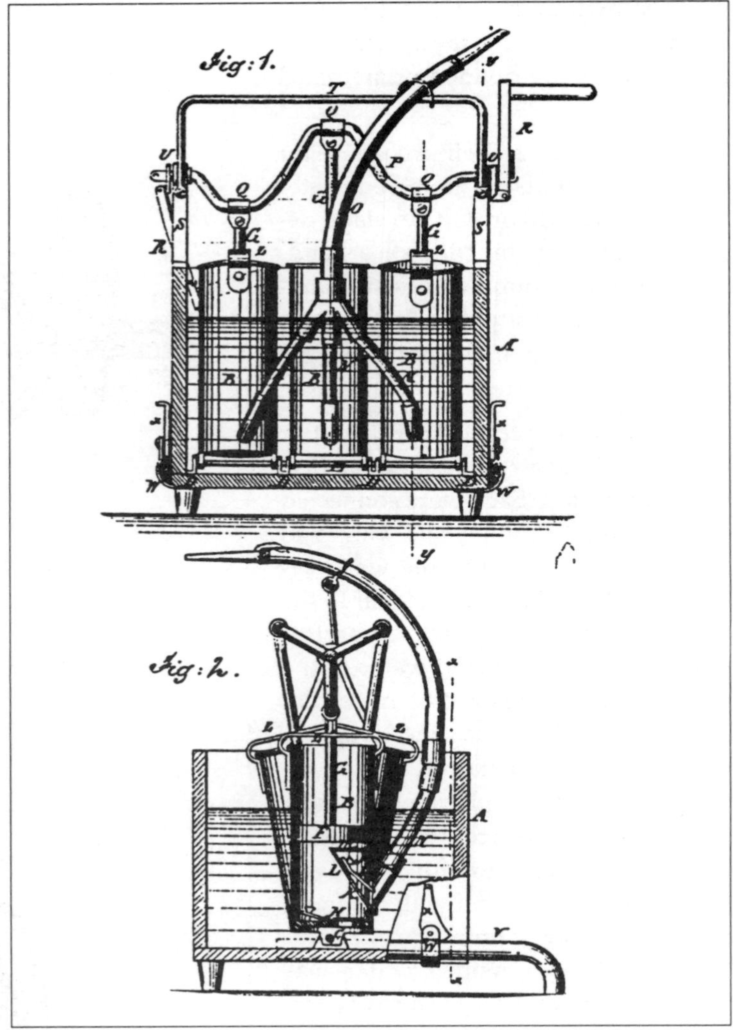

FIGURE 17

The Taylor-Hall Pumper

The thought of carrying such devices, the weight of which must have been considerable, leads to some intriguing speculations about the muscular strength of the firemen of yesteryear. There was, for example, this claim for a combined hand-powered pump and hose carriage, that, "Two men can readily draw it over all common grades. At present the manufacturers are building one size only, the entire equipment weighing only 500 pounds." And when about to use it at a fire, "Two men can lift the engine from the carriage, as it weighs alone only 325 pounds."

This particular fire engine was invented by Levi Taylor (patent number 195,183 dated September 11, 1877), and it is seen in FIGURE 18. The publicity blurb which acclaimed that this 325 pound device was a featherweight and, "so light that it can be lifted and set over a well, cistern, or reservoir," was a direct quotation from the inventor's patent writeup. Mr. Taylor must have sold his patent rights to Mr. A.M. Hall, since the latter was credited as the patent owner in a Scientific American article (FIGURE 19) in the August 26, 1876 issue. Notice that the manufacturer of record was the S.C. Forsaith Company of Manchester, New Hampshire. The distributor was the Fred J. Miller Fire Apparatus and Fire Department Supplies Company, 72 Maiden Lane, New York. Miller's 1884 catalogue listed this engine at $350, just a couple of pennies more than one dollar a pound!

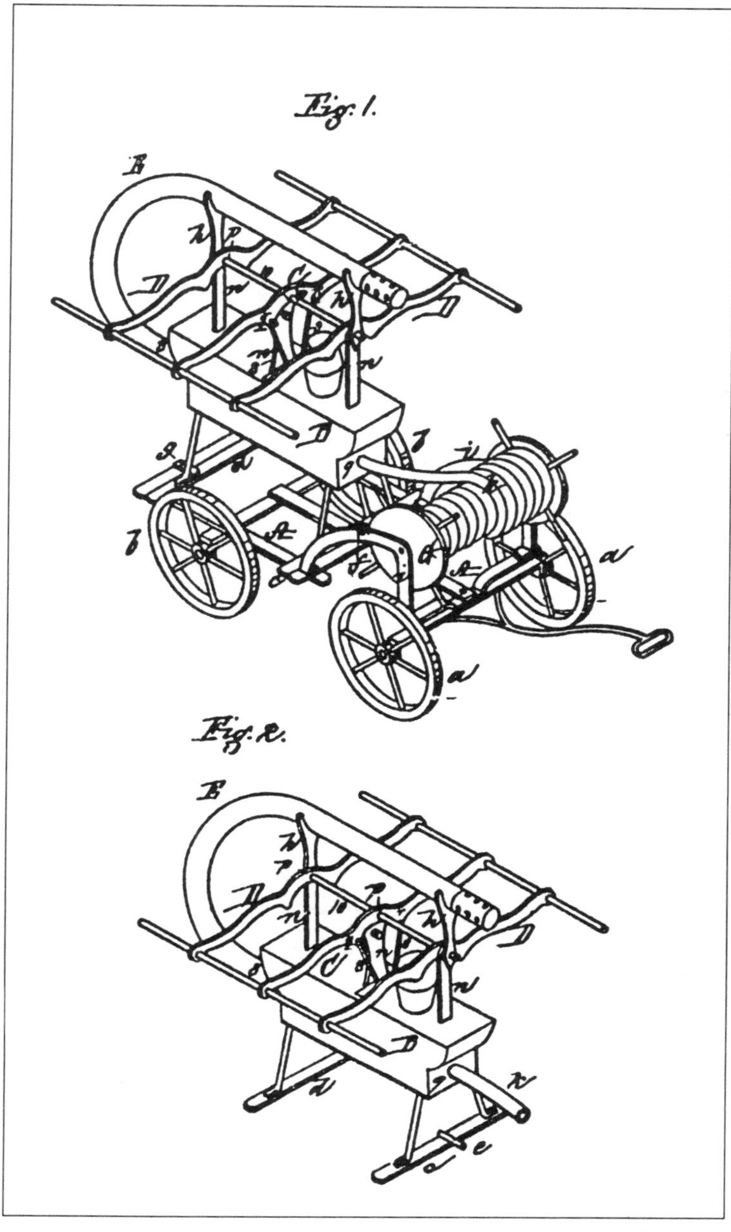

FIGURE 18

In the next illustration, FIGURE 20, can be seen Brooklyn, New York inventor John W. Stanton's patent 202,068 dated April 2, 1878. This small, very portable, hand-pulled and operated pump could be set up to supply either water or chemical solutions. Inventor Stanton provided paired axle-mounted hose reels to facilitate simultaneous fire suppression and replenishment of the on-board liquid reservoir. Also shown in the drawing was his method for bracing and stabilizing the rig whenever the pump handles were being worked.

Another small pumping engine was the one invented by Hale P. Kauffer, patent number 258,233 dated May 23, 1882 (FIGURE 21). This inventor said that his interests and intentions were mostly domestic, e.g., watering gardens, showering lawns, and washing windows. And, by the way, as he so noted, perhaps the machine could also extinguish minor conflagrations (a strange word selection, since the standard definition of "conflagration" is a calamitous fire).

Inventions From the American Mid-West

Even though hand-powered pumps of one sort or another had been used in firefighting since antiquity, the time for them to be retired to the revered status of parade pieces and muster apparatus was approaching. Nonetheless during these remaining years of the nineteenth century inventors continued to churn out new design ideas, and the U.S. Patent Office had, by law, no choice but to process them in the usual manner. One interesting aspect of this last group of patents, and probably no coincidence, was that almost all of these particular inventors were from towns of the American mid-west.

FIGURE 19

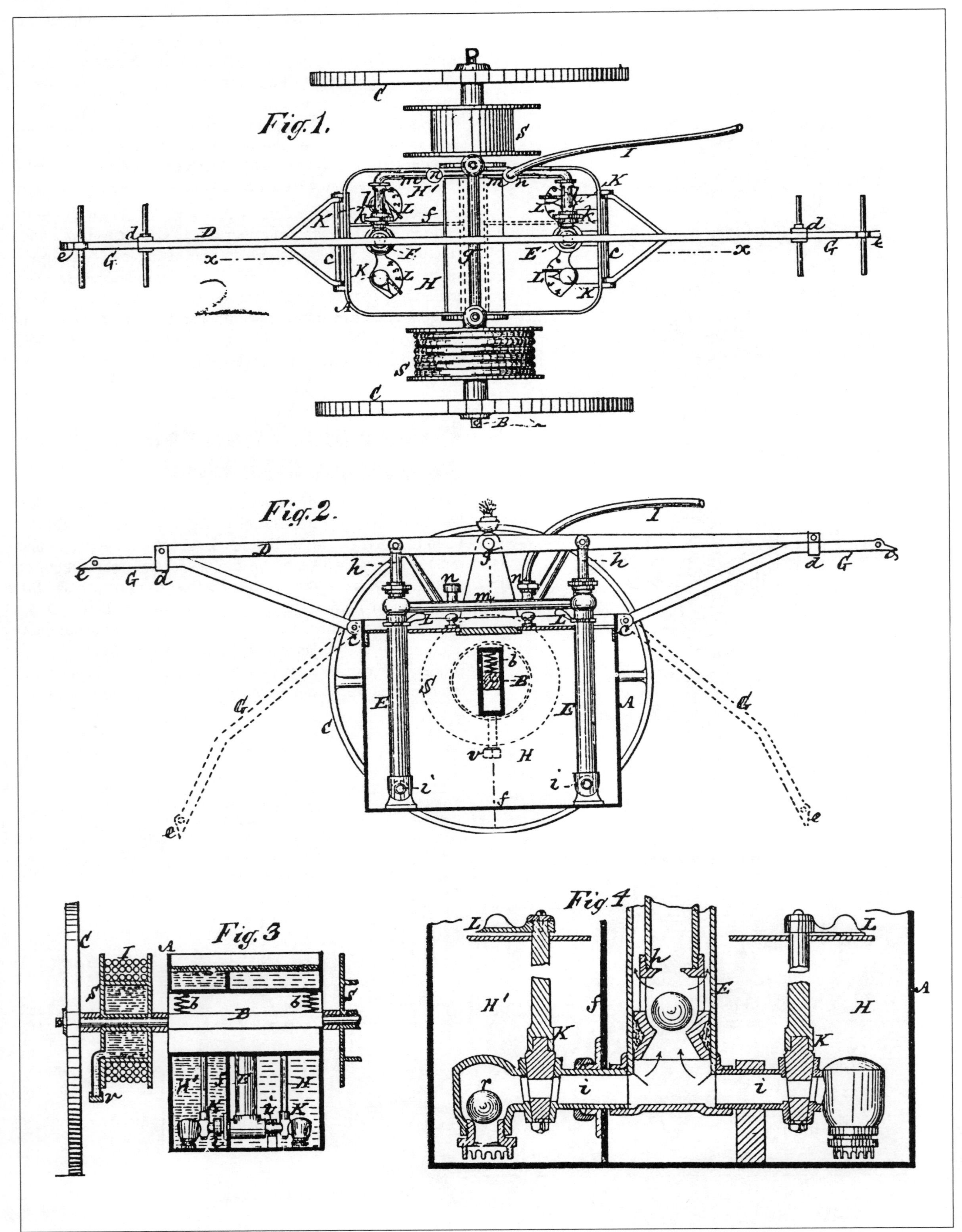

FIGURE 20

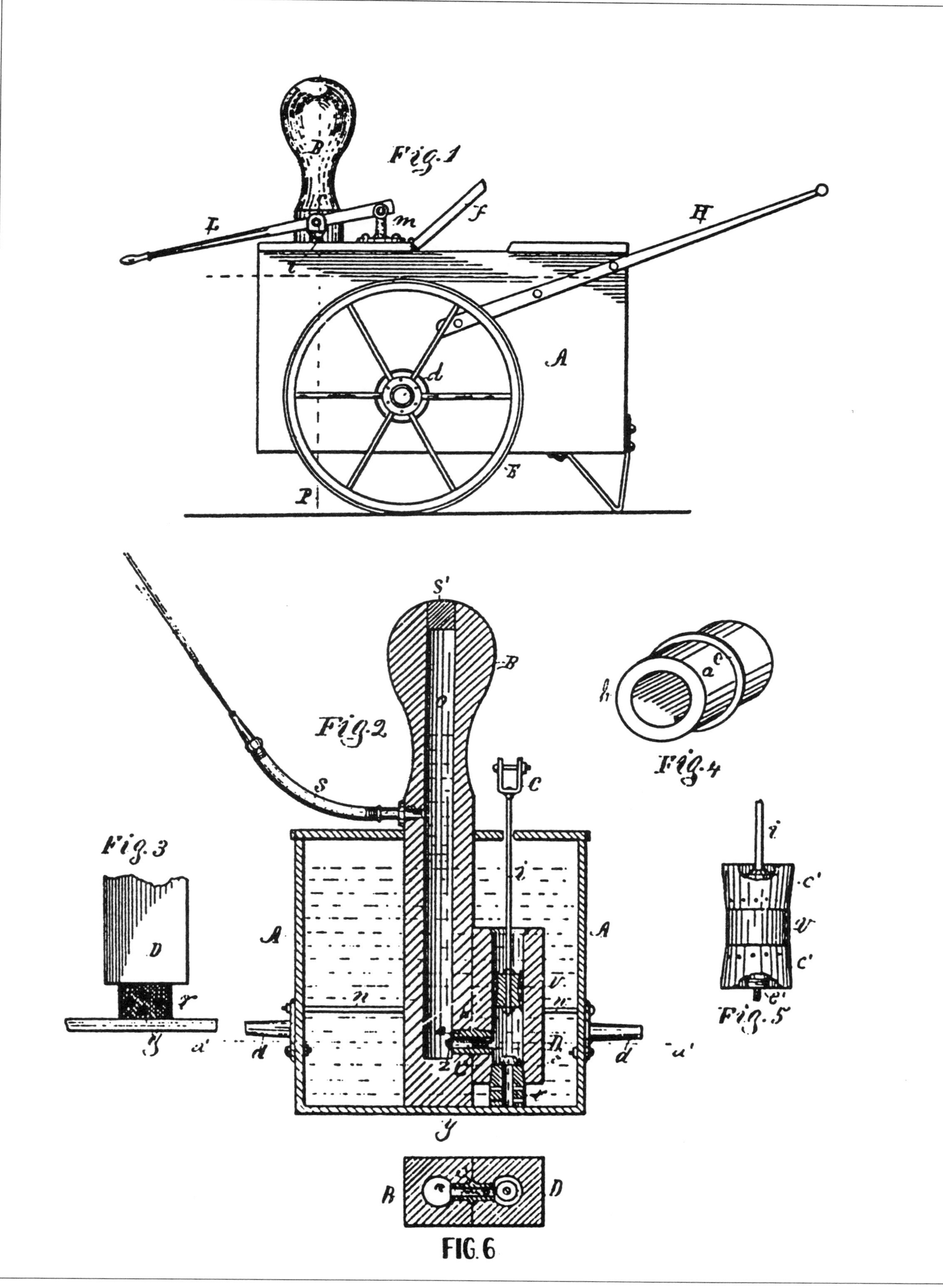

FIGURE 21

Patent number 268,295 dated November 28, 1882 is seen in FIGURE 22. The inventor, Lyman H. Ziegler, was from Redkey, Indiana, a village situated about 15 miles to the northeast of Muncie. This was another small fire pumper, but one which incorporated a new and useful design contribution: it had a U-shaped axle, and this enabled the apparatus to rest on the ground while being used, thus steadying it against the vibrations of active pumping.

The next illustration, FIGURE 23, shows a fire engine invented by Henry Losse, a gentleman who lived in a small town in Illinois called Pinckneyville, located in the southern part of the state about 25 miles south of Interstate Route 64 and 25 miles west of Interstate Route 57. Losse's device, which was assigned patent number 321,510 and was dated July 7, 1885, was especially unusual for two reasons. First, there was no air chamber, continuous water flow thus depending upon the action of a pair of piston pumps set for operation in opposition one to another. And second, the motive force for working these pistons was derived from the circular motion of a crank and its handle, and this has never been an optimum way to use muscle power.

Robert Morrell was a prolific inventor of hand-operated fire pumpers during the final two decades of the last century. In September, 1883 he was the recipient of patent 285,055 for a simple pump (FIGURE 24). And his "improved" water-barrel fire engine, for which he was granted patent number 337,424 dated March 9, 1886, is seen in FIGURE 25. Anyone who has read "Triumph and Tradition, Firefighting in Prince George's County, Maryland 1887-1990" will recognize in this drawing the prototype for the water-barrel fire pump which was illustrated as the frontispiece of that fine volume. Morrell's next patent was number 371,209, and it is shown in FIGURE 26. This 1887 invention involves a hose reel and a force pump mounted on a ladder truck. Morrell commented that by combining all of this equipment in one machine, a rather unusual practice at that time, it enabled, "all

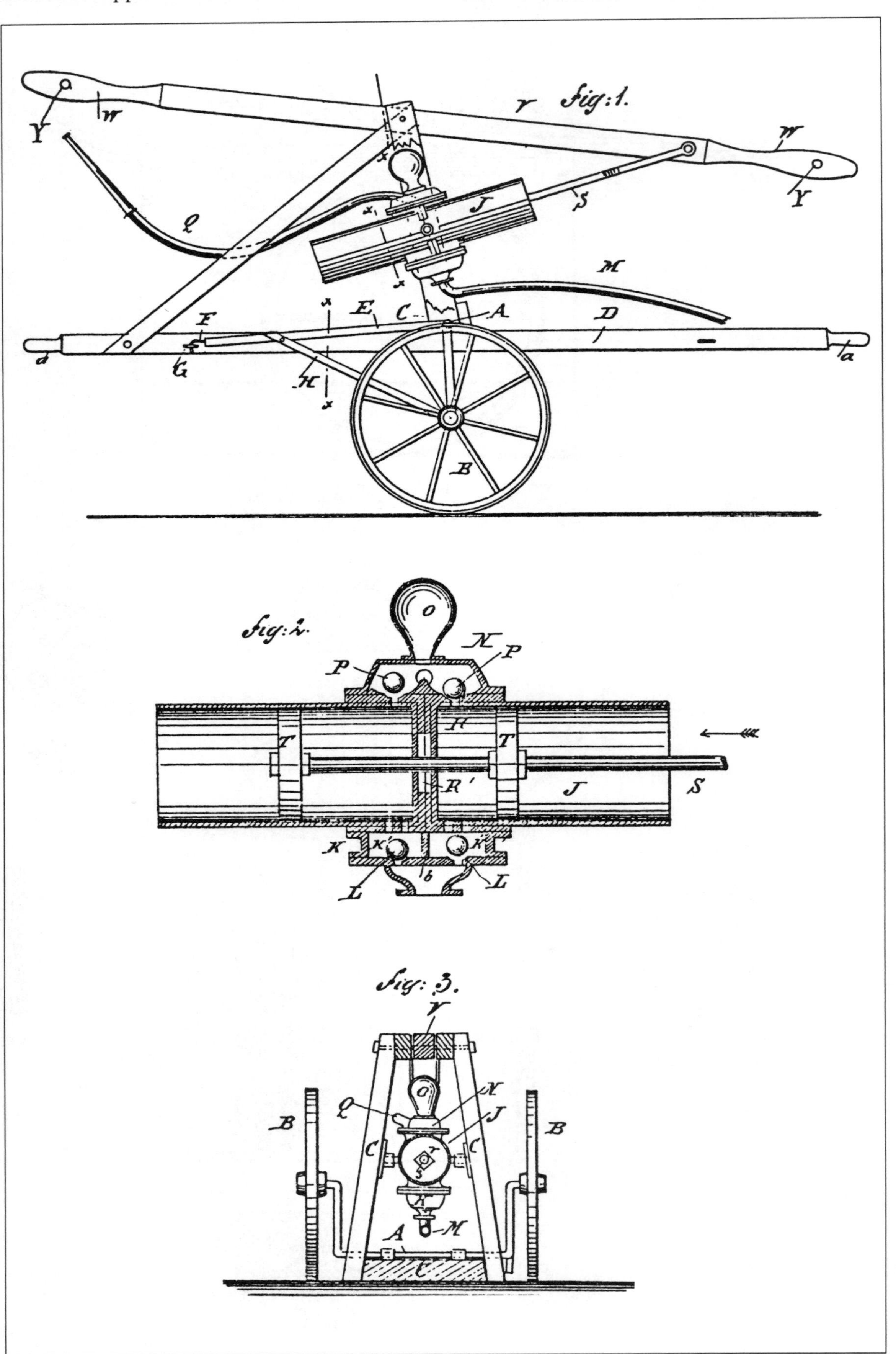

FIGURE 22

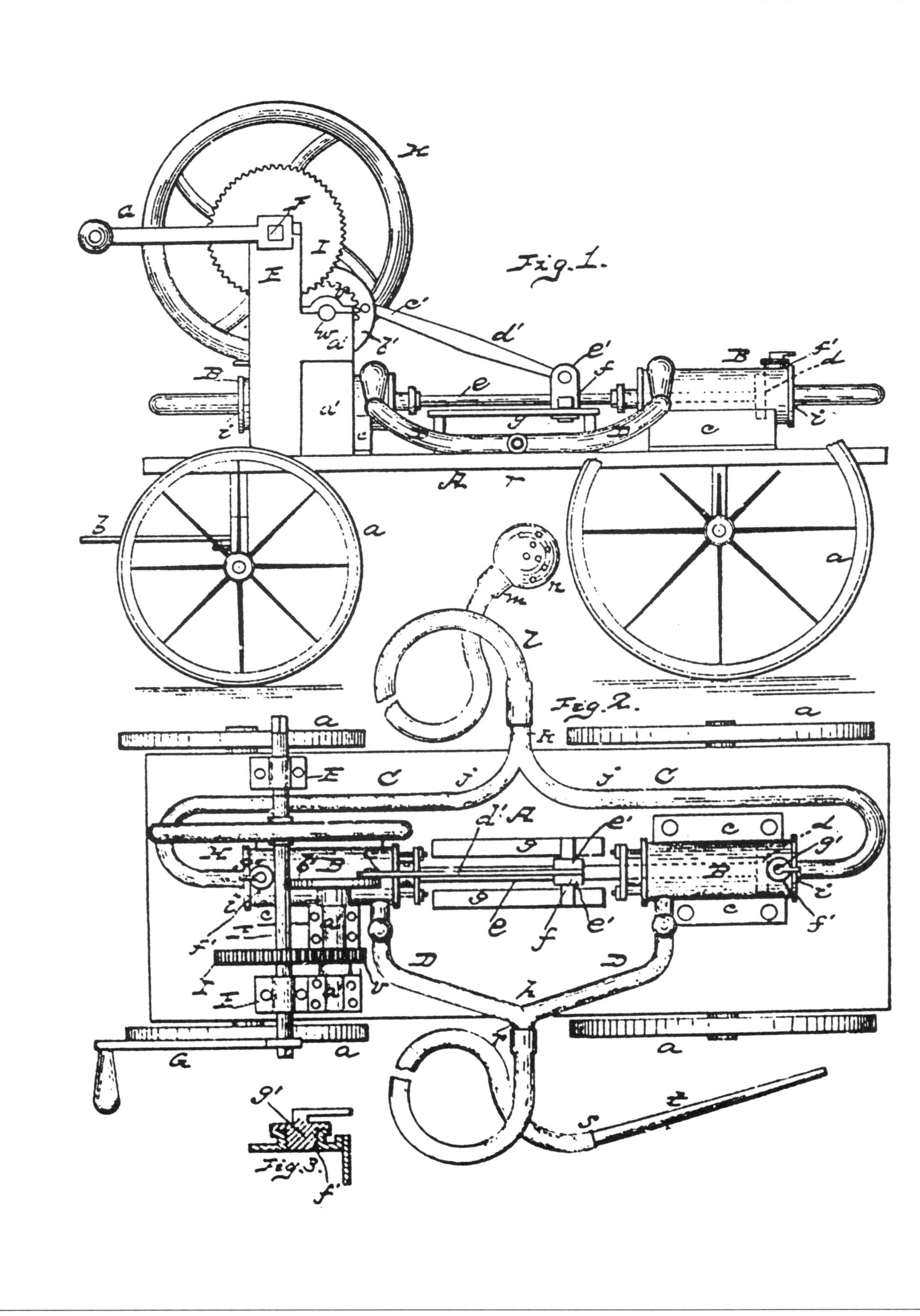

FIGURE 23

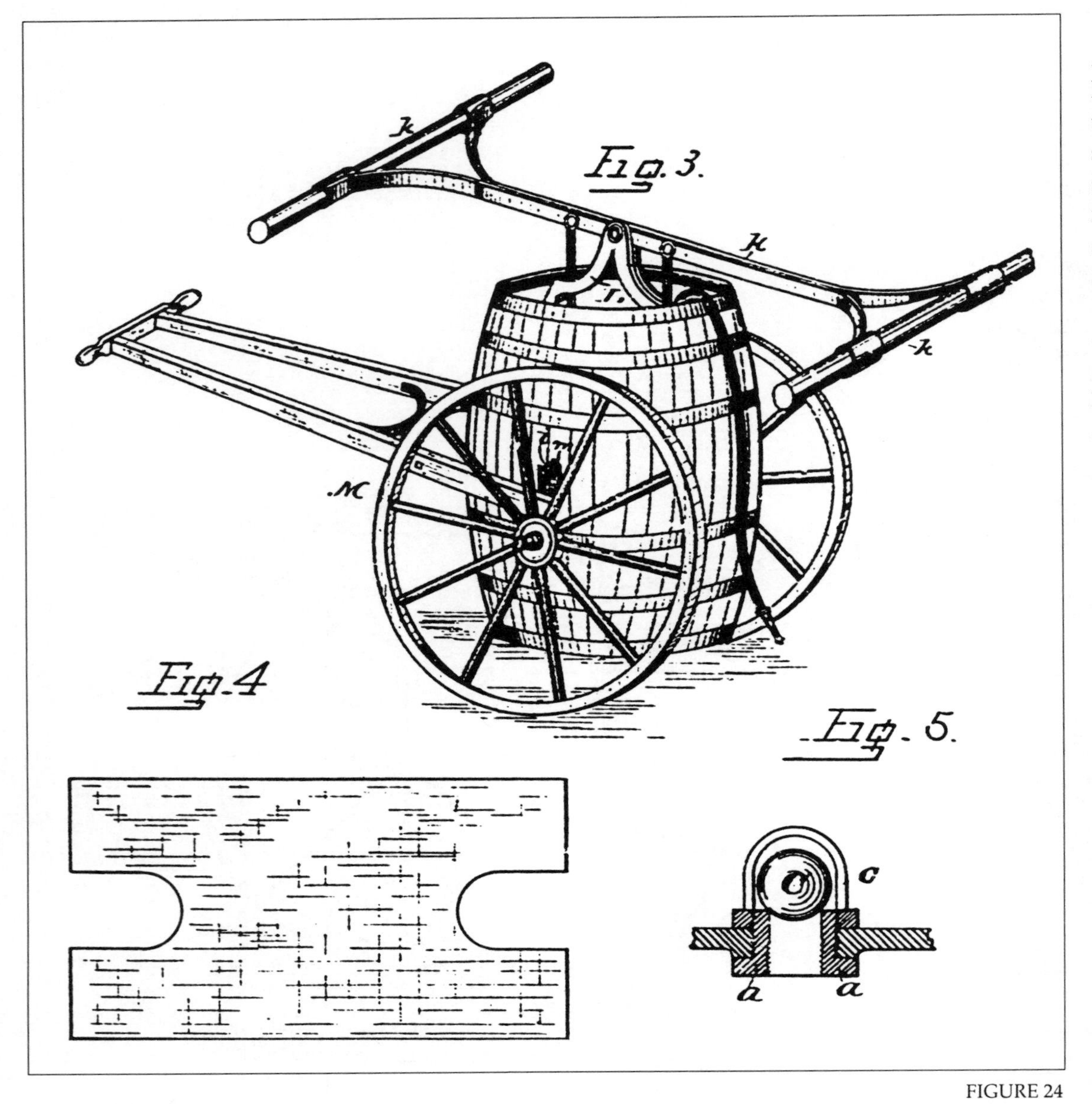

FIGURE 24

the requisite appliances for firefighting to be available in one portable apparatus for the convenience of small communities where it is difficult to collect help for hauling more than one machine and also difficult to raise the money to buy the common fire apparatus."

Notice that his patent drawing shows a hose reel mounted on the rear axle, with the ladders resting on the chassis and the pumps situated within a large water tank. Also seen are the pump levers, the adjustable legs for bracing the vehicle during pumping, and a forward-projected, large-diameter pipe which was supposed to enable the bucket brigade to maintain water in the tank without getting in the way of the pump operators.

Joshua G. Chancellor, Elias C. Henry, and August T. Simon all had in common that they resided in Bloomington, Illinois, that they were fire engine inventors, and that they received patents during the year 1886. Chancellor and Henry, patent 345,571 dated July 13, 1886, provided a pump which was easily carried between a pair of wheels and which could be lowered to the ground for use (FIGURE 27). Mr. Simon's patent was number 346,872, and it was dated August 3, 1886. The novelty of this invention was the method by which the frame carried the pump. It is illustrated by FIGURE 28.

Another mid-westerner was A. Dayton Elliott of Viola, an Illinois town located at the intersection of U.S. Route 67 and state road 17 on the western side of the state. Elliott's remarkable contraption was the subject of patent number 375,317 dated December 20, 1887 (FIGURE 29). If this machine was ever built as described and then operated as built it certainly would have exhausted one platoon of firemen after another. This apparatus was designed with twelve force pumps, set out in two rows of six each. The inventor claimed that his fire pumper could be operated by less than twelve men, but he expected the energy expenditure per man to increase as their numbers dwindled. The direct up-and-down style of muscular exertion of the kind required was obviously quite far from being ergonomically economical.

One of the really great names among all of the inventors and builders of fire engines of the late nineteenth century was Indianapolis, Indiana resident Benjamin J.C. Howe, the founding father of a dynasty as well as a famous company. Howe was another prolific inventor of firefighting gear but it is his hand-powered chemical fire pumper, patent number 443,983, dated on the 30th of December, 1890 which is pertinent right here. The illustration is FIGURE 30. Notice the crane neck and the hose reel. The pump levers were said to be workable with but a pair of operators. Even though Howe had given his home address as Indianapolis, the Howe Fire Apparatus Company had its factory in the city of Anderson, Indiana, about 25 miles to the northeast of present-day Interstate Route 465, a circumferential beltway around the state capital city.

FIGURE 31 is the drawing for patent number 539,850 dated May 25, 1895, Stephen Banfill inventor. This was another device which was alleged to be useful with either chemical solutions or with water, the tank being designed to hold either. Banfill also made provisions for pumping water directly from any convenient supply source, probably having in mind the many small lakes and streams of Palo Alto County in Iowa, where his Ayrshire home was located along state road 314.

The last patents directed to hand-powered pumper inventions just before the turn of the century were the ones received

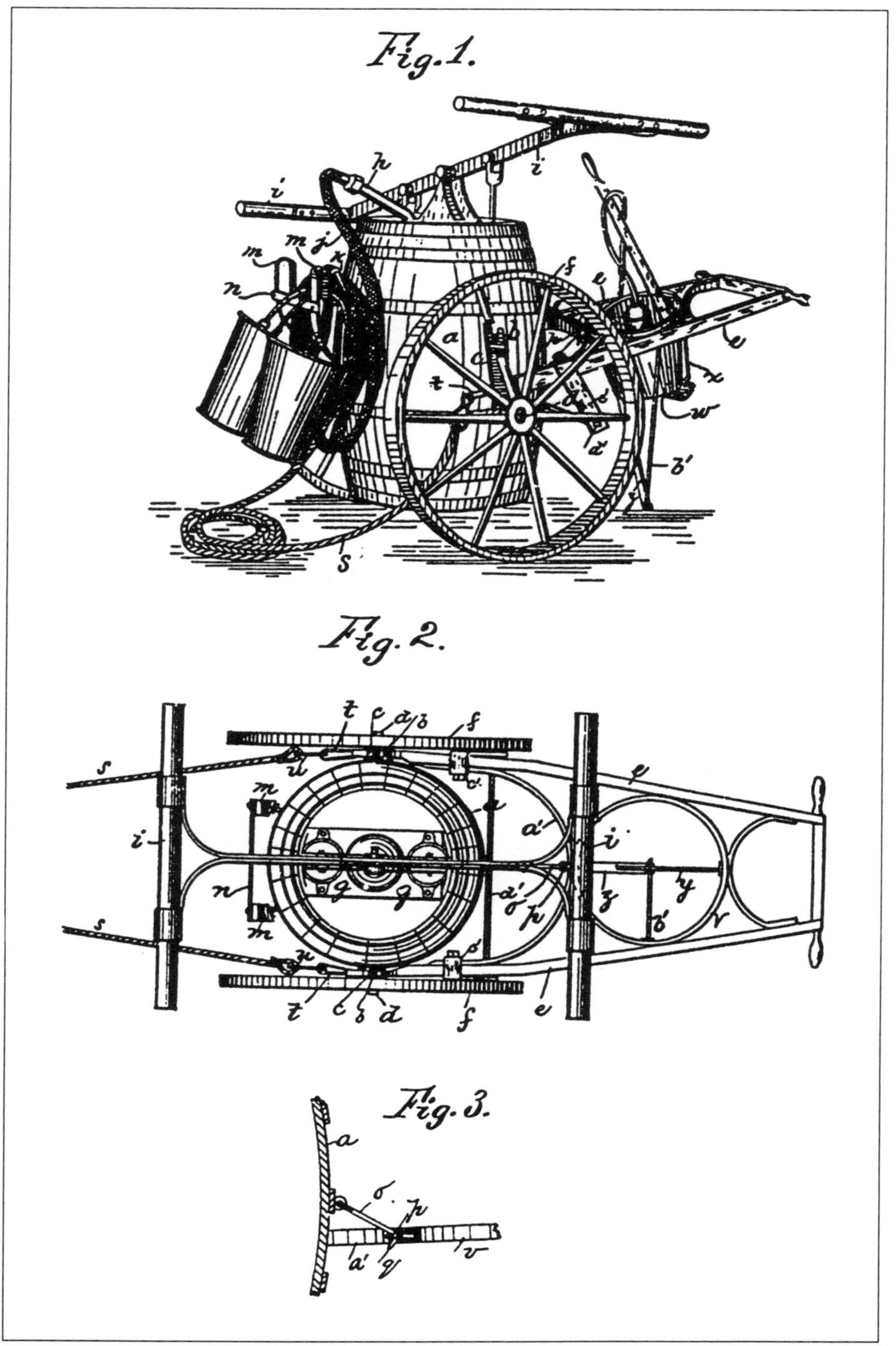

FIGURE 25

by Henry C. Atkinson and William Miller, both from Louisville, Kentucky. In their own words what they had intended was, "to provide a convenient machine for extinguishing fires in villages and towns of small population that are without waterworks and are unable to maintain a regular fire department."

FIGURES 32, 33, and 34 are their illustrations for patents 547,871 and 572,096 dated, respectively, October 15, 1895 and December 1, 1896. The second invention was presented by Atkinson and Miller as an improved version of the former one. In these final two patent drawings the very first figure is a side view of their machine, ready to be moved from the fire station, while the third figure shows the apparatus ready for use at a fire. Sharp-pointed feet have been driven into the ground to steady the machine, and the ladders, the buckets, and the axe have all been removed, and the hose has been unreeled and the water reservoir made ready for bucket-brigade filling. The pump mechanism was fairly conventional, there being four pistons which were actuated by two firemen who rotated the carriage wheels using handles fitted for that purpose. Turning these wheels revolved the axle to which were attached the piston rods. Notice that the bulk of the pump machinery rested within the large water reservoir or tank.

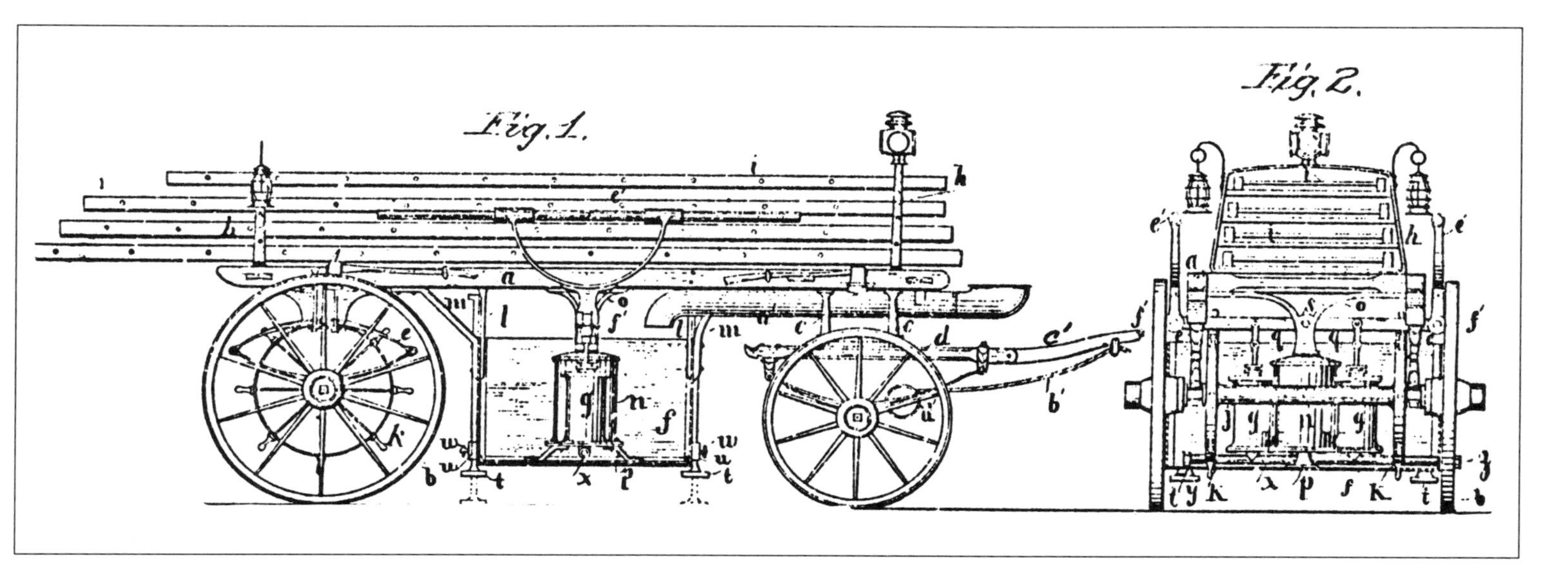

FIGURE 26

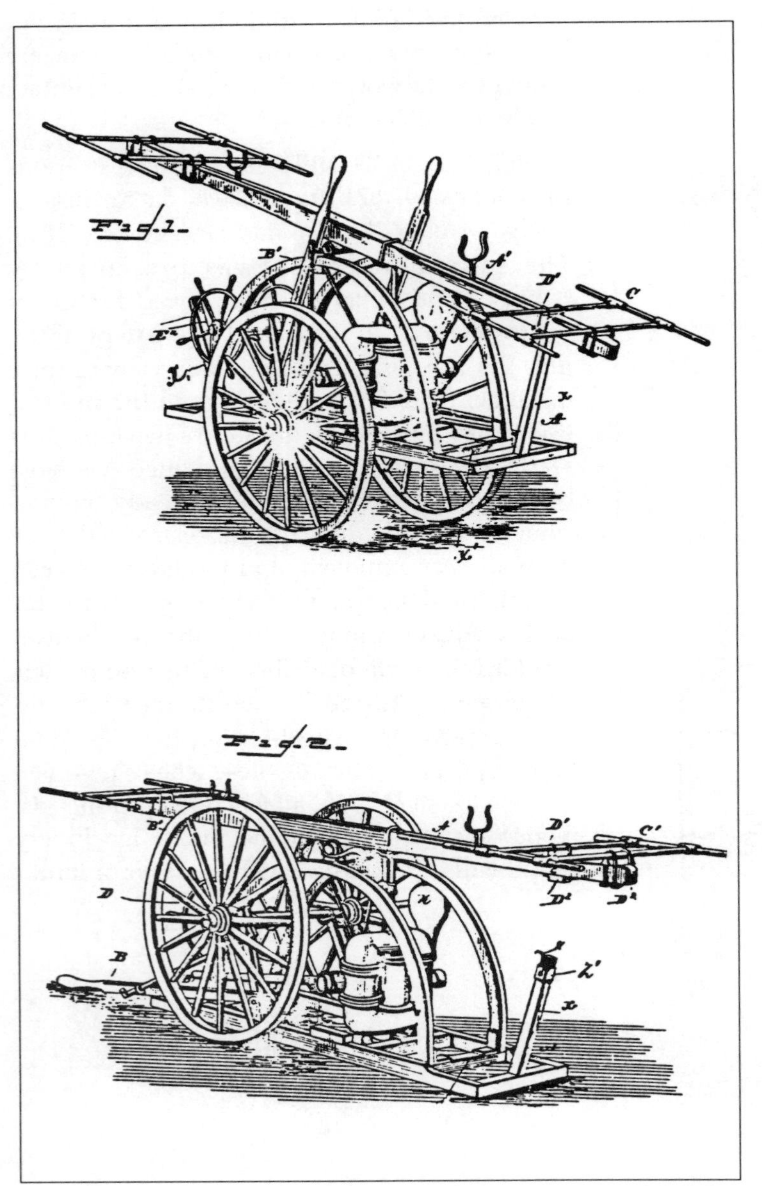

FIGURE 27

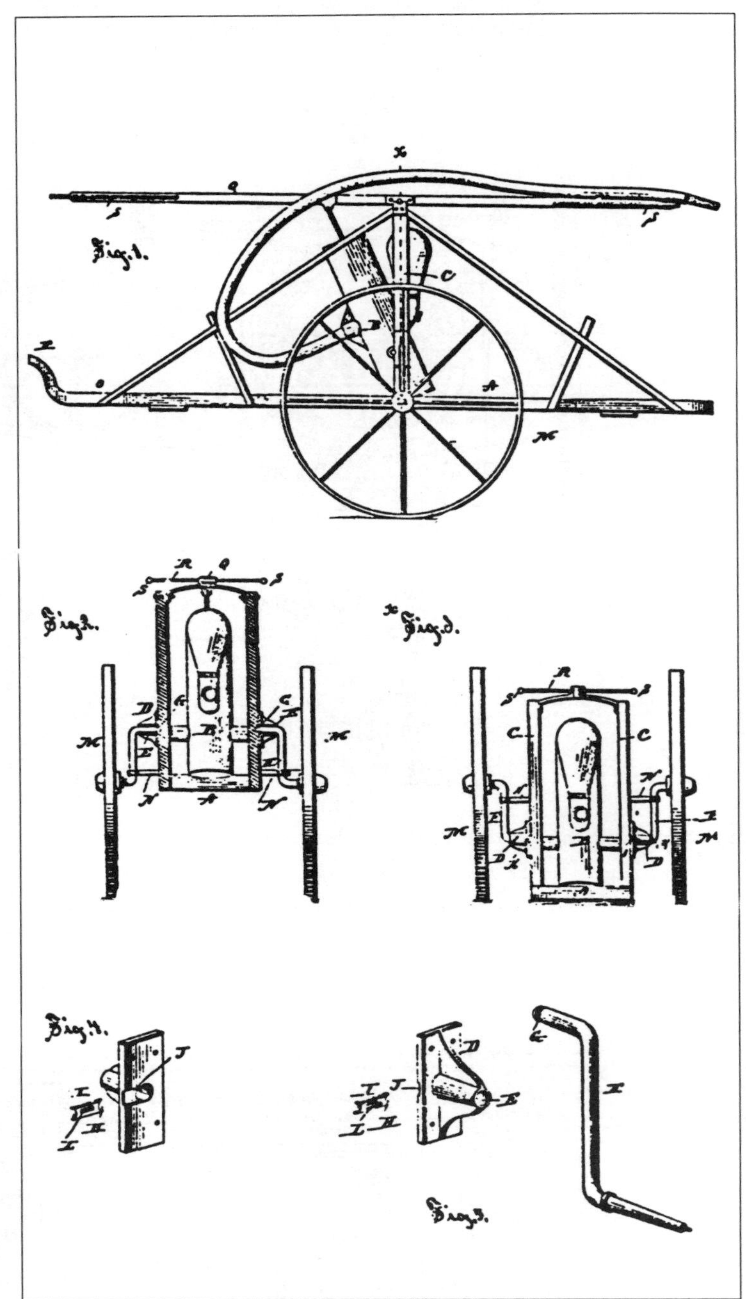

FIGURE 28

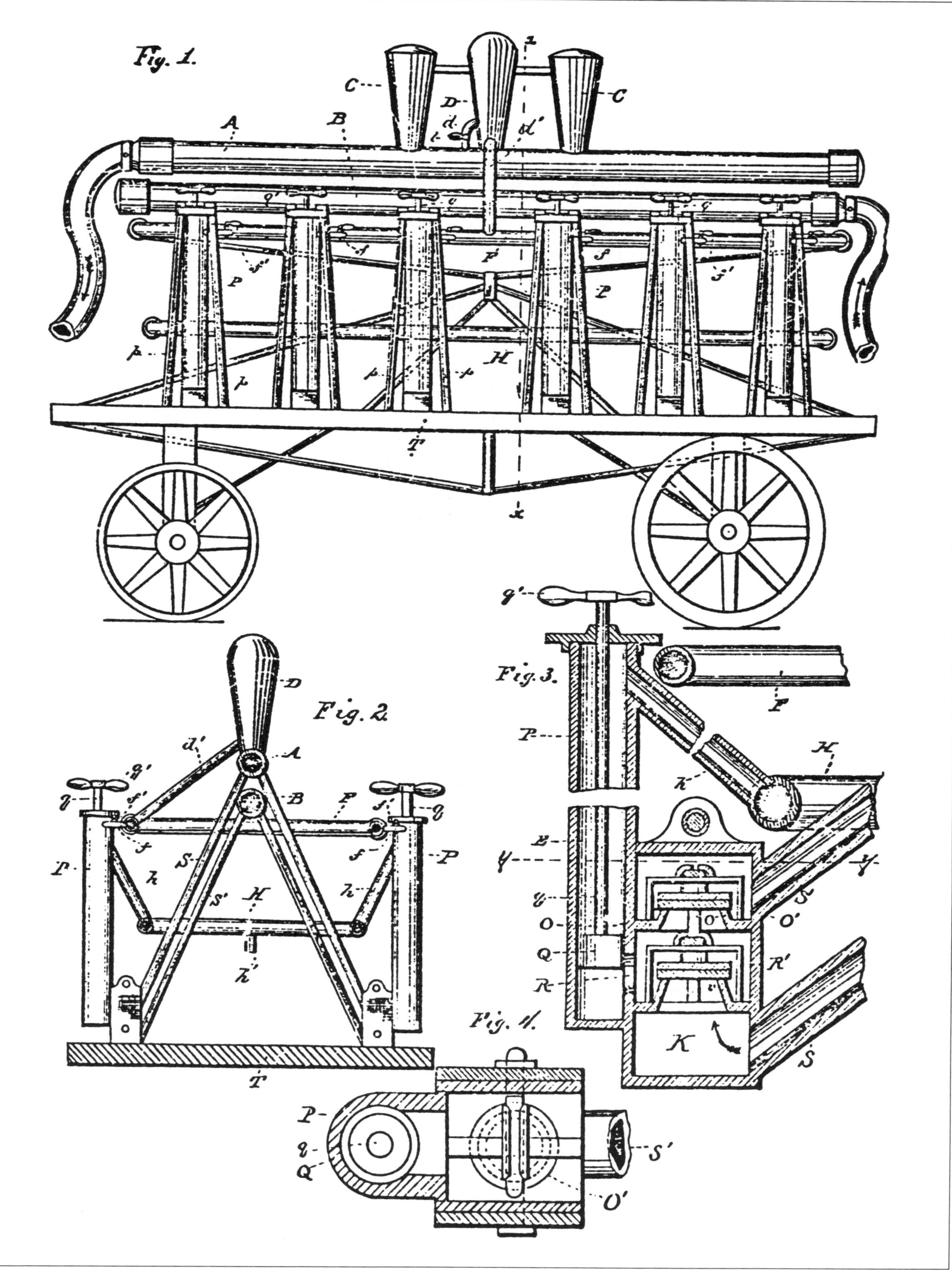

FIGURE 29

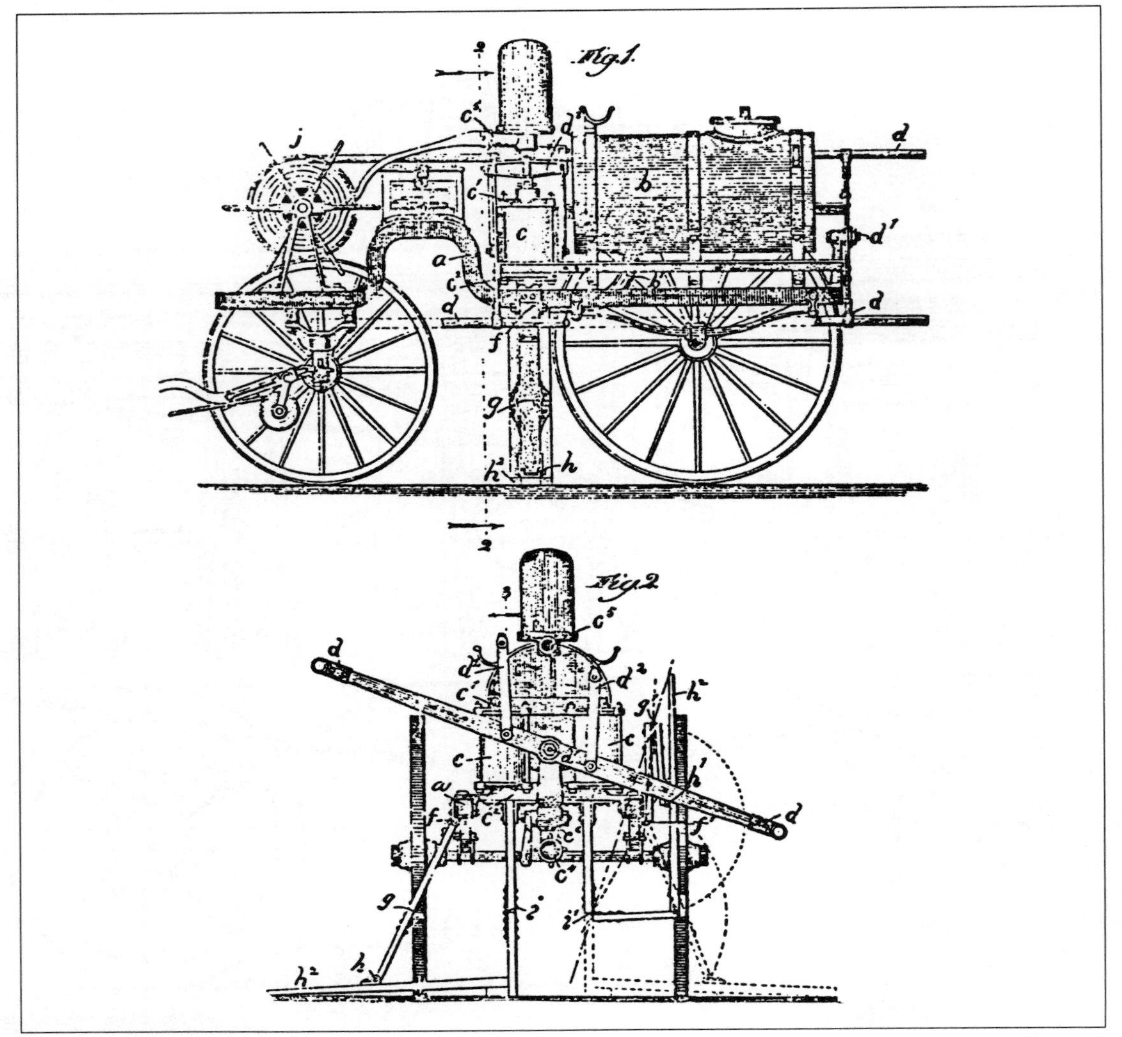

FIGURE 30

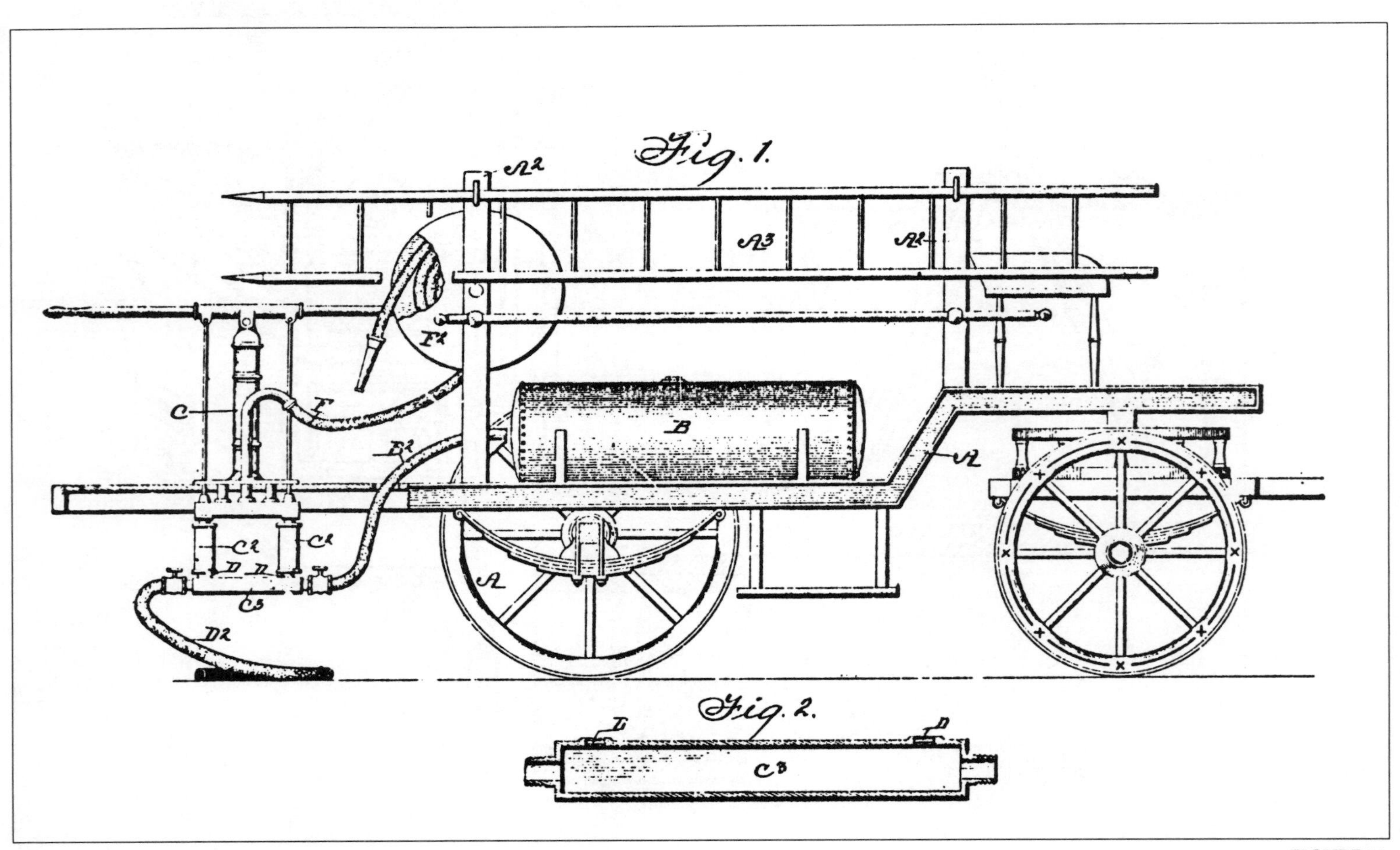

FIGURE 31

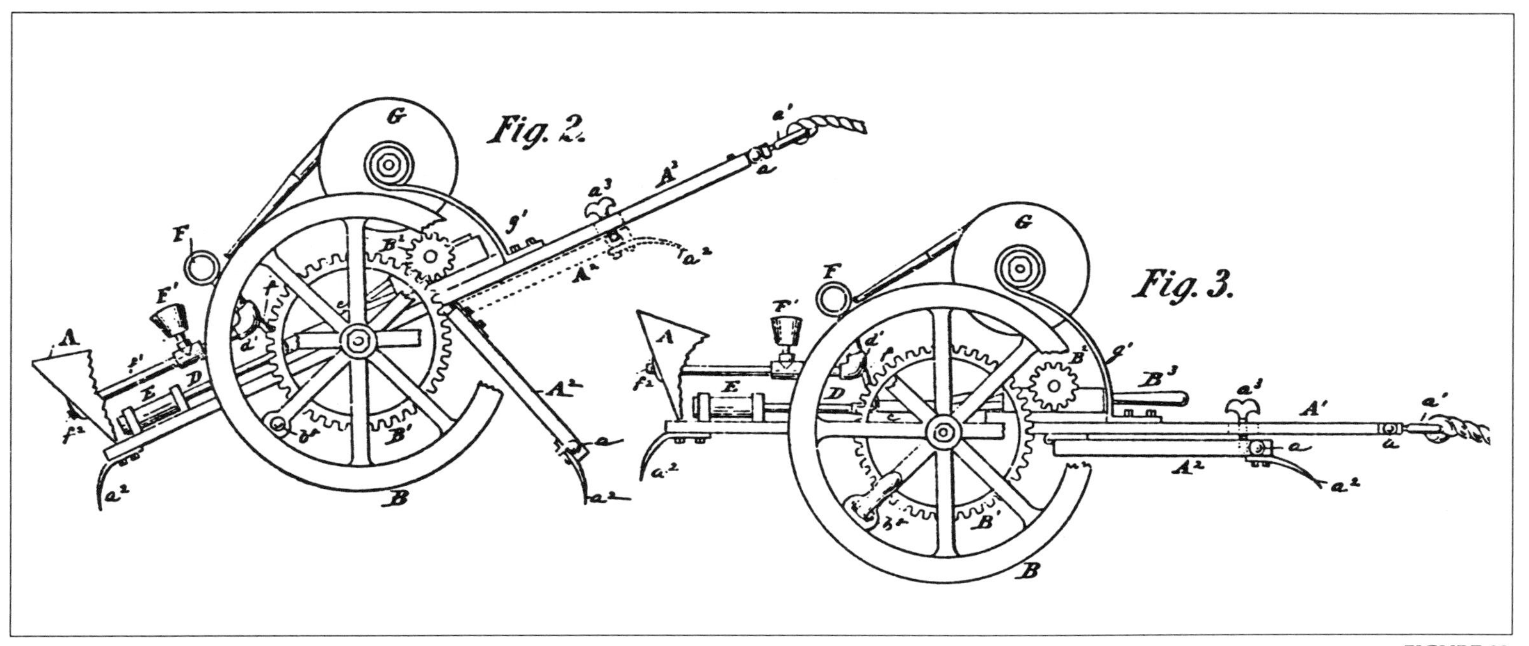

FIGURE 32

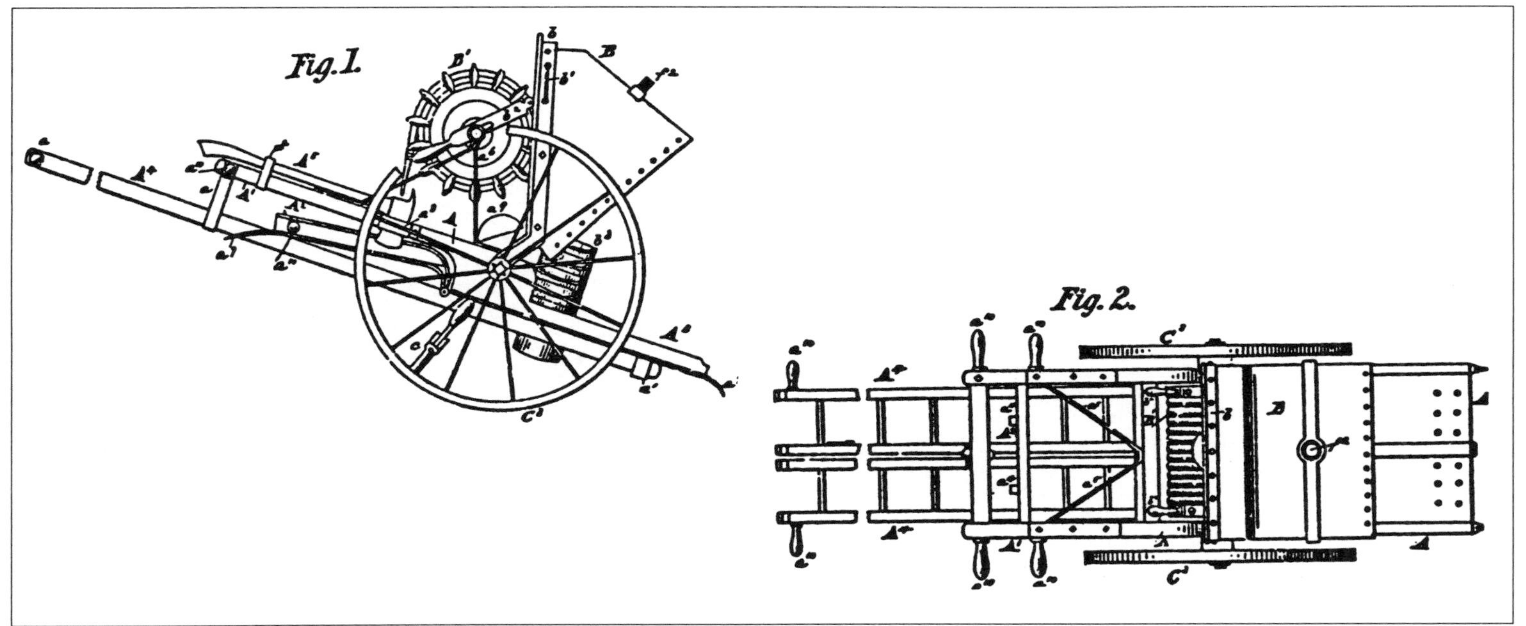

FIGURE 33

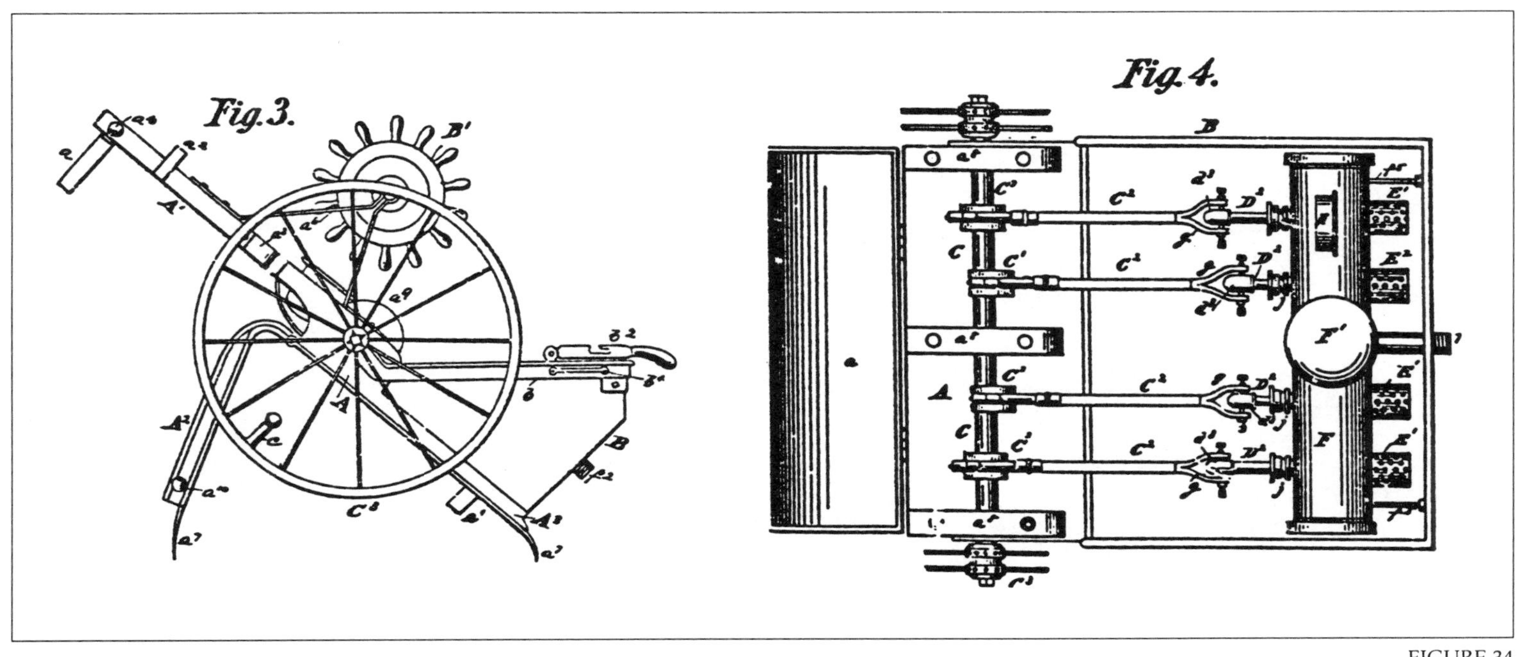

FIGURE 34

Hand-Powered Pumper Photo Album

In bringing to a close this discussion of hand-powered fire pumpers, an album of photographs of such machines is presented. FIGURE 35 is an end-stroke pump built by Philip Mason in 1795 and rebuilt in 1856 by John Agnew. It is owned by the Clayton Volunteer Fire Company of Clayton, New Jersey. Two well-preserved antique rigs owned by the Singerly Fire Company of Elkton, Maryland are seen in FIGURES 36 and 37. The "Hydraulion" built in 1818 was probably constructed by Richard Sellers of Philadelphia. This was a combination pumper and hose wagon, and it was not a successful apparatus. The second Singerly hand pumper was built in 1859 by John Rogers. Shown mounted on a trailer for parade purposes, FIGURE 38, is an 1855 side-stroke pumper built by the Rochester, New York firm of Wright Brothers, and owned by the Penn Yan Fire Department, whose members rebuilt and reconditioned it. Sometimes these invaluable antique fire engines are transported within specially modified or even custom-built trailers. FIGURE 39 is just such an example, and the vehicle thus carried is seen in FIGURE 40. (The patent drawing for this Howe hand pumper was shown as FIGURE 30.) An 1892 Rumsey side-stroke hand pumper, property of the Rescue Fire Company of Dallastown, Pennsylvania is illustrated in FIGURE 41. And in FIGURE 42 there is shown a very peculiar hybrid: a hand-powered pump mounted on a gasoline-powered vehicle. This rig, from Jacobus, Pennsylvania, is housed at the Fire Museum of York County, located in the old Royal Fire Company hall in the city of York, Pennsylvania. It is a 1919 Ford Model T with body by American-LaFrance, equipped with a Gould hand pump.

FIGURE 36

FIGURE 37

FIGURE 35

FIGURE 38

FIGURE 39

FIGURE 40

FIGURE 42

FIGURE 41

Fire Pumps Powered By Animals

INTRODUCTION

In many of the older cities of the United States the fire stations were located in accordance with certain criteria. These would of course include such factors as the value of the surrounding properties and buildings (and no doubt sometimes the greed of public officials anxious to unload some real estate for hard cash). But fire chiefs also preferred that stations be situated upon the highest ground available so that the horses could run downhill when responding to an alarm. And the intervals between fire stations were often set by the distance that a team of straining, running horses could haul a heavy steam fire engine.

Out in the countryside, however, in the rural areas which formerly encompassed so much of this land, a steamer often provided un-needed pumping capacity at unacceptable cost. Horses there were, and in abundance, and there was no scarcity of inventors ready to design horse-powered fire engines. So patents were granted for fire pumps powered by animals, meaning as a rule horses, although oxen were also used. These animals were thus required to pull the relatively light apparatus to fires, and upon arrival they were hitched up to the pump.

McCarthy's Invention

FIGURE 43 is an engraving which was used to illustrate a Scientific American magazine article in the issue of December 25, 1869, a time which seemed to mark the beginning of a twenty-year period in which animals were so employed in firefighting. This artist's conception illustrated John C. McCarthy's November 2, 1869 patent. In this invention the horses had to walk upon a treadmill, with the power they generated being transmitted to a force pump by crank motion actuated by a pair of bevel gears. It was said by the inventor that the weight of the entire apparatus, including treadmill, gears, pump, hose and reel, suction hose, and the wagon itself, was less than 2,500 pounds.

McCARTHY'S HORSE-POWER ENGINE FOR EXTINGUISHING FIRES.

FIGURE 43

In the lengthy article about this fire engine the SA reporter made note that, "For rural towns this engine possesses many advantages, coming, as it does, between the hand engine and the expensive steam fire engine. It is light enough to be rapidly drawn to a fire, and the cost of fuel is saved. Its cost is much less than a steam engine, and its efficiency is greater than that of a hand engine, as the number of horses is not limited to two. One of the defects of hand engines now avoided is the generally admitted demoralizing tendency of volunteer fire company organizations upon the youth who, for the most part, compose them. The machine may be placed in charge of some responsible person in small towns, and when required, two or three men may effectually operate it. Where the water has to be raised only a short distance through the suction pipe it is claimed that two horses will throw a three-quarter inch stream to a height of seventy feet, through two hundred feet of hose."

"We think this machine peculiarly adapted to the wants of far-western towns. In such cases it might be placed in the care of the postmaster, merchant, or other responsible party centrally located, and would be an important safeguard against those disastrous conflagrations which have so frequently ravaged our border settlements."

Underwood's Track

Louisville, Kentucky resident William T. Underwood was another inventor who yearned to design an ideal, cheap, and efficient fire engine for towns and villages. FIGURE 44 is the drawing for his patent, number 209,202 dated October 22, 1878, a pumper which could be pulled as well as operated by oxen or horses. These beasts were required to walk upon the circular track shown in the illustration. It could be folded up into segments for transport and, of course, deployed on the ground for use. According to inventor Underwood, "The track and carriage were fastened together, firmly, so that one could not be moved without moving the other. The feet of the animals would tend to drive the track in one direction, while the force of their draft would move the carriage in the opposite direction. Therefore the two forces would counteract one another, and the whole would remain stationary. This is the great value of the portable track." Water was moved by the action of a pair of simple piston pumps, actuated by vertical cog wheels which were turned by a horizontal wheel hitched to the animals by long levers.

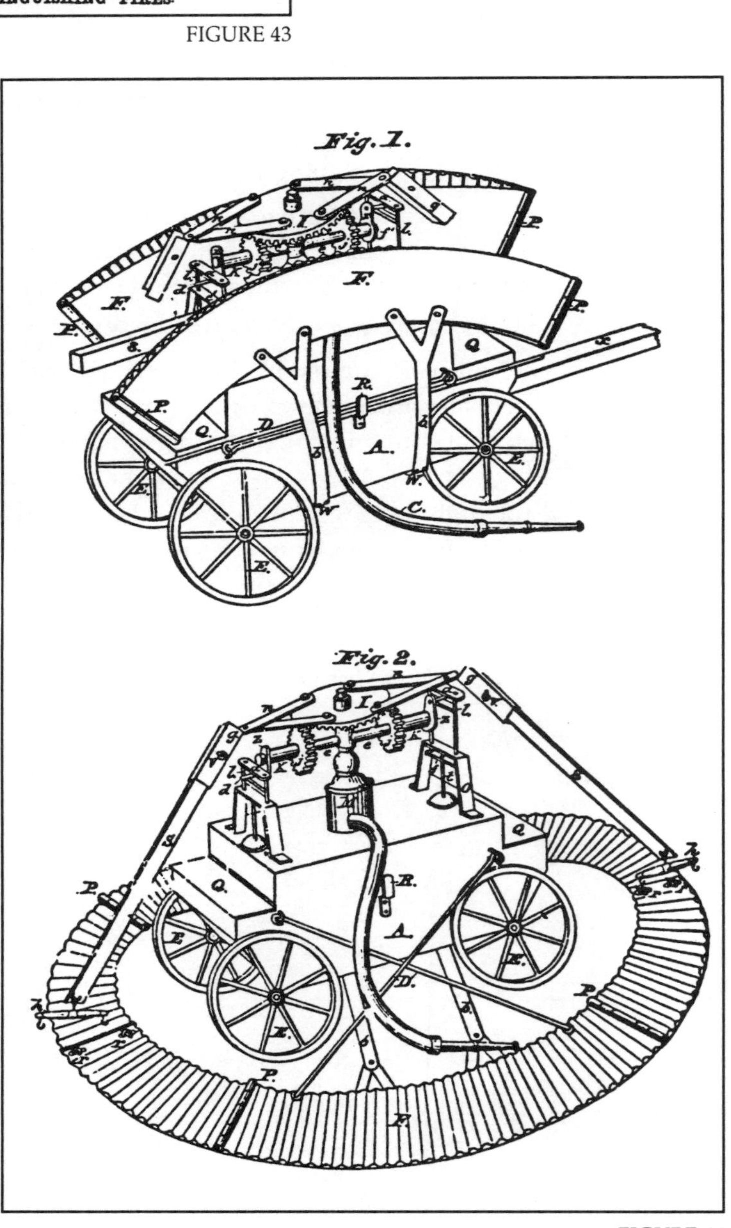

FIGURE 44

B.J.C. Howe

The "master inventor," so to speak, of animal-powered fire pumpers during the decade of the 1880s was Benjamin J.C. Howe, who we have already met in these pages. FIGURES 45 and 46 which represent, respectively, a top view and a side view of a fire engine, were illustrations for his patent number 244,131, which he received on July 12, 1881. This document must have been, from Howe's point of view, a kind of basic patent, as he made reference to it, subsequently, several times. Essentially it was a simple device, fitted with three piston pump chambers. Most of the vital working parts are clearly shown.

On August 1, 1882 Howe received two additional patents, numbers 262,195 and 262,196. These inventions are illustrated by FIGURES 47, 48, 49, 50, and 51. Notice that the trio of piston pumps were now set directly alongside of one another, and that a "fire chamber" for heating the water chambers had also been built into the apparatus. The major design change for the second of these patents was the substitution of rotary pumps for the usual piston pumps.

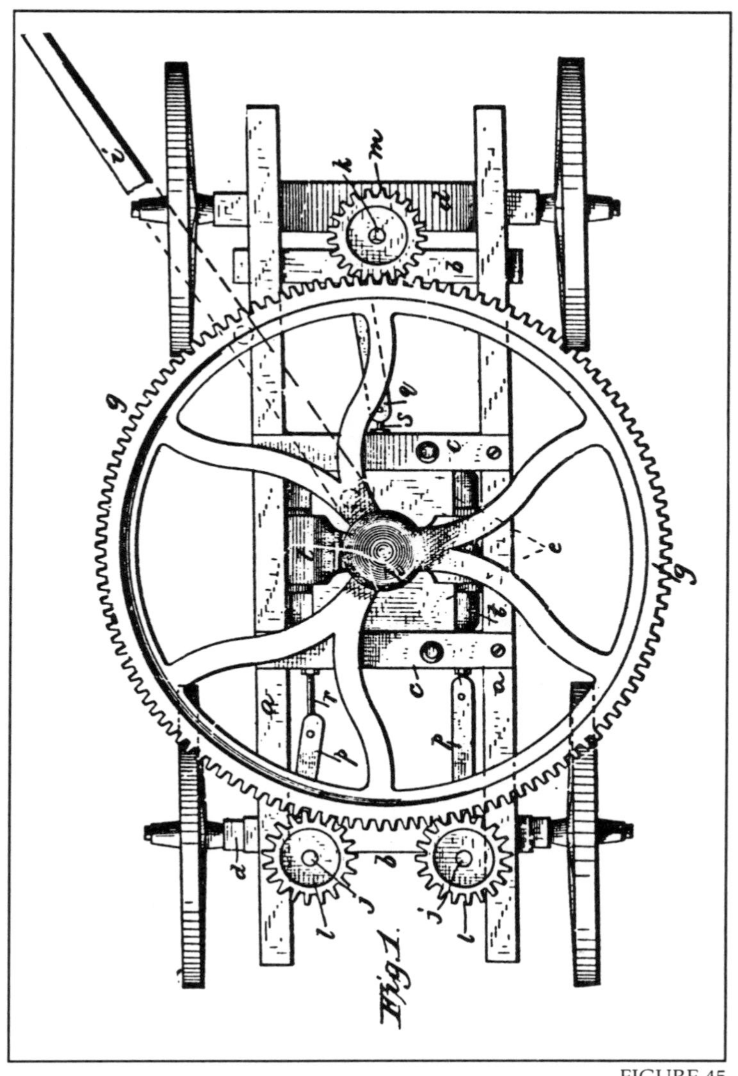

FIGURE 45

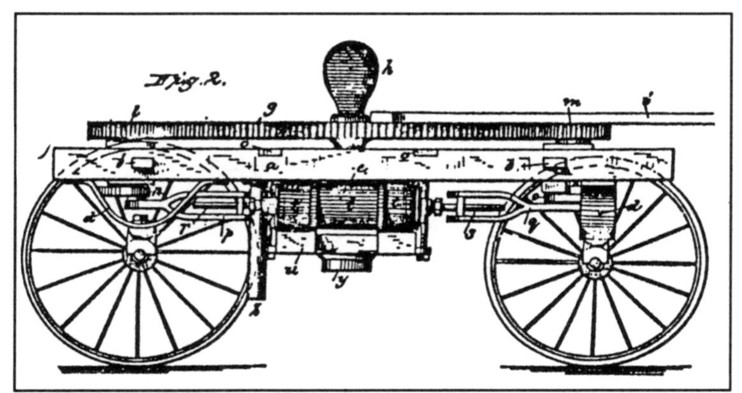

FIGURE 46

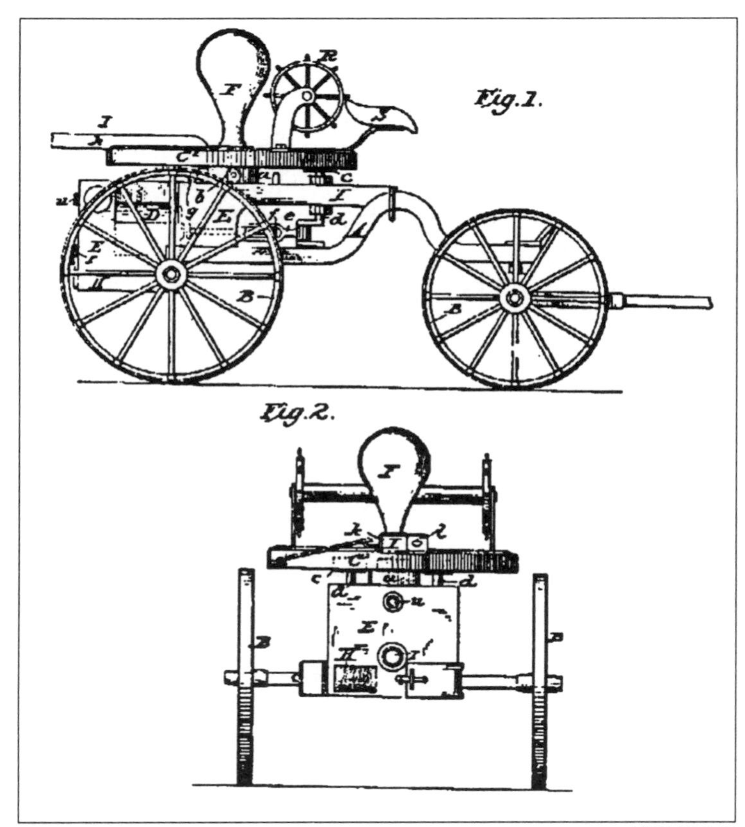

FIGURE 47

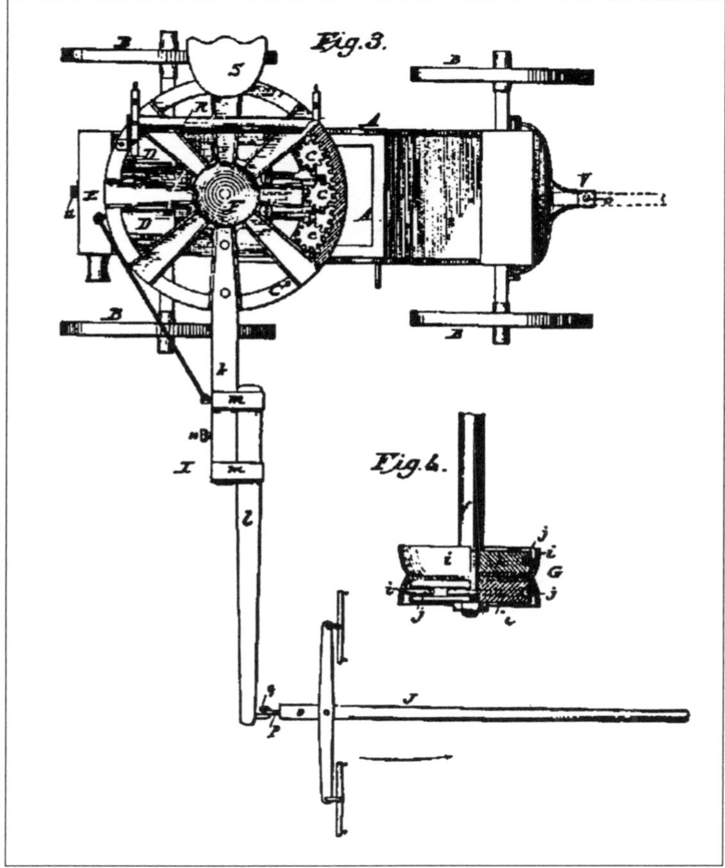

FIGURE 48

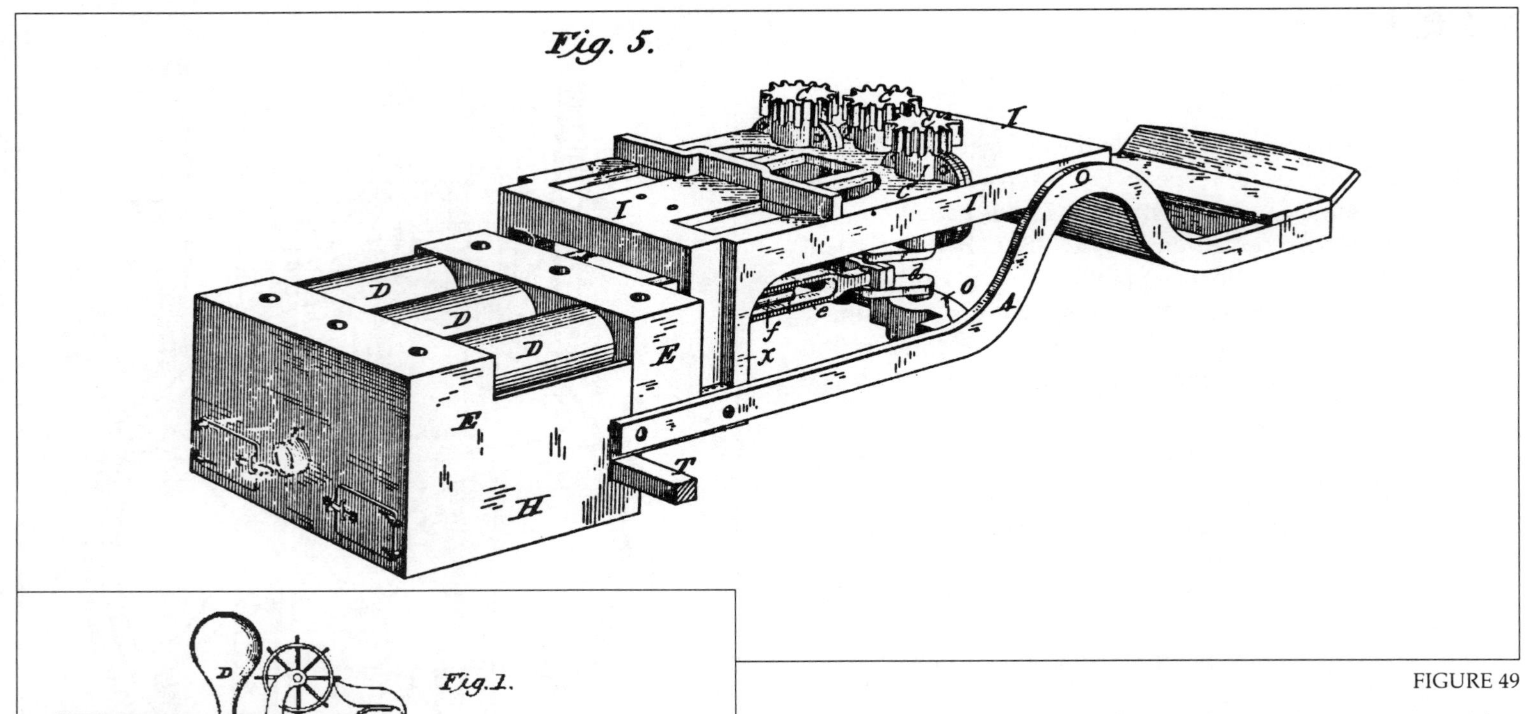

FIGURE 49

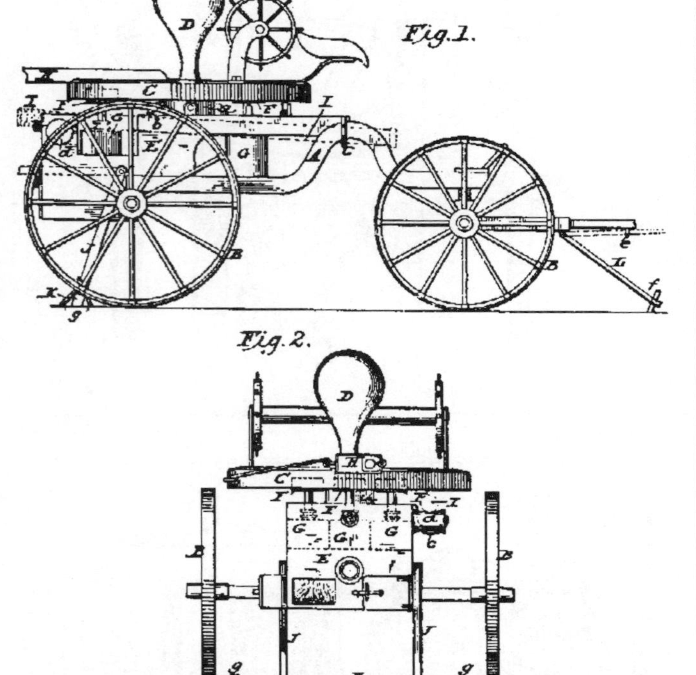

FIGURE 50

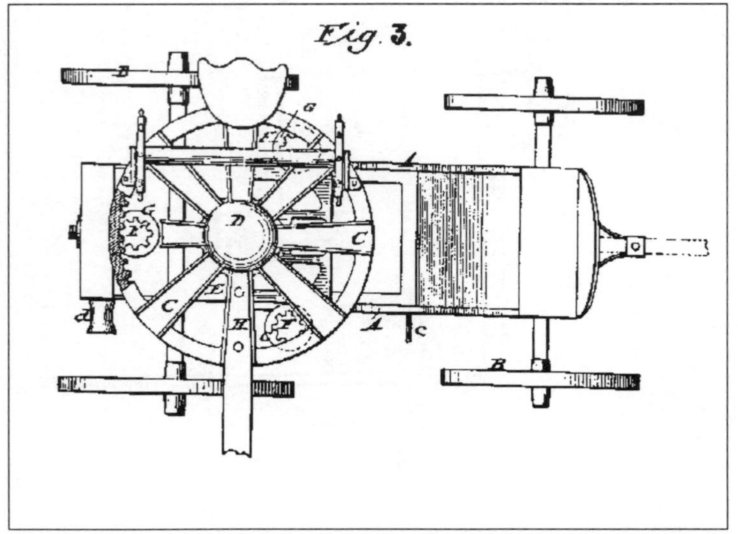

FIGURE 51

One of the builders and sellers of Howe horse-powered fire engines was the Remington Agricultural Company, whose factory was located near Utica, New York in the town of Ilion. One of these rigs was illustrated (FIGURE 52) and described in the March 15, 1884 issue of Scientific American, and the following lines have been exerpted from that article:

> "It is made entirely of metal except for the levers, and it weighs about 3,000 pounds. The driving wheel is placed horizontally on top and is furnished with sockets for eight levers, all of which can be used when the engine is worked by men. Only two levers are used when horses are available. There are three double acting pumps, driven by pinions which mesh into teeth on the driving wheel, and so arranged that their combined action produces a continuous pressure and even flow of water. The pump cylinders are five and three-quarter inches in diameter, and the stroke is eight inches. The capacity is 200 gallons per minute. The horses travel in a convenient circle at the ordinary walk of a work team, about the same as in plowing. During operations the engine is held in position by iron braces on each side, fastened to the street with steel pins. When operated by a pair of horses, as described, a horizontal stream can be forced to 160 feet through a seven-eighth inch nozzle."

This fire engine is also illustrated in FIGURE 53, which is from the 1884 catalogue of the Fred J. Miller Fire Apparatus and Fire Department Supplies Company.

Seemingly the last of this genre of fire engine invention was Matthew Morton's patent, number 356,778, which he received on February 1, 1887 (FIGURE 54). This must have been a rather lightweight rig, because, "The whole mechanism may be readily transported by hand, if desired, and if horses have not arrived a number of persons may jump upon the chain and operate the pumps."

THE REMINGTON HORSE POWER FIRE ENGINE.

FIGURE 52

FIGURE 53

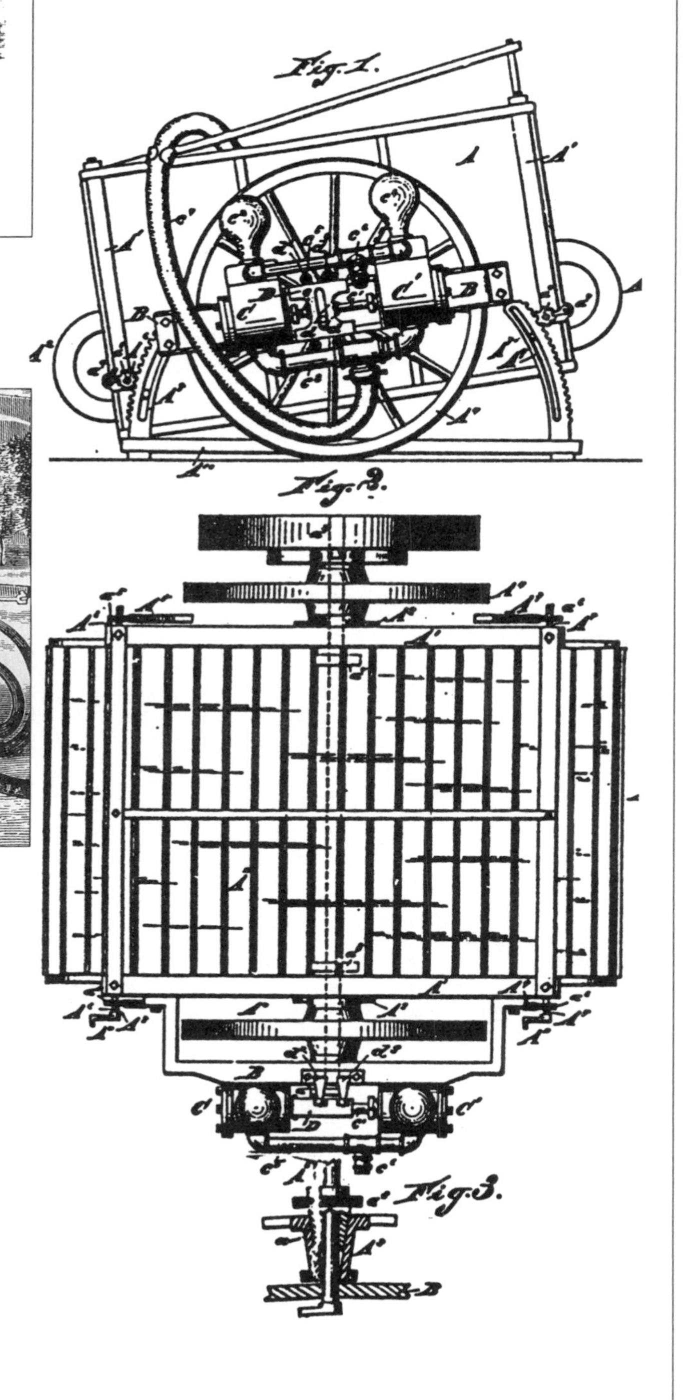

FIGURE 54

The Horsepower

And finally, just before closing this discussion about horse-powered pumps, the term, "horsepower," as it has come to be used deserves a few words with respect to its origin and definition. James Watt (1736-1819) deserves the credit for originating the concept of rating a steam engine in terms of the number of horses that it replaced, although he could not foresee that he was introducing a unit destined to become applied universally. And, as a matter of fact, for many years it was not generally known by engineers that the term "horsepower" had definite dimensions of pounds, feet, and minutes. In 1814 Watt wrote that, "Horses being the power generally employed to move the machinery in the great breweries and distilleries of the metropolis, where my engines first came into demand, the power of a mill-horse was considered to afford an obvious and concise standard of comparison, and one sufficiently definite for the purpose in view. A horse going at the rate of two and one-half miles an hour raises a weight of 150 pounds by a rope passing over a pulley, which is equal to raising 33,000 pounds one foot high in a minute. This was considered the horse's power."

FIRE PUMPS POWERED BY STEAM

INTRODUCTION

To this day there endures within Westminster Abbey, London, a statue of James Watt. The epitaph inscribed upon it expresses the tribute of mankind to Watt the man, the genius, and the inventor of the steam engine: "Not to perpetuate a Name, which must endure while the Peaceful Arts flourish, but to show that Mankind have learned to honor those who best deserve their gratitude, the King, his Ministers, and many of the Nobles and Commons of the Realm, raised this monument to James Watt, who directing the force of an Original Genius, early exercised in Philosophic Research to the improvement of the Steam Engine, enlarged the resources of his Country, increased the power of Man, and rose to eminent place among the most illustrious Followers of Science and real Benefactors of the World."

Steam fire engines were among the favored topics of *Scientific American* magazine reporters during the last half of the last century (FIGURES 55, 56, 57, and 58). And yet the entire era of steam fire engine manufacturing in the United States of America lasted only about 65 years, having its genesis in 1853 and its exodus about 1915. During these years there were 75 or more firms which built steamers, but there were only seven companies which actually built and sold 100 or more of them. These seven organizations were the Ahrens-Fox Fire Engine Company, the Ahrens Manufacturing Company, the Amoskeag Manufacturing Company, the Button Fire Engine Works, the Clapp and Jones Company, the LaFrance Manufacturing Company, and the Silsby Manufacturing Company. Altogether these seven builders produced more than 3,700 steam fire engines, or about 80 percent of all such apparatus sold in this country.

THE ADVOCATE OF INDUSTRY, AND JOURNAL OF SCIENTIFIC, MECHANICAL AND OTHER IMPROVEMENTS.

VOL. XIII. NEW YORK, MAY 29, 1858. NO. 38.

THE
SCIENTIFIC AMERICAN,
PUBLISHED WEEKLY
At No. 128 Fulton street, (Sun Buildings,) New York,
BY MUNN & CO.

O. D. MUNN, S. H. WALES, A. E. BEACH.

Responsible Agents may also be found in all the principal cities and towns in the United States.

Sampson Low, Son & Co., the American Booksellers, 47 Ludgate Hill, London, Eng., are the British Agents to receive subscriptions for the SCIENTIFIC AMERICAN.

Single copies of the paper are on sale at the office of publication and at all the periodical stores in this city, Brooklyn and Jersey City.

TERMS—Two Dollars per annum.—One Dollar in advance, and the remainder in six months.

☞ See Prospectus on last page. No Traveling Agents employed.

Steam Fire Engine.

There are very few persons in a community who have not, more or less, partaken of the intense excitement and sympathy always consequent upon fires taking place in their midst; and it is not strange, therefore, that a large amount of thought should be expended toward their prevention and extinction. In view of their frequent occurrence in this country, and the vast amount of property annually destroyed in this manner, it becomes the duty of every good citizen to exercise his mind and skill, and endeavor to avert and remedy this terrible evil, and to encourage the efforts of others to these ends. Of late years the attention of inventors has been directed to the application of steam to the suction and forcing of water in fire engines; and many powerful fire engines worked exclusively by this force are in successful operation in many of our western cities.

Our engraving represents a side elevation of a compact form of steam fire engine, manufactured by Silsby, Mynderse & Co., which is a modification of the plan and arrangement of the parts of the one previously manufactured by them, and which was illustrated in No. 10, Vol. XII, SCIENTIFIC AMERICAN.

A represents the steam boiler, provided with three hundred $1\frac{1}{4}$ inch upright tubes, and connected to the supply water tank, B, by a suitable pipe. C is the smoke pipe, in the lower part of which is placed a blower or fan, D, which receives its motion by a band passing around a wheel on its shaft, and around another wheel, secured to one of the hind wheels. E is a rotary engine, constructed on the plan of Holly's patent, as shown in our former illustration. F is a rotary pump, also constructed after the plan of Holly's patent. G is a take-off for a hose for conducting water to the fire to be extinguished. H is a $4\frac{1}{2}$ inch suction opening, for attachment of the suction pipe or hose usually carried with fire engines of this description. It can be used to draw water from cisterns, rivers, and other places, or attached to the ordinary hydrant. I is the pump for supplying the boiler with water, geared to and worked by the shaft on which the rotary engine, E, and rotary pump, F, are secured. J is a rotary donkey pump and engine combined, constructed and operating in every respect similar to the engine, E, and pump, F. This additional supply pump is for the purpose of supplying the boiler with water when the machine is not in operation, and cannot, therefore, receive a supply of water from the pump, I. K is the steam supply pipe. L is the exhaust steam pipe. M M are india rubber springs, on which the machine rests. N is the driver's seat. O is a tongue, to which the horses for drawing the machine are attached. This tongue is made to disconnect by means of a lever under the control of the driver. P is a steam whistle, Q a steam gage, and R the platform for firemen. S is the heater for feed or supply water. T is a safety valve, and U the throttle valve.

We think this a simple and convenient form of steam fire engine, and admirably adapted to the object for which it is designed. Its weight is from 4,500 to 5,000 pounds, with capacity to force two 1 inch streams 175 feet, or one $1\frac{1}{4}$ inch stream the same distance; and

SILSBY, MYNDERSE & CO.'S STEAM FIRE ENGINE

the pressure of steam required to produce these effects will only range from 40 to 60 pounds per square inch. A working pressure of steam can be generated in the boiler in from eight to ten minutes, and this can be maintained constantly to force one or two streams of water of the sizes mentioned. Its main working parts are constructed under Birdsill Holly's patent, issued in 1855, by Messrs. Silsby, Mynderse & Co., Island Works, Seneca Falls, N. Y., who will furnish any additional information desired.

HOADLEY'S FEED-WATER HEATER FOR BOILERS.

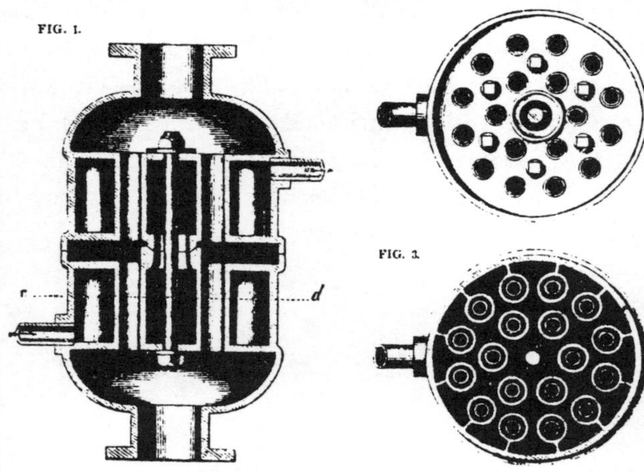

It is a well-known fact that a great saving is effected by conducting the exhaust steam of a steam engine through a feed-water heater after performing its function in the cylinder, and thus imparting through this agency a great portion of its heat to the cold water before it enters the boiler; and hence its follows that a heater, to be thoroughly effective, should be so constructed as to present as much surface as possible to the action of the steam, and contact of the supply water.

The accompanying engravings show an improved honey-comb heater, which accomplishes this desideratum in a marked degree, in which Fig. 1 represents a vertical section, Fig. 2 a top view of the under section, and Fig. 3 a horizontal section of the same at the dotted line, *c d*, of Fig. 1.

This heater receives its name on account of its resemblance to a honey-comb, and consists of two cylindrical sections containing a series of tubes cast with them, and communicating above and below, with spaces at the center part, where they are secured together by a screw bolt at the upper and lower ends of the heater, the spaces at these last mentioned parts being provided with nozzles for the entrance and exit of the exhaust steam to and therefrom. These tubes and spaces cause the steam to be displayed and brought in contact with a large area of heating surface, which receives a great portion of its heat, and in turn constantly imparts it to the supply water passing through the several water spaces between and around, and above and below the upright tubes and spaces, as it ascends after entering the lower part of the heater, through the horizontal pipe, and before it is discharged from the upper end through the corresponding pipe conducting it to the boiler in a heated state.

This multitubular heater differs from all others in use in being formed as described in the claim upon which the patent was issued, "with the tubes, tube sheet or heads, and case, all of one piece of metal, without joints uniting these parts." It is at once cheap, durable and efficient, characteristics which commend it to all owners of steam engines.

Further information may be had by addressing J. C. Hoadley, agent, Lawrence, Mass.

FIGURE 55

A JOURNAL OF PRACTICAL INFORMATION IN ART, SCIENCE, MECHANICS, AGRICULTURE, CHEMISTRY, AND MANUFACTURES.

VOL. II.—No. 9. NEW YORK, FEBRUARY 25, 1860. NEW SERIES.

IMPROVED STEAM FIRE ENGINE.

The superiority of a rotary motion for a steam engine is so manifest that it is not strange that many attempts have been made to overcome the practical difficulties to which it is subject. One of these difficulties—indeed, the principal one—has been the packing of the part which performs the office of the piston in the straight cylinder. Robert Stephenson expressed the opinion a few years ago, that a rotary engine would never be made to work profitably on account of the difficulty of packing. For our own part, though we have cautioned our readers that the field had been gone over many times by men of rare genius, our experience has so impressed us with the fertility of resource among our inventors that we have always entertained a lingering hope of seeing the defects in the rotary engine all removed, and its great advantages rendered available. The most palpable of these advantages are the reduction in the size of the engine in proportion to the power (resulting from the greater velocity of the piston) avoidance of the tremendous strain, especially in propelling ships, and finally a great saving of power which is expended in the reciprocating engine in overcoming inertia, in changing the direction of the motions. This last subject is forcing itself more and more upon the attention of mechanicians, and its importance is not yet by any means fully appreciated. These advantages adapt the rotary engine in an especial manner to the driving of a locomotive fire engine; and when our readers are told that this application has been made, that all difficulties have been surmounted, and that the packing, after 18 months' trial, has been found perfectly tight, they will sympathize with us in our interest in these statements. They are made by Silsby, Mynderse & Co., in relation to the engines which they manufacture on the plan invented by Birdsill Holly, and patented in 1855. We have already given two engravings of engines made on this plan, but the important modifications which have been made in the mode of constructing them, combined with the great interest felt both in rotary engines and in steam fire engines, induces us to give this third illustration in order to keep our readers informed in regard to the progress of improvement in both these machines.

Fig. 1 is a view of the whole machine, E being the boiler, B the engine, and C the pump. The construction of this engine is illustrated in Fig. 2. The steam enters at A, and passes out at F, turning the two revolvers, c and d, in its passage. The sides of these revolvers are packed as represented, by the blocks of metal inserted into the grooves and pressed out by the elastic springs. The ends of the revolvers are ground to the ends of the cylinders in which they turn; and we are assured that, after eighteen months' use, these ends still fit absolutely steam-tight. The pump is made precisely like the engine; the revolvers, being carried around by gears upon the outside of the cylinder, are worn very little where they run together.

The principal improvement in this fire engine, made since our last illustration, is in the boiler, of which Fig. 3 is a vertical, and Fig. 4 a horizontal section. The fireplace is represented at M, Fig. 4, with the vertical water-pipes, *i i i*, passing directly through the fire. These pipes are closed at the bottom, and open at the top where they pass through a water-tight plate, *g*, and communicate with the water in the boiler which rises to the level of *f*. They are represented in section at the sides, *k k* being the external, and *l l* the internal tubes, both open at the top, and the internal tubes having openings at the bottom. This arrangement causes a constant current, the water rising on the outside of the tube, *l*, as it is heated, and its place being supplied by a current flowing downward through the tube from the boiler. The smoke and flame pass among the tubes, *i i i*, and up through the flues, *h h h*, which are represented in section at the sides, *j j*. It seems to be now the pretty general opinion that steam can be generated more rapidly in vertical tubes than by any other plan yet tried.

SILSBY, MYNDERSE & CO.'S STEAM FIRE ENGINE.

The inventor of this improvement in boilers is M. R. Clapp, who has assigned his interest to Messrs. Silsby, Mynderse & Co., of Seneca Falls, N. Y., to whom persons desiring further information in relation to these boilers or engines will please address.

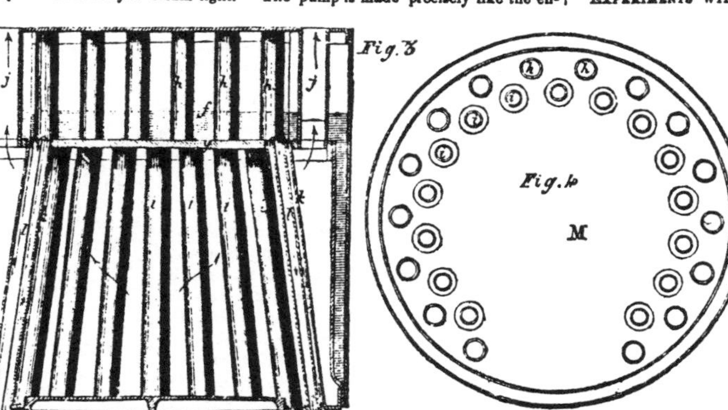

EXPERIMENTS WITH WATER WHEELS AT PHILADELPHIA.—As we have had many inquiries regarding the experiments with water wheels at Philadelphia, under the supervision of Chief-engineer Birkenbine, we would state, for the general information of all, that they are not yet quite finished. Two or three wheels have yet to be tested, but it is expected that these operations will be completed this month, and that some time during the month of March a report on the entire subject will be made out. We shall endeavor to present the same to our readers at as early a date as possible. We know that a very great interest is felt in the subject by our hydraulic engineers, mill-wrights, and mill-owners, becaus it is conceded by all those who have been witnesses of the experiments that they are conducted in a most fair and trustworthy manner.

FIGURE 56

A JOURNAL OF PRACTICAL INFORMATION IN ART, SCIENCE, MECHANICS, AGRICULTURE, CHEMISTRY, AND MANUFACTURES.

Vol. II.—No. 15. NEW YORK, APRIL 7, 1860. New Series.

A LIGHT STEAM FIRE-ENGINE.

We present herewith an illustration of the steam fire-engines built by Messrs. Lee & Larned, of this city, at the Novelty Iron Works. These engines are fitted to be drawn by hand, being intended especially for the use of engine and hose companies; so that villages and small cities may now avail themselves of the superior and untiring power of steam, for fire-engine purposes, with no change in existing organizations, and without the expense of a horse establishment. The engine from which the view is taken has been on duty for several months, in the hands of the Valley Forge Hose Company, stationed in Thirty-seventh-street, this city, and it has rendered signal service on several occasions. It is about 10 feet in length, exclusive of the pole, and weighs 3,700 pounds; which weight, we understand, will be reduced at least 200 pounds, in engines of the same style to be hereafter built. Having large wheels and sensitive springs, it runs as easily as an ordinary fire-engine of 500 or 600 pounds less weight, and easier than the average of first-class hand engines. Its best single stream, for distance, is one inch diameter; for quantity, 1⅛; but for ordinary fire duty, it will handle, with good effect, *two* one-inch streams, drawing its own water. This it did, for ten consecutive hours at the fire on the ship *John J. Boyd*, in January last.

The steam power is derived from one of Lee & Larned's patent annular boilers, of 125 feet of heating surface, with which steam can be raised to working pressure in from six to eight minutes. The pump, which is of brass, and highly finished, is Cary's patent rotary, driven by a single reciprocating engine, of 7 inches bore and 8½ inches stroke, with a pair of light balance wheels to carry it over the centers. It is intended to make from 200 to 400 revolutions per minute. A flange-disk, cast on the pump shell, makes one of the heads of the steam cylinder; the two, thus combined, forming a steam pump, of novel form and unequaled simplicity and compactness; occupying, indeed, so small a space (only 27 inches in length) that they are hardly seen in the engraving. The piston rod, passing out through the opposite head, acts on a cross-head of such length as to allow a connecting rod from each end of it to pass the cylinder and take hold of cranks on the pump shaft. The valve movement is obtained by means of a rockshaft, actuated by an eccentric rod from the main shaft. The boiler is supplied from an independent feed pump, but has also a connection with the main pump, which may be used at pleasure. The carriage frame is, in front, simply a horizontal bed plate of iron, of less than a foot in breadth, expanding, behind, into a ring, to the inside of which is bolted an upright open cylinder of thin, but stiff, sheet iron, strengthened at the bottom by an angle-iron ring, the whole forming at once a seat and a casing for the boiler, which is placed within it. This end of the bed or frame is hung on platform springs, arranged like those of an omnibus, by means of tension rods and braces, taking hold of the angle-iron ring. The center of weight is directly over the hinder axle, which opens into a hoop allowing the boiler to hang within it. The springs are plates of steel, one or more to each, of uniform thickness, but tapering in width from the middle towards either end. In front, two springs of this form are used, placed one above the other, in line with and directly under the bed, receiving the weight of the machinery at the middle or widest part. These serve the two-fold purpose of spring and reach, taking hold in front, by means of forked ends, on swivel-boxes at each end of a short vertical shaft, forming a universal joint with the front axle; giving thus a single point of front suspension, annihilating the tendency of the bed to wring and twist under its load in traveling over rough roads, saving all the weight of metal needed under the ordinary arrangement to counteract that tendency and secure the necessary stiffness, protecting the machinery perfectly against the concussions of travel, and dispensing with the complication and friction of a fifth wheel.

These engines are built of several different sizes, the one we have described being the smallest. The next size larger, weighing 5,200 pounds, is also a hand engine (though either can be fitted to be drawn by a horse or horses, if required), and being of proportionally greater power, it is to be preferred where the condition of the streets is favorable, in respect to surface and grades, and the company is strong enough in numbers to manage it. This engine has thrown a 1¼-inch stream 260 feet, a 1⅜-inch 228 feet, and for fire duty not unfrequently plays a 1¼-inch stream with great effect. The *Manhattan* engine, which, in the hands of Manhattan Company, No. VIII., of this city, has done such admirable service at the severe fires of the past winter, and has been, according to the estimate of competent authorities, the means of saving property to the amount of at least a hundred times its cost, is of this size.

For any further information address Lee & Larned, at the Novelty Iron Works, this city.

LEE & LARNED'S STEAM FIRE-ENGINE.

VELOCITY OF ELECTRICITY.

Messrs. Guillemin and Burnouf have been making numerous experiments on the transmission of electricity by telegraphic wires, with a view of discovering some law which governs this transmission. They conclude from their researches that the electric fluid is not propagated like the waves or undulations of light, and that it has not a constant and uniform velocity. They find it necessary to fall back upon the idea of Ohm, expressed in 1827, that electricity is propagated through wires, in virtue of the same kind of laws which govern the propagation of heat in a metallic bar. To determine experimentally which of these two opinions ought to prevail—that is, whether electricity is propagated with a constant and uniform velocity, or whether it is transmitted like heat—the authors disposed an apparatus, showing the intensity of the electric current in a certain point of a conducting wire, at different instants of its propagation. The first or the second opinion would then be justified, according as the current acquired suddenly in this point its definite intensity, or arrived at this intensity gradually. The authors found that the current at the point in question began with a very feeble intensity (the galvano-meter marking 0° 50′) which augmented gradually, and soon attained a maximum which it did not surpass, however long the contact of the pile with the conducting wire was continued. This maximum or permanent state was obtained in 0·024 of a second of time (the galvanometer then marking 19° 50′) in four lines of different lengths. The experiments were made during very fine weather, from 10 to 12 o'clock at night, from the 4th to the 6th of October, on a telegraph circuit of 104 leagues in length, passing from Nancy to Strasbourg, Mulhouse and Vesul, back to Nancy.

Poisoned Perfumes.—A Paris correspondent of the *New York Daily Times* says that the police of Paris have been for some months engaged in the examination of a variety of falsifications, and among the rest that of perfumery. Several actresses have been suffering from the effects of poison absorbed from the face, without suspecting that their sufferings came from this source. The quantity of corrosive sublimate, arsenic, verdigris, vitriol, and other poisonous substances daily absorbed in Paris must in effect be immense, and the reform did not commence too soon. The investigation was instigated by an actress of the Varieties Theater against a perfumer for damages for indisposition attributed to his cosmetics. At the same time the Academy of Sciences is occupying itself with the question of lucifer matches, and the reform necessary in their mode of fabrication. Several deaths from poisoning in the use of these agents directed this learned body to the subject.

Our thanks are hereby tendered to the Hon. John Cochrane, member of Congress from this city, for a full set of the Patent Office Reports for 1858.

FIGURE 57

A JOURNAL OF PRACTICAL INFORMATION IN ART, SCIENCE, MECHANICS, AGRICULTURE, CHEMISTRY, AND MANUFACTURES.

VOL. II.—NO. 26. NEW YORK, JUNE 23, 1860. NEW SERIES.

STEAM FIRE-ENGINE FOR RUSSIA.

Owing to the great number of wooden structures which were originally erected in the cities and villages of our country when it was new and timber so abundant, their combustible character naturally led to frequent and extensive conflagrations. To prevent and mitigate such evils, the energies of our people were aroused, and their natural mechanical genius was excited; and as a consequence, they became the inventors and builders of the most efficient hand fire-engines in the world. These were divided into several classes, and with some variations, generally consisted of two upright pumps, situated opposite one another, at the rear of a long water box placed on a carriage, and connected with a suction pipe behind and a discharge pipe before leading through an air-chamber to the hose and nozzle. The pumps had two valves—an inlet and discharging one. Some of these were furnished with springs, but the most common kind were simple flap valves. The suction valve opened into the cylinder, the discharge one opened outwards, thus forming a force pump. The two pumps were operated by a long horizontal lever or brake at each side, united by cross arms to an oscillating central shaft supported in bearings. The engine therefore was a large double force pump, by which a continuous stream of water was forced through the hose upon a fire by a row of men at each side working the brakes up and down as most of our readers have no doubt seen. Engines of this character, and of various capacities, are still made and used; but in cities and large towns, they are being rapidly superseded by steam fire-engines, in the construction of which some of our mechanical engineers have already obtained a world-wide celebrity. The annexed engraving represents the one which has been recently built for Messrs. Winans, Harrison & Winans, of St. Petersburg, Russia, by Messrs. Ettenger & Edmond, of Richmond, Va., and wherever it goes it will carry with it credit to the builders and to our country. The workmanship of it appears to be first-class as regards strength, beauty of finish, and efficiency. The boiler is a vertical-tubular, with an extensive heating surface, so as to generate steam rapidly. The entire machinery is secured on a strong four-wheeled spring truck; and the whole arrangement embraces great simplicity and compactness. Pressure and water gages, and every other device necessary to secure safety, convenience and efficiency, have been provided.

This engine was built from designs made by Mr. Alexander McCausland, and has proved itself equal in every respect if not superior to any engine of its dimensions heretofore built. The boiler, while running at an ordinary rate, is about twenty horse power; it is easily managed, and has the advantage of working very dry steam, not raising its water as most engines of the same class do. There is one 9-inch steam cylinder, with the steam chest beneath, and the valve is so arranged that, in case of any water working over from the boiler, it will work its own way out without having to open the cylinder cocks. The valve is worked by an eccentric on the fly-wheel shaft, in connection with a rock-arm, as on ordinary engines. The stroke of the engine is 15 inches. The pumps are of gun-metal, and set one above the other, and are reciprocating in their action, the cross-head of the engine being made in such a way that the piston rod from the steam cylinder is fastened in the center; the two pump pistons take hold, one above the other, below the steam piston, while the side or connecting rods take hold of the ends which project over the sides of the frame, and give motion to the fly-wheels. One of the pumps—the lower one—is cast solid in the vacuum chamber, so that, no matter how much the engine is jolted over the streets, the vacuum chamber will never leak. This arrangement of pump gives a chance for the valves (suction valves) to be placed between the pumps, so that the instant the engine changes its direction the water is taken off the valves, leaving them free to act without any dead water on them. A patent has been applied for, to secure this pump to the inventor of the arrangement. The pump valves are ordinary clack valves, and so arranged that one valve will fit in the place of another as well as it does in its own. The two pumps are each $8\frac{1}{4}$ inches in diameter, same stroke as the engine, and are equal in area to a 5-inch diameter pump. The engine is placed low down on straight axles, and cannot turn over, no matter how fast it may be going while turning corners; the body of the engine rests on six *semi-elliptic springs*, and rides very easy. It weighs (all complete with wood and water) 6,500 lbs. The back wheels are 4 feet 9 inches high, and the front ones 4 feet 6 inches, and it is easily managed with two horses. The engine is furnished with a lamp; also, with a horse and a man-tongue. The boiler is handsomely jacketed with Russia iron, with heavy brass bands; a fly-wheel is on each side of the air chamber, which is tall and handsomely shaped. This engine, altogether, was thought to be equal in looks to any yet made, and the Chief-engineer of the Philadelphia Fire Department, in speaking of it, makes use of the following language:—"I consider it, on account of its size, *one of the best I have ever seen perform*; having thrown water 250 feet, which distance was measured in my presence by George Eckfield, engineer of the United States Mint." At a trial in Richmond, this engine was drawn up to a fire-plug on Main-street, and the fire lighted; in 10 minutes, an abundance of steam to start was made. The steam was then turned on, and a 1-inch stream was thrown high above the eagle on the American Hotel, and then along Main-street, a distance of 240 feet. It was then taken to the canal, and, while raising its own water, threw a $1\frac{1}{8}$-inch stream 220 feet, a $1\frac{1}{8}$-inch stream 143 feet, and two $\frac{7}{8}$-inch streams 183 feet each. This trial was made when everything was perfectly new, and the boiler foaming from the grease or oil used in making it. It was pronounced by the Chief-engineer of the Richmond Fire Department worth the whole department put together at a fire. At a subsequent trial in Philadelphia, it threw the $1\frac{1}{8}$-inch stream 260 feet, which distance the builders and those present at the trial think cannot be beaten by any 5-inch diameter pump made; and the builders are willing to put their engine alongside of any other of the same capacity for a trial at any time. Steam is guaranteed in *eight minutes*, and the boiler to maintain any pressure required for hours at a time; in fact to blow-off at 100 lbs. pressure all the time.

ETTENGER & EDMOND'S STEAM FIRE-ENGINE.

It is somewhat surprising that the power of steam was not applied at an earlier date to operate fire-engines in our large cities, as it is only a very few years since the first successful one was built and put into practical use. To the city of Cincinnati does the credit of first introducing the steam fire-engine belong. But if our mechanics have been tardy in applying, and some of our cities rather conservative in adopting steam fire-engines, a spirit is now abroad to redeem our credit, and make amends for past neglects. In Cincinnati, where they were first adopted, no other fire-extinguishers are employed; and St. Louis, Chicago, Philadelphia, Baltimore and New York, are each furnished with several, and the time is not far distant when they will be used exclusively. Sinews of iron and steel never tire, and if the boiler is furnished with food and water, and the joints lubricated with oil, the fire-horse will obey the behests of his masters and spurt copious streams by day and night, and stream defiance at weariness and sleep. It is much to the credit of Messrs. Ettenger & Edmond, and it affords evidence of their abilities and facilities in building engines, that this one was completed in 70 days from the date when the first line of it was drawn.

FIGURE 58

Rating and Configuration of Steam Fire Engines

Steam fire engines were classified or rated by their builders according to their pumping capacity, using a nomenclature which stratified them by size. Pump output data and size parameters are shown, by way of example, in the following table, which describes the apparatus of four manufacturers:

Size	Ahrens	Amoskeag	LaFrance	Silsby
Double extra first	1,000	1,350	–	–
Extra first size	900	1,100	1,100	1,100
First size	800	900	900	950
Second size	700	700	750	800
Third size	600	550	650	700
Fourth size	500	350	550	600
Fifth size	400	–	450	500
Sixth size	300	–	375	400
(Pump output in gallons of water per minute)				

Troubled by such discrepancies in the rating scales for steam fire engines, the National Fire Protection Association finally set up a committee to study apparatus standards. The chairman, Fire Captain Greeley S. Curtis, presented this committee's report during the tenth NFPA Annual Meeting, held in Chicago that year, 1906, on May 22, 23, and 24.

According to Captain Curtis and his committee there was clearly a need for reforming the way in which fire engines were being rated: "Uniform standard ratings should be established for future engines and for those already in service. The present system whereby any manufacturer rates his engines as first, second or third size, or extra first and double extra first size without much regard to the action of his fellow manufacturers leads to constant confusion."

"The engines in a certain city were tested under like conditions by the chairman of the committee. In the list were one first size engine stationed in the valuable business district, and one fourth size engine of different make stationed in the suburbs. When tested for capacity the little fourth size engine exceeded the discharge of the first size engine by more than 200 gallons per minute, even with the handicap of being run by the crew of another engine."

"A few manufacturers have a habit of rating their engine at capacities which cannot be attained in actual practice, and as prices have hitherto been based largely on rated instead of actual capacity, injustice has been done both to the purchasing city and to competing manufacturers of full capacity engines."

"After a study of the engine sizes in use throughout the country, and after conferences with manufacturers, it seems advisable to limit the standard ratings for the present at least to three sizes, viz, 600 gallon, 800 gallon, and 1,000 gallon per minute engines. No smaller capacity than 600 gallons per minute is given consideration for the reason that smaller engines after a few years' service could not be relied upon with ordinary handling to furnish two efficient streams. The upper limit is set at 1,000 gallons per minute, as experience shows that but few engines now in service could meet the proposed requirements of this rating."

"The objectionable multiplicity of existing sizes is cut down to three. It is believed that city and town requirements can be as well filled by these three sizes as by the eight or ten hitherto in vogue." And the report then went on and presented specifically and in detail the methods for rating steam fire engines, the standard methods for testing steam fire engines, and the features which indicated efficiency of steam fire engines.

But, as the following tables shows, there were differences other than pumping capacity which could be used to describe these fire engines:

Size	Weight (lb.)		Length (ft.)		Width (ft.)		Height (ft.)	
	Amoskeag	LaFrance	Amoskeag	LaFrance	Amoskeag	LaFrance	Amoskeag	LaFrance
Double Extra First	17,000	–	16.5	–	7.25	–	10	–
Extra First	9,000	8,500	24.75	25	6.5	6.25	10	9.5
FirstSize	8,000	7,300	24.5	24.5	6.1	6	9.1	9.5
Second Size	7,000	6,500	24.25	24.5	6	6	9	9.5
Third Size	6,000	6,000	24	24.25	6	6	8.7	9.3
Fourth Size	4,200	5,500	20.25	24	5.8	6	8.25	9.1
Fifth Size	–	4,700	–	22	–	6	–	9
Sixth Size	–	4,000	–	22	–	6	–	9

It is also somewhat instructive and of interest to compare the performance of hand-powered pumpers to steamers. In the tabulation which follows it has been assumed that the operators are maintaining a sixty times per minute stroke rate:

Hand-powered Fire Pumper Output (GPM)	first size	second size	third size	fourth size	fifth size
	160	130	115	60	50

With numbers such as these it can readily be seen that hand engines were at best only about 15-18 percent as efficient as steam pumpers. Of course the fallacy of any such estimate is that the efficiency as calculated really applies only to an instant in time. But men do become tired and fatigued, men do get thirsty and hungry, and some men do imbibe spirits. The steamer, however, just continues to pump water. No wonder that Miles Greenwood, first chief of the Cincinnati Fire Department, remarked that, "Their only drawback is that they can't vote."

Now there were a few other characteristics, in addition to pumping capacity, weight, and so forth, with which steam fire engines could be described and classified. These included the type of boiler, the type and the alignment of the pump, and the type of supporting framework or chassis.

Briefly, the boilers were of two kinds, although there were a number of variations for each. The "water tube boiler" carried water, within tubes, through the fire-box. The "fire tube boiler" carried heated gases up through the water and out via the stack. The pumps were either of the horizontal or the vertical piston variety, and either single or double, or else a horizontal rotary pump was used. Finally, the frames and chassis were one or another of five kinds. There was the straight-frame steamer which provided very little space for angular movement of the front wheels, and therefore had no capability for making sharp turns. In distinction, the crane-neck vehicle had its frame sufficiently well-arched for the front wheels to turn under the body. The barrel tank steamer had a feed water heater which extended forward from the boiler. Then there was the U-tank design which replaced the barrel tank type. It was used with some vertical piston steamers, and this frame has been described as looking a bit like a horizontal letter "U." Finally, there was the harp tank steam fire engine, characterized by its curved, light, harp-like frame.

Just like steam railroad locomotives steam fire engines have continued to exert a very special appeal, partially intellectual and mechanical and mostly emotional, for many people. This subject area has been well and adequately reported in pictures and in words, during recent years, by Peckham and by Hass. There is much detail concerning size and construction specifications of steam fire engines in their books. The thrust of our present narrative emphasizes patent information, and therefore is not at all similar to the information which was so well presented in these other books.

The Hodge Machine

During the long, cold winter of 1839-1840 the frequency and the extent of devastating fires was alarming to the citizens of the city of New York, and financially threatening to the fire insurance companies which carried fiscal responsibility for the burned-out structures. And so a machinist named Paul Hodge was urged by the underwriters to build a steam fire engine, which he did. Mr. Hodge's machine received its first public test in front of the City Hall on a fine and clear day in March, 1841. Some spectators remarked that the eight-ton steam fire engine looked very much like a railway locomotive (FIGURE 59). It was given a subsequent trial by the Pearl Hose Company, Number 28, but it was judged to be not successful and it was sold off to a maker of packing boxes to be used as a stationary engine.

The Matteawan Machine

There is another tale which surfaces whenever historians search about for the first American steam fire engine, and again the same approximate date, 1841-1842, turns up. An organization by the name of the Matteawan Company constructed a steamer under contract to the fire insurance companies of New York. Apparently it was maintained by Matteawan, and was

FIGURE 59

used at a number of fires. These occasions for its use were strictly limited however, and required the permission of New York fire officers. Finally the insurance companies realized that the hostility of the firemen would ultimately cause their financial losses to increase well beyond any savings which could be realized by using the steam pumper. And so, it was withdrawn.

As a matter of fact the general introduction of steam fire engines to the city of New York was delayed for a number of years by the opposition of the volunteer firemen whose foresight was sufficient to recognize such a formidable opponent. After all, if three or four men could handle and deploy such a machine there would no longer be any need for 50 or 60 men to work the brakes of a hand pumper.

John Ericsson

More or less simultaneously with the Hodge and the Matteawan machine developments, the Board of Directors of the Mechanics Institute of the City of New York was offering its highest honor, its Gold Medal, as a reward for a working steam fire engine. The Committee on Arts and Sciences of the Institute reviewed the several specifications and drawings which were submitted, and they reported in favor of John Ericsson. Not yet the renowned and venerable engineer and inventor of later years he did, at least, have some prior experience in England with steam fire pumps (FIGURE 60). Ultimately he developed a successful "caloric engine" type of hot air machine, and he invented a screw mechanism for ship propulsion. No doubt he remains most famous as the designer and builder of the Civil War gunboat U.S.S. Monitor. His interests even extended to solar heating, which he began to study after crewmen of the Monitor complained that the sun made the flat deck of that ship so hot that they were unable to enter the forecastle below.

FIGURE 60

The Committee on Arts and Sciences in its review of Ericsson's design submission remarked that, "The points of excellence as thus narrowed down were found to belong in a superior degree to an engine weighing less than two and one-half tons that, with the lowest estimate of speed, has a power of 108 men, and will throw 3,000 pounds of water per minute to a height of 105 feet through a nozzle one and one-half inches in diameter. By increasing the speed to the greatest limit easily and safely attainable, the quantity of water thrown may be much augmented."

Ericsson's drawings and specifications were sent in with a covering letter dated July 1, 1840. His drawings were published in the October, 1840 issue of American Repertory of Arts and Science, and they were again presented in *Family Magazine*, Volume 8, pages 224-226, 1840. An engraving of the Ericsson machine is shown in FIGURE 61.

FIGURE 61

The Lay Machine

In 1851 William L. Lay of Philadelphia, Pennsylvania designed and secured a patent for his steam fire engine. This unique apparatus was to have been self-propelled, and it employed a rotary pump. Lay was unable to obtain sufficient financial backing, and thus no working model was ever built. This invention is pictured in FIGURE 62.

Alexander Bonner Latta

It was the steam fire engines of Cincinnati, Ohio, and the names Abel Shawk, Miles Greenwood, Joseph S. Ross, Richard G. Bray, and Alexander Bonner Latta which esthetes of the subject proclaim as being the first successful machines and their inventor-builders. And while this is not the case literally, it is in the practical sense true. FIGURE 63 is the patent drawing for Shawk's water tube boiler (U.S. Patent 10,041 dated September 20, 1853), and FIGURE 64 is A.B. Latta's patent number 11,025, June 6, 1854, another water tube boiler.

FIGURE 62

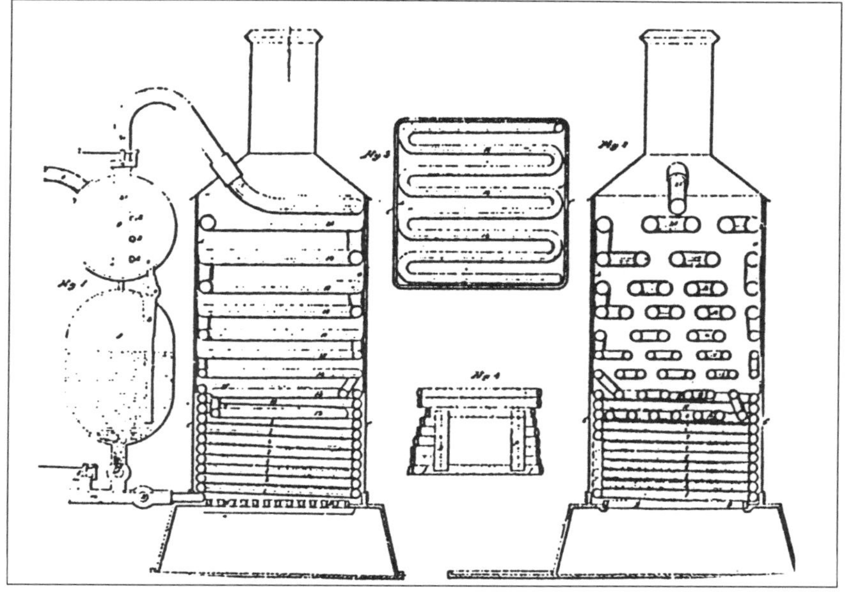

FIGURE 63

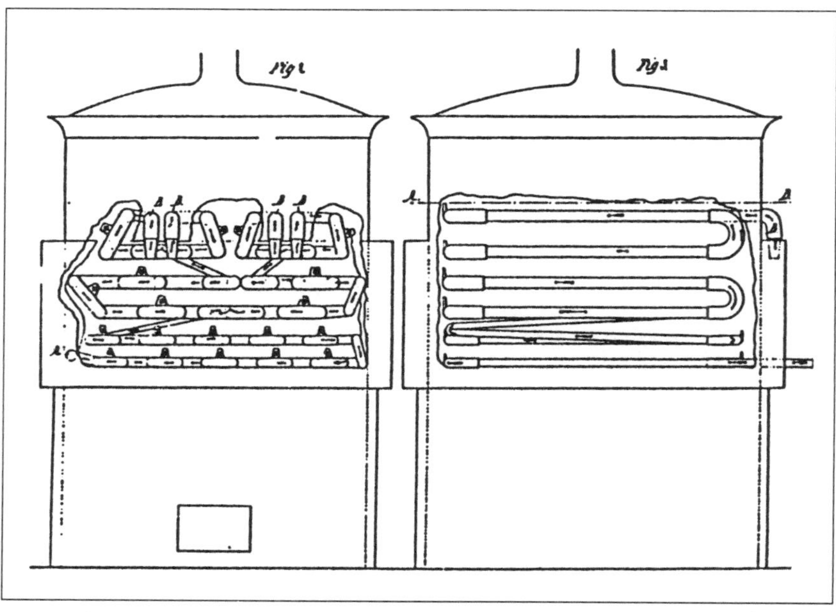

FIGURE 64

Alexander Bonner Latta received at least seven patents during the five-year period 1854-1859. One of his most historically-important inventions was his "Carriage for Steam Fire Engines," patent number 12,912 of May 22, 1855. FIGURES 65, 66, 67, and 68 were the drawings for this document. Within these papers Latta addressed a number of the peculiar aspects typical of self-propulsion, and following are some exerpted comments:

"Carriages of steam fire engines must often be driven with great rapidity over paved streets, causing violent and continual shocks and therefore requiring, in consequence of the great weight upon them a yielding lateral support of the boxes of the wheel in the line of motion of the wheel, or line of momentum, which is not necessary on steam carriages for railroads–although it would be useful there. As the fire engines are required to turn short curves in the streets, and that with rapidity, it is necessary that the guide wheel should be capable of turning 'short' while the bracing is preserved nearly in the direction of motion of that wheel."

"After a fire, or in moving from one fire to another, the arrangement for giving locomotion to the machine will be found very useful. The engine house should, if possible, be located on the highest ground so that the engine may be easily drawn by horses to any point where a fire may occur while steam is being raised, and, in returning up the elevation, the aid of the engine will be found exceedingly useful."

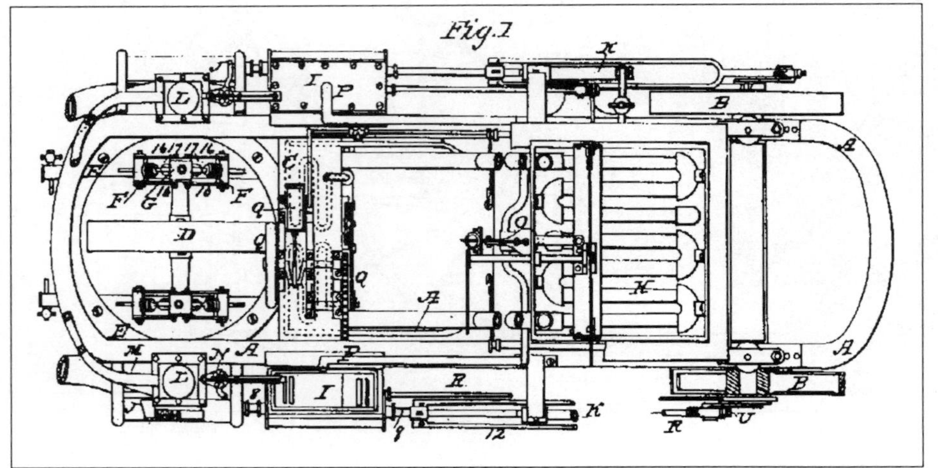

FIGURE 65

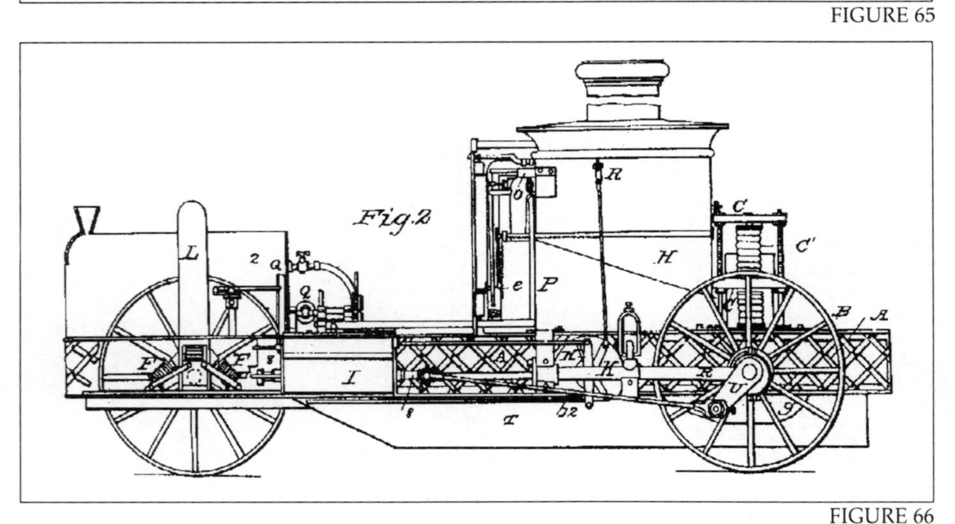

FIGURE 66

It was in 1852, allegedly, that an unidentified writer for the old *Chicago Herald* newspaper made a visit to the shop and foundry of John McGowan, on Ohio Avenue in Cincinnati during which he met and talked with A.B. Latta. The reporter first recounted the story of this meeting twelve years later:

"Having to wait awhile I was allowed the privilege of going into the work room where the steam fire engine inventor was at work. It was a long, high room, the walls on the East side being hung with drawings of the engine. Beneath the drawings ran a long work bench, and at this stood a very diminutive specimen of a man, short and spare, stoop-shouldered even to deformity. He had a square white cap on his head, and was busy measuring something while I looked at him. I saw that his head redeemed his poor body, for it was massive, and the eyes had in them the light of genius. In a moment he turned to me and asked, 'Did Mr. Greenwood give you permission to come in here?' "

" 'He did, Sir; he told me to come and see how the steam fire engine was getting on, so I could report its progress to Mr. Probasco' " (of the great hardware house of Tyler Davidson & Co.).

"'Ah, very well,' said the inventor, 'very well. My name is Latta, Moses Latta, and Mr. Probasco knows me well, and as you came from him, you shall see what few

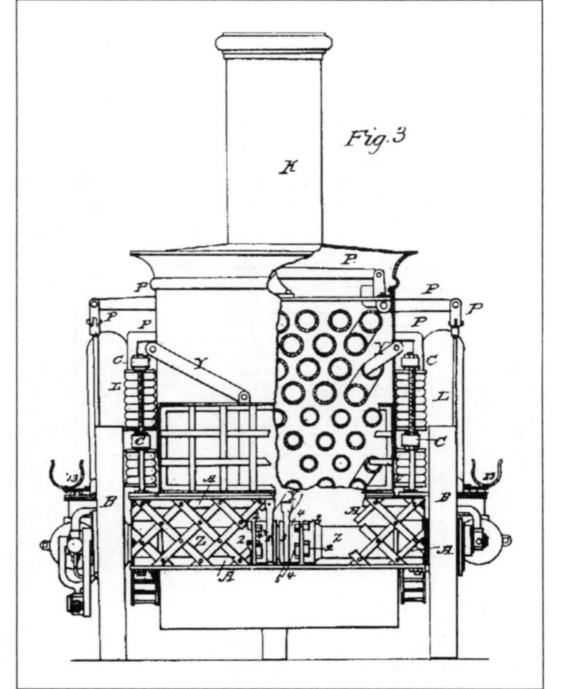

FIGURE 67

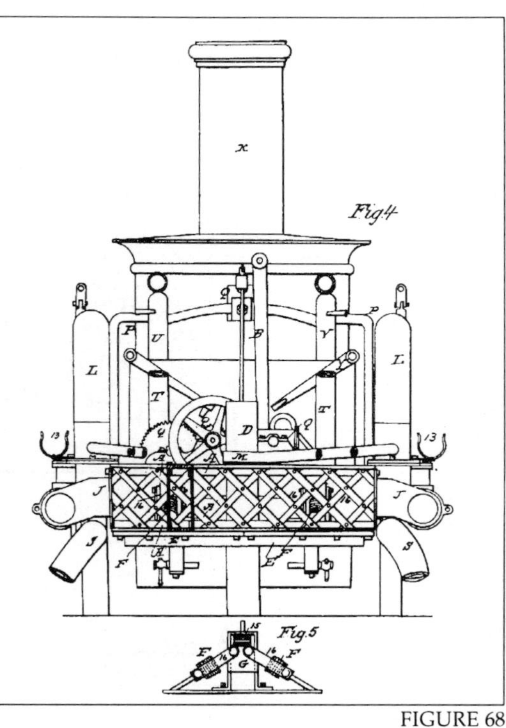

FIGURE 68

see. Can you in any way or to any extent understand the drawing on the wall?' I confessed that I could not. 'Well, it is very simple. Let me explain. The engine is intended to throw at any time eight streams of water–four from each side–and whenever the water can be obtained in sufficient quantity for the eight streams, there will be no trouble in supplying it to the eight hose lines. It is intended, of course, to take the engine to the scene of the fire with horses–four horses. As the engine starts out the furnace is fired up, and ordinarily, by the time we shall arrive at the fire, steam will be up and the engine ready for service. Eight of these large streams focused out onto a fire with the pressure we shall be able to command will drown any fire; even four of them, well-directed, will be of wonderful value. But,' added Mr. Latta, 'The trouble is that there is no certainty that this or any other steam fire engine will ever run to a fire. You are not aware, probably, how bitter the feeling of the volunteer firemen is against this engine. They say it shall never throw a stream on a fire in this city. The recent riots here show what a mob can do in our city, and I fear sometimes that I shall never live to see this grand idea brought into the service of the world. My steps are dogged; spies are continually on my track; I am worried with all sorts of anonymous communications, threatening me with all sorts of ills and evils unless I drop work on this engine and pronounce myself a failure.'"

"The old man's eyes flashed as he said, 'I'll never give it up! I'll build it, and there are men enough in this city to see that it has a fair trial, and I shall have it. When it is finished it will be heard from at the first fire, and woe to those who stand in its way.' "

"With that we separated. As the time approached for the public trial of the engine, the volunteer firemen were in a ferment. It would never do to destroy the engine before it had a trial, and to destroy it after a successful exhibit of its powers was made equally useless, so it was understood that no demonstration, pro or con, would be made on it until it should come to a fire; then it was to be rendered useless, and all who had a hand in its working were to be rendered useless also."

"The public trial came off. The engine far exceeded in efficiency anything that had been claimed for it by its inventor or by his backers, and a feeling of satisfaction swept over the city at knowledge that such a great auxiliary power was with them to fight fire. Still it was known, or believed generally, that its first appearance at a fire would be the signal for as bloody a riot as had ever disgraced the city. The volunteer fire department was there, as everywhere else, a political ring, far more efficient, under ordinary circumstances at the polls than at a fire, and its members were selected for their influence at the voting precincts and for their ability to make the contents of the ballot box when it was emptied show by a large majority their man ahead, no matter what kind of ballots had gone into it. Then, if this steamer was of any account it would ruin and break up not only the companies, but their friends and backers, and the manufacturers who built hand engines."

"One night an alarm rang out from some great warehouse on Third Street, near Main. A minute or two elapsed to the listeners on Main Street, above Fourth, and then down came the great steam fire engine, four mammoth gray horses in front of it at a gallop, the smoke streaming from its stack, the fire flashing from its grates, its ponderous wheels grinding the cobble stones into powder as they struck them, and as the great monster went down the hill, people woke as out of a trance and started after it."

"The engine was brought in front of the block, and soon stream after stream shot from it. The warehouses were among the most valuable in the city, and were stored with costly goods. The time had come, the engine was there, four streams had been gotten on, when the cry, 'The hose is cut,' rang out. Then the melee began, but the citizens were stronger than the volunteer firemen, and after a struggle the steamer drowned the fire and was taken home."

"The next morning Moses Latta awoke to find himself famous, and the action of the appreciative citizens of Cincinnati soon put him in a position where his genius was made more available to the world. The steamer of today has little in it outside of the fact that it is built to effect the same purpose as was Latta's engine, but that was the germ of all those which now at the tap of the electric bell seem to hitch themselves to the horses and tear down our streets when an alarm is struck."

This wordy, old accounting is so lyrical, so much "good guys in triumph over the bad guys," that upon reading it one hopes, and fervently so, that this was really how it all happened.

Nehemiah Bean and Amoskeag

The Amoskeag Manufacturing Company, already cited as a builder of steam fire engines also produced self-propelled versions of these machines beginning in 1869. Nehemiah S. Bean was the most prolific inventor associated with this company and its successor, the Manchester Locomotive Works. Bean, who resided in Manchester, New Hampshire, received a number of patents during the decade of the 1860s, among which was number 31,138 dated January 15, 1861, a U-tank design frame together with a vertical piston pump. This arrangement can be visualized from the drawing, FIGURE 69. Of much greater importance both to Bean and to his company was patent number 75,348 of March 10, 1868. This document and drawing (FIGURE 70) described the original double extra first size, self-propelled steam fire engine. The Hartford, Connecticut Fire Department took delivery of the first one of these double crane neck frame monsters in July, 1889.

In June, 1897 and again in January, 1898 the Boston Fire Department took delivery of self-propelled steamers, and assigned them to Engine Companies 38 and 35, where each served until 1926. FIGURE 71 is an engraving of one of these fire engines from the February 13, 1897 issue of Scientific American, in which the following descriptive text also appeared: "There is now being constructed for use by the Boston Fire Department a horseless steam engine, of great size and power, having a contract capacity of 1,350 gallons per minute, but the builders, in view of recent tests, are confident that this engine will throw 1,850 gallons of water per minute. For some time past the fire commissioners of nearly all the great cities have had under consideration the question of adopting a specially powerful steam fire engine for use in portions of the city in

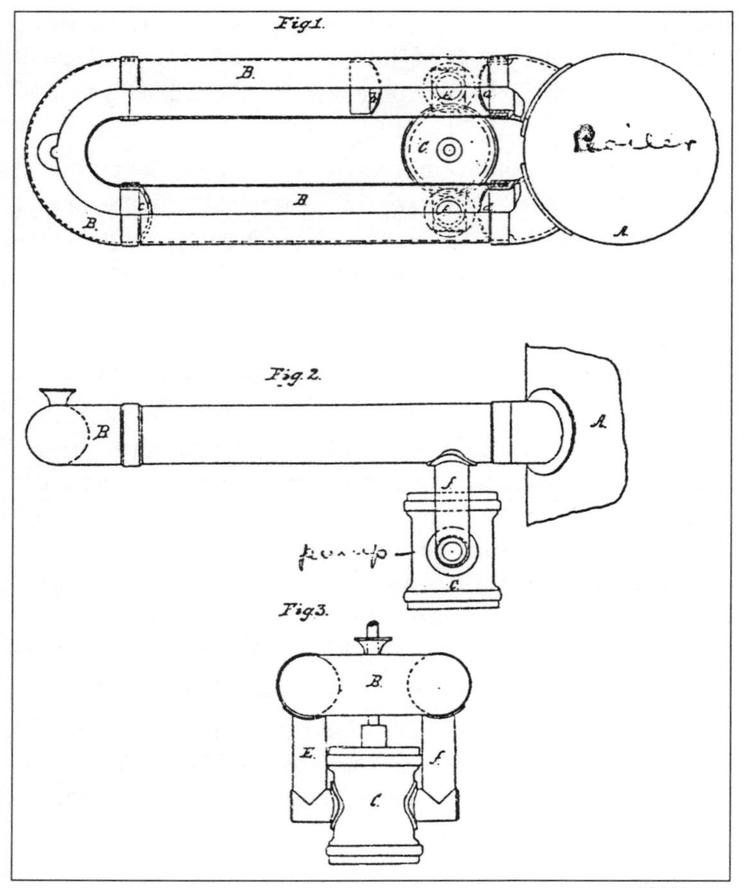

FIGURE 69

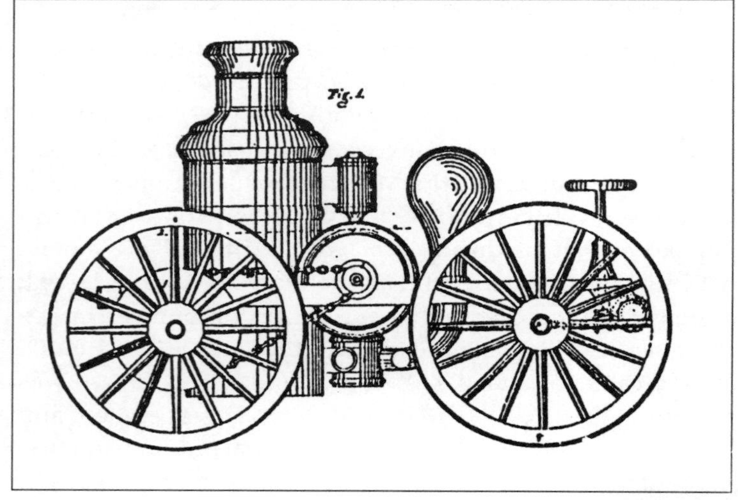
FIGURE 70

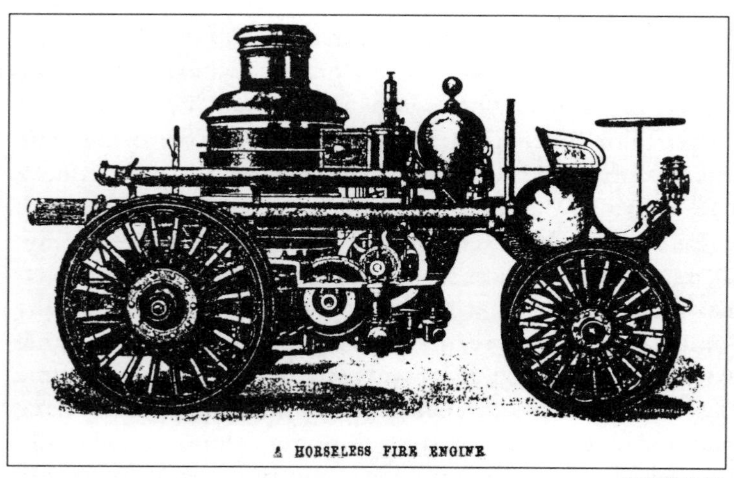

FIGURE 71

which the great office buildings are located. One of the options for avoiding disastrous fires in these tall buildings is the development of powerful engines capable of projecting higher streams of water."

"The heaviest fire engine for horses weighs about 10,000 pounds, and requires three horses to get it about the city. It has a guaranteed capacity of 1,100 gallons per minute. A heavier engine would be almost unmanageable, if horses were used as a means of moving it about from place to place, and in the narrow streets of Boston and lower New York it is even difficult for an engine with three horses to make rapid progress, and the liability of some of the horses becoming injured is also very great."

"The double extra first size self propeller is of the Amoskeag type, built by the Manchester Locomotive Works. It is ten feet tall, sixteen feet, six inches long, and seven feet, three inches wide. Equipped for service it weights 17,000 pounds. The boiler is upright and tubular in style, with a submerged smoke box, and is expanded at the lower end to increase the grate surface. It is made of the best quality of steel plate, with seamless copper tubes, and is thoroughly riveted and stayed. It is jacketed with asbestos and has a lagging of wood which supports the metallic jacket. The connections with the steam cylinders are simple and have the advantage of being entirely unexposed to the air. The steam cylinders are cast in one piece. They are firmly secured to the boiler and framing and are covered with a lagging of wood, with a metallic jacket on the outside. The main shell of the pump is in one solid casting. It is a double acting and vertical pump and its valves are vertical in their action. The pump is arranged for receiving suction hose on either side and has outlets on either side for receiving the leading hose. The connection between the steam cylinders and the water cylinders or pumps may be made by the old and familiar link motion and link block, or the equally-familiar cross-head and connecting rod plan, both giving excellent results for ordinary steam fire engines; but in the self propelling engine, where the engine power is transmitted to the driving wheel through the main crank shaft, which is not the case when this power is transmitted to the pumps, the cross-head and connecting rod plan has many advantages and is therefore adopted for self propelling engines."

"The manner of handling the self propellers is very simple. The chief engineer rides on the fire box of the engine and has directly under his hand the various levers and wheels which start, stop, and regulate the speed of the machine. The assistant engineer rides on the driver's seat, and by means of the large steering wheel he steers the machine in exactly the same manner as the rear wheels of the long ladder trucks are governed, through a system of bevel and worm gearing. Very little machinery in addition to the ordinary mechanism of a fire engine is required to operate the self propellers. The road driving power is applied from one end of the main crank shaft to an equalizing compound, the two endless chains running over sprocket wheels on each of the main rear wheels permit these rear wheels to be driven at varying speeds when turning corners. The driving power is made reversible, so that the engine may be driven forward or backward at the will of the operator. When it is not necessary to use the power of the engine for driving purposes, the driving mechanism can be discontinued by the removal of a key, so that the pumps may be worked with the engine standing still. An extra water tank is carried at the rear of these engines to supply the boiler until connections can be made with a hydrant. The self propeller can travel on a fair level road at a maximum rate of twelve miles an hour. It can climb any ordinary grade; in fact, any one that a team of horses can climb with a heavy load."

M.R. Clapp

Another one of the prominent inventors of the seventh decade of the last century was M.R. Clapp of Seneca Falls in New York. Earlier on Clapp often assigned his interests to Edward Mynderse as well as to himself, later to Silsby, Mynderse and Company, and later still to the Clapp and Jones Company. Clapp's January 7, 1862 patent , number 34,087, is seen in FIGURE 72. According to the inventor it was the ceaseless motion of the pistons, during pumping, which subjected steam fire engines to much undesirable vibrating, which was, in turn, the primary reason for serious wear and deterioration of parts of the machinery. Clapp obviated this design defect by arranging two cylinders in such relation one to another that their strokes were simultaneous as well as in opposite directions.

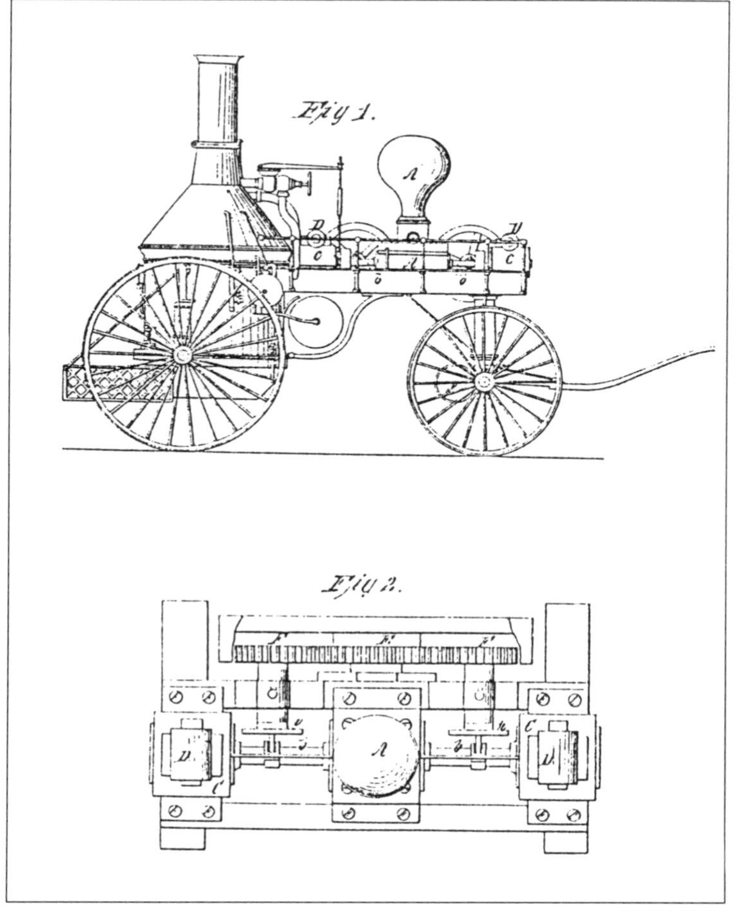

FIGURE 72

The Knibbs Valve

U.S. Patent 42,920 dated May 24, 1864 described a most important device. The inventor was James Knibbs of Troy, New York, and the patent document concerned the "Knibbs relief valve." Illustrated in FIGURE 73, the function of this valve on a steam fire engine was to return excessive water from the discharge side to the suction side of a pump. The New York Fire Department adopted this valve and installed it on practically every one of their steamers, and since the patent claims were so comprehensive, so all-embracing, the royalties — and they were substantial, became the object of a prolonged legal dispute.

After a time Knibbs assigned his interests to Christopher J. Campbell, who ultimately filed a suit against the city of New York. In 1877 the litigation was transferred to the U.S. Circuit Court, and finally in 1897 judgment in favor of Campbell was upheld. The City was ordered to pay him $818,074.32!

As one might expect it was structural modifications and the development of accessories, often quite minute, but patentable, which continued to engage the efforts of inventors during the last third, more or less, of the last century.

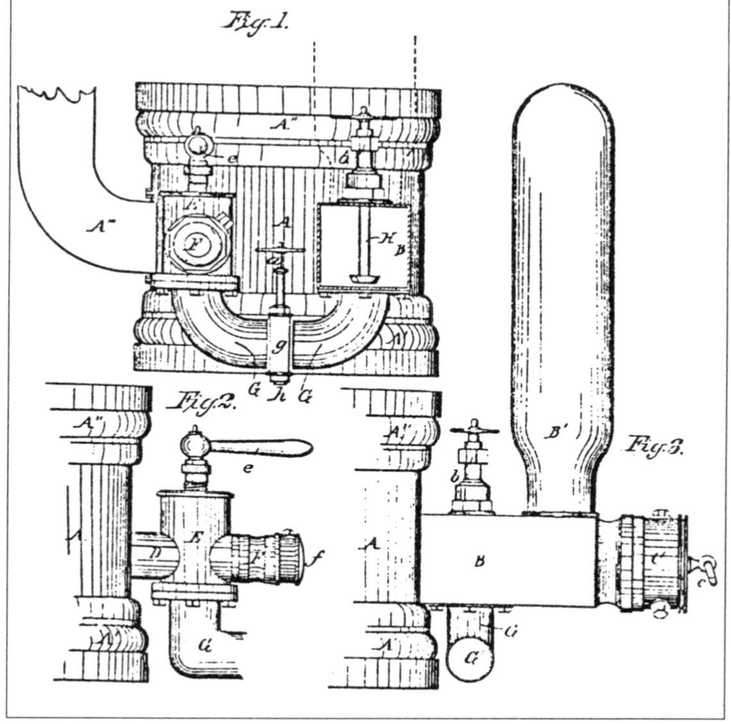

FIGURE 73

The Coles, Chris Ahrens, and the Buttons

The Coles of Pawtucket, Rhode Island, Edward R. and Henry S., received patent number 79,730 for their 1868 partial redesign alterations which minimized the vibrations and the pounding of vertical piston pumps. Their patent drawing is seen in FIGURE 74. Another patent issued during the same year was for the "Improved Steam Device for Washing Windows," invented by Parisian Charles Nivert. The illustration, FIGURE 75, shows that with only minimal alterations it could be converted into a fire engine. This was patent number 84,132

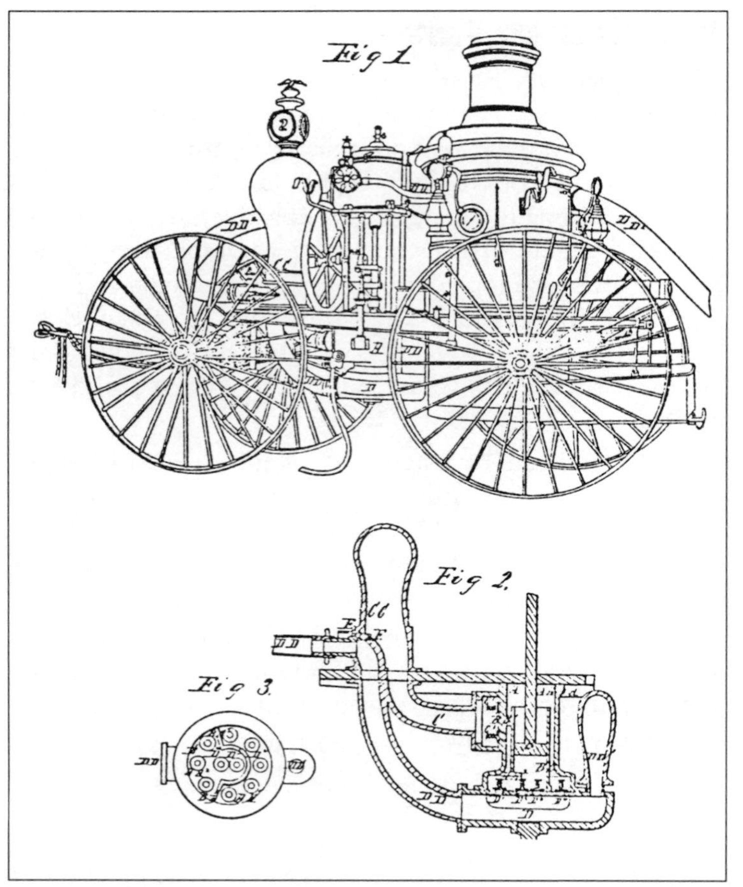

FIGURE 74

dated November 17, 1868. And again in 1870 an invention by the Pawtucket Coles was recognized with a patent, this time number 108,452 (FIGURE 76). This was a hollow bed-plate used as a part of the circulating water pathway. U.S. Patent 111,298 dated January 31, 1871 was Christopher Ahrens significant water-tube boiler invention, illustrated in FIGURE 77. Ahrens and his co-inventor Frank Kamman provided a design which permitted easy access and replacement of tubes, and during actual operations quick verification of steam flow through each tube section. FIGURE 78 is from patent number 116,161 which was received on June 20, 1871 by father and son inventors Lysander Button and Theodore Button. Their contribution was the re-arranged placement of components which thereby improved upon the turning radius of steam fire engines.

An 1872 patent (number 132,426) by Albert F. Allen of Providence, Rhode Island was received by him for another bed-plate modification. He hollowed out this component of his crane neck steamer, using it, again, as part of the water conduit. FIGURE 79 is the illustration. And during the following year William C. Davol, a resident of Fall River, Massachusetts was granted patent number 143,750 (FIGURE 80) for his design of a pan for catching ashes and cinders. A device such as this was of no little importance when one considers the risks of ignition of the wooden wheels of the vehicle, as well as nearby grasses and shrubs.

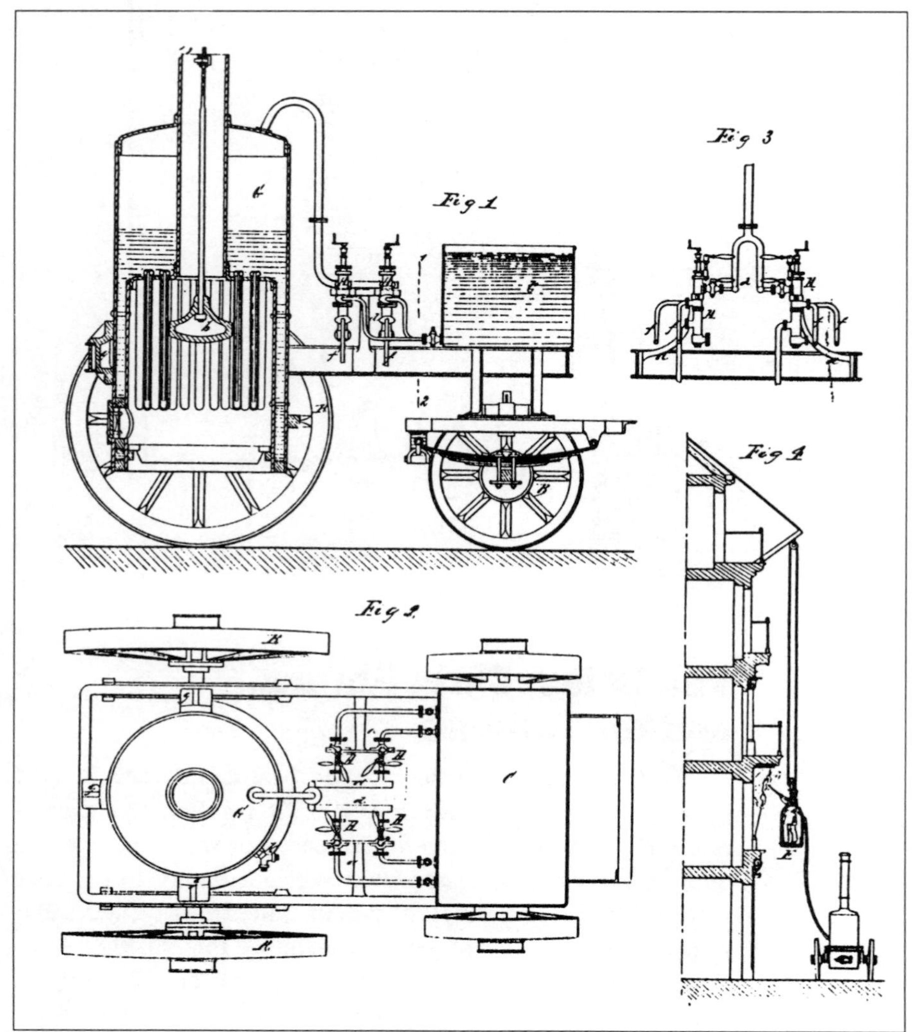

FIGURE 75

The Gould Steam Fire Engine

Just as many of us today hold vivid memories of the United States Bicentennial celebration of 1976, the centennial year, 1876, must have had particularly strong meaning for the people of that time, perhaps Civil War veterans most of all. There were many observances and commemorations, and one of the greatest was the International Exhibition in Philadelphia. Along with hundreds and hundreds of manufactured products and artifacts on display were steam fire engines. One of these was the "Gould" fire engine made by B.S. Nichols & Company of Burlington, Vermont. An un-named correspondent for the British journal "Engineering" composed a description of this rig, and it is of more than passing interest because of the comparisons made to contemporary English steam fire engines:

"The general appearance of the engine differs in several points from those of English makers, the fore wheels being much larger than are usually employed here, while there are no seats for the firemen, beyond that for the driver, and the boxes and mountings for hose, ladders, tools, etc, which add to the weight of our engines being absent. The framing of the engine is simple, consisting of a pair of bars, or "crane necks," which connect the boiler with the fore-carriage, the latter being provided with a central spiral spring. Similar springs are also applied to the hind axle." (FIGURE 81).

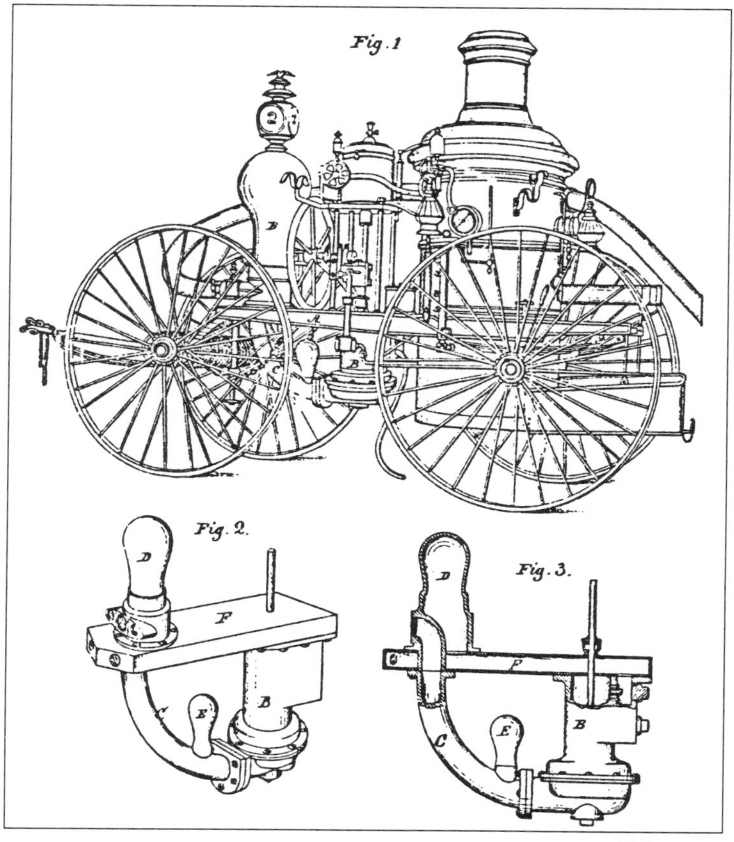

FIGURE 76

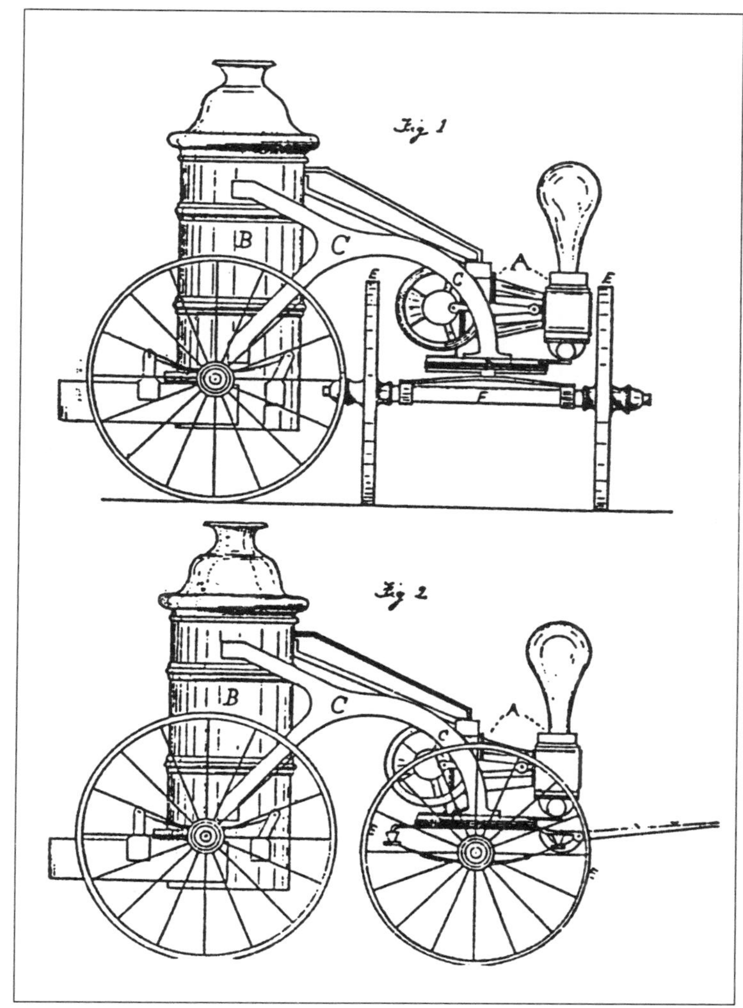

FIGURE 78

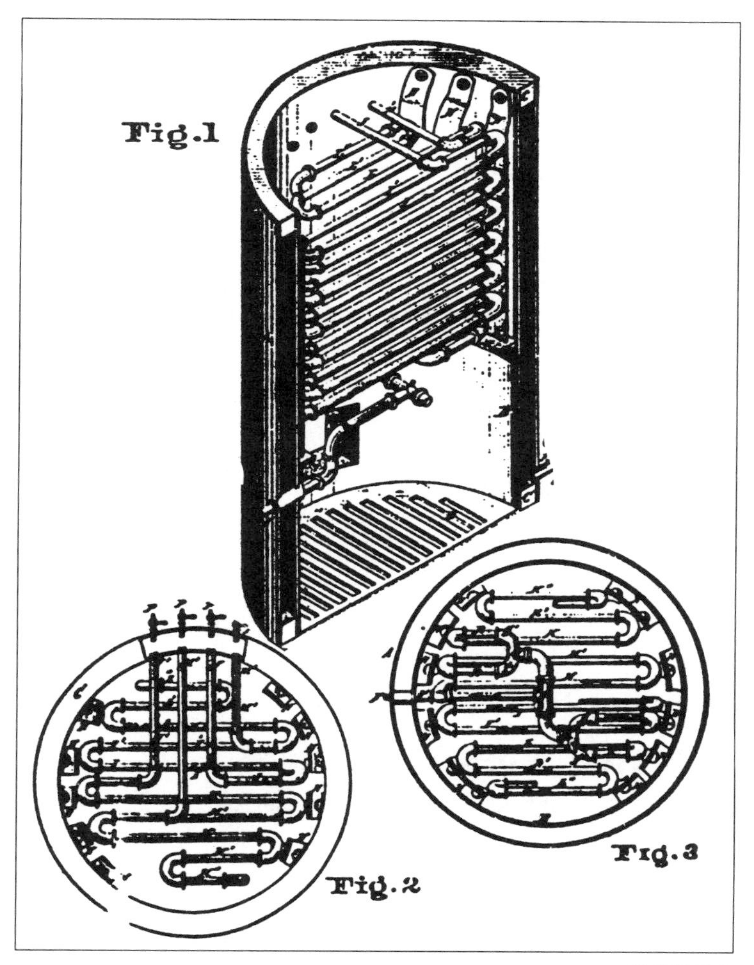

FIGURE 77

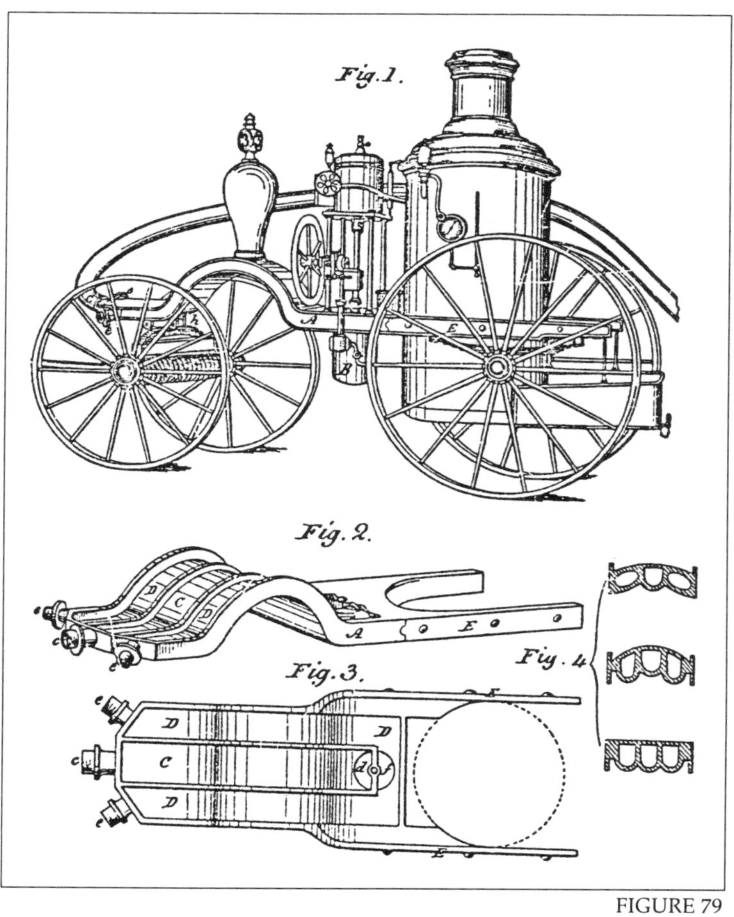

FIGURE 79

FIGURE 80

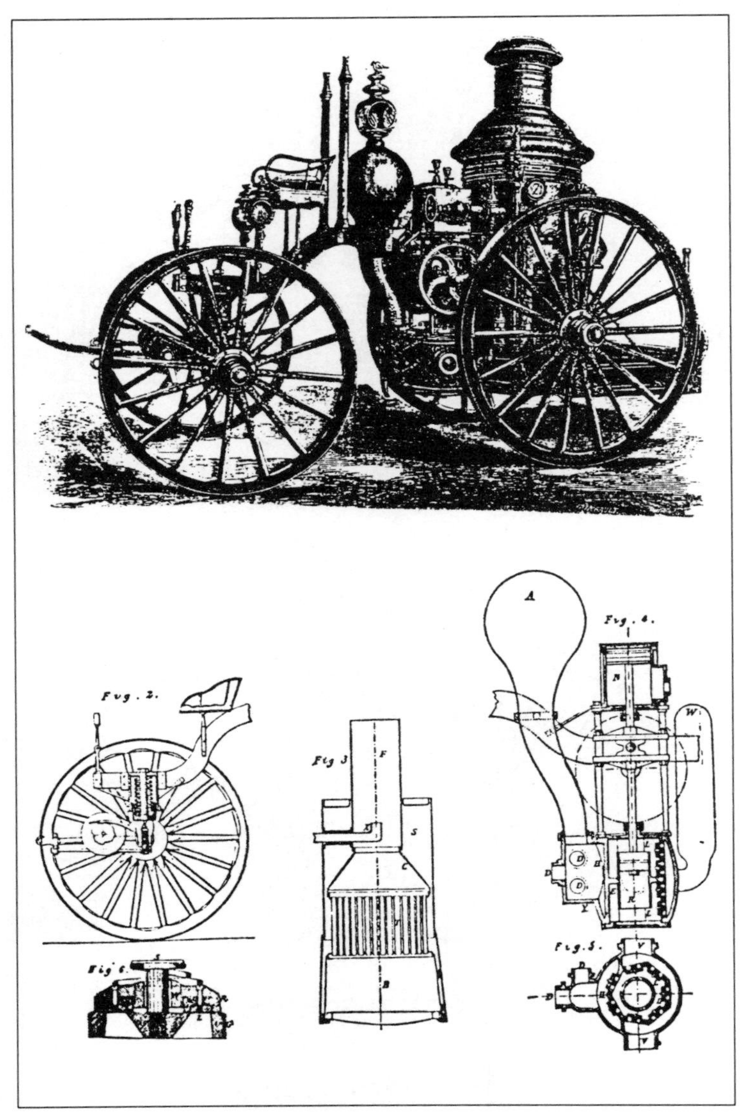

FIGURE 81

"The boiler in the engine is of the vertical tubular type, with a submerged smoke-box. The lower part of the shell is conical, the fire box being also conical, so as to obtain a larger fire-grate. The taper of the fire box is rather more rapid than that of the shell, so that the water spaces increase in width toward the top. The submerged smoke-box has also a conical top, a form which makes a good connection with the chimney, and enables stays to be dispensed with. The tubes are of copper and are 14 inches long by one and one-fourth inches in diameter. With the exception of the tubes, the boiler, which is tested to a pressure of 200 pounds per square inch, is made of steel."

"The steam cylinders and pumps are fixed vertically between the frames just in front of the boiler. Messrs. Nichols & Co.'s first and second class engines having a pair of cylinders, and their third, fourth and fifth class engines a single cylinder. The valve-chest opens at the side in the single engines and at the front in the double engines, the slide valve being in either case readily accessible. The cylinder and pump are connected by wrought-iron columns, which are bolted firmly to the crane necks, the cylinders being braced to both the boiler and the crane necks, and the pump casing being braced to the boiler. The exhaust is made variable by a simple arrangement."

"The pump consists of an outer shell and the pump-barrel and valve-plate, the barrel and valve-plate being in one casting. Partitions divide the two ends of the casing, and other partitions separate the suction from discharge. By this arrangement a large area of valve-plate is obtained, and in all cases the area of the suction valves for each end is slightly greater than that of the pump piston, while the delivery valve area is correspondingly large. With such ample waterways the pump works with great ease, and Messrs. Nichols & Co. inform us that it is found to take its full contents of water at all speeds. The piston is packed with a couple of cupped leathers placed back to back."

"By taking off the bottom cover the pump-barrel and valve-plate can be dropped out of the shell, thus at once exposing all the valves so that they may be readily examined or replaced. This arrangement for giving ready access to the valves is a special feature of the pump and a valuable one. Messrs. B.S. Nichols & Co. have a high reputation as steam fire engine builders, and they take great care to secure strength and efficiency with as little weight as possible, while the general design of their engines is very neat. A 'first class' steam fire engine of the type we have described has thrown a stream from a one and one-half inch jet a distance of 354 and one-half feet."

Another Gould steamer is shown in FIGURE 106.

Miscellaneous Appliances and Hair-Brained Schemes

Turning to an overview of some of the various appliances and impedimenta associated with steam fire engines, perhaps it was the idea of placing elevators in fire stations to accommodate additional steamers which was the most hair-brained. It seems that this suggestion was studied by officials of the New York Fire Department during the period 1884-1887, when the city was rapidly expanding and the cost of real estate was inflating. The fire department needed some double engine companies, but found that most of their stations were too small to house a pair of steamers. So it was proposed to introduce elevators sufficiently powerful to raise and lower an engine or tender from floor to floor. Sections of the cellar floor and the first floor would be made moveable and connected by heavy stantions, so as to preserve an invariable distance from each other. When the lower platform sinking into a depression in the cellar floor came to be level with it the upper platform was to have been flush with the working floor. After the regular engine company was called out the platforms would be raised until the lower one was even with the first floor. A locking device would be provided. By counterpoising, the relative weight to be raised would be almost nothing. Although a steam fire engine might weigh 10,000 pounds, the elevator could be counterpoised within a few pounds of this load, or it might even be overloaded so that upon release of a catch the platform would rise automatically (FIGURE 82).

FIGURE 82

Thinking about this does, at the least, provide us all with some reassurance that the politicians and bureaucrats of 100 years ago were neither any brighter nor more innovative than those of our own day.

Such extreme proposals aside, it was a fact that the cellar of a city of New York fire station was usually neither empty nor unused. Upon receiving a box alarm, the operators at fire headquarters in turn alerted the fire stations. The first alarm received sounded a small gong, the "joker." The electric current through the circuit activated an electromagnet which caused the release of a small weight sliding upon a rod adjacent to the gong. The falling weight then would strike a lever arm, in turn permitting a heavier weight, located in the cellar, to fall, withdrawing the harness bolts, and thus freeing the horses and enabling them to dash forward to their places under the harness rigs. The boilers of steam fire engines were directly connected with pipe coils of stoves situated within the basement. Low down upon the rear of each engine were pipes which telescopically joined pipes leading upward from the cellar stove coil. Just prior to leaving, a pair of valves were closed so that no water could escape from the boiler. This was accomplished by the throw of a single lever, and an adjacent rod pressing down through a hole in the floor operated four other valves. Two of these closed the pipes leading upward through the floor, while two others opened pipes from a small water tank up near to the ceiling. In this way the heating coils were supplied with water when the fire engine was not there.

Harnessing the horses was potentially the most time-consuming part of the entire evolution. The horses were kept bridled, and in stalls as close as possible to the engine. The harness was attached to the engine and was raised high enough that a horse could pass under it, to his place. It was suspended from a kind of Y-shaped frame, having a downward curved hook at each end which held the harness. By yanking on the reins these hooks released the harness so that it fell down and upon the horses. The collars were hinged in the center, and one end was provided with a bolt which entered a socket in the other end, and was held there by a spring catch. Some of these fire companies were regularly able to roll out the door within only a few seconds after receipt of an alarm.

U.S. Patent 285,453 dated September 25, 1883 was awarded to Henry M. Young for his invention of a device which enabled steam fire engines to come to action rapidly, to pump water. The device patented by this Minneapolis, Minnesota resident called for the provision of small tanks of gasoline and a "pilot light" type of burner. FIGURE 83 is the patent drawing. Known as the "Paragon Heater" it was manufactured and distributed in three sizes during the years of its widespread use, and FIGURE 84 is a reproduction of an advertisement for it from the 1888-1889 catalogue of the D.A. Woodhouse Manufacturing Company. Another invention of a related kind is shown in FIGURE 85. This was patent number 315,721, received by Loudon Campbell of Alexandria, Virginia on April 14, 1885. This device used a very low flow of illuminating gas as a heat source for the boiler. Upon receipt of a fire alarm and as the steamer exited the station a gas-fed flame suddenly blasted the fire box to ignite the kindling, following which it was automatically shut off.

Edwin Medden of Seneca Falls, New York, an employee of the Silsby Manufacturing Company received patent number 376,330 dated January 10, 1888 for his invention of a heater for steam fire engines. This patent is illustrated in FIGURES 86 and 87. The first one of these drawings shows the heater unit, while the second one demonstrates the patent-holder's concept of using it to heat the firehouse and the hose-drying tower as well as the fire engine boiler. As with each of the other, competing systems there was some simple gadgetry to ensure clean separation and automatic sealing of the heater lines as the engine left the station for a call.

Truckson LaFrance

No discussion of steam fire engines and their component parts can be considered at all scholarly without some notice being given to the work of Truckson LaFrance. The young T.S. LaFrance moved to Elmira, New York sometime during the decade of the 1860s, and there became interested in steam and

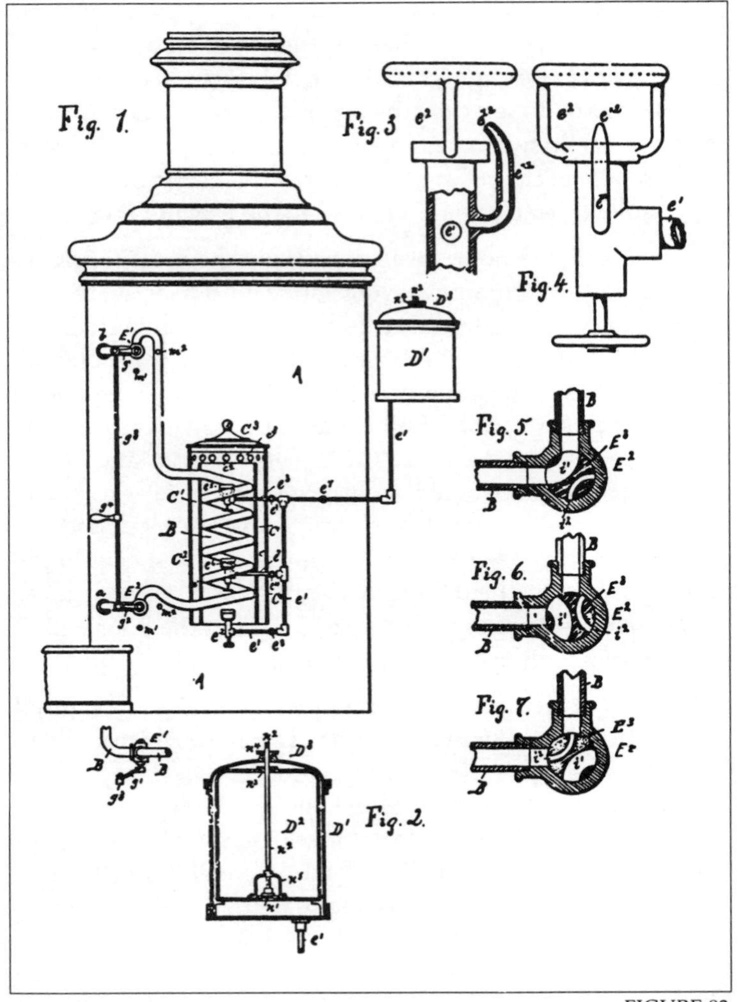

FIGURE 83

FIGURE 84

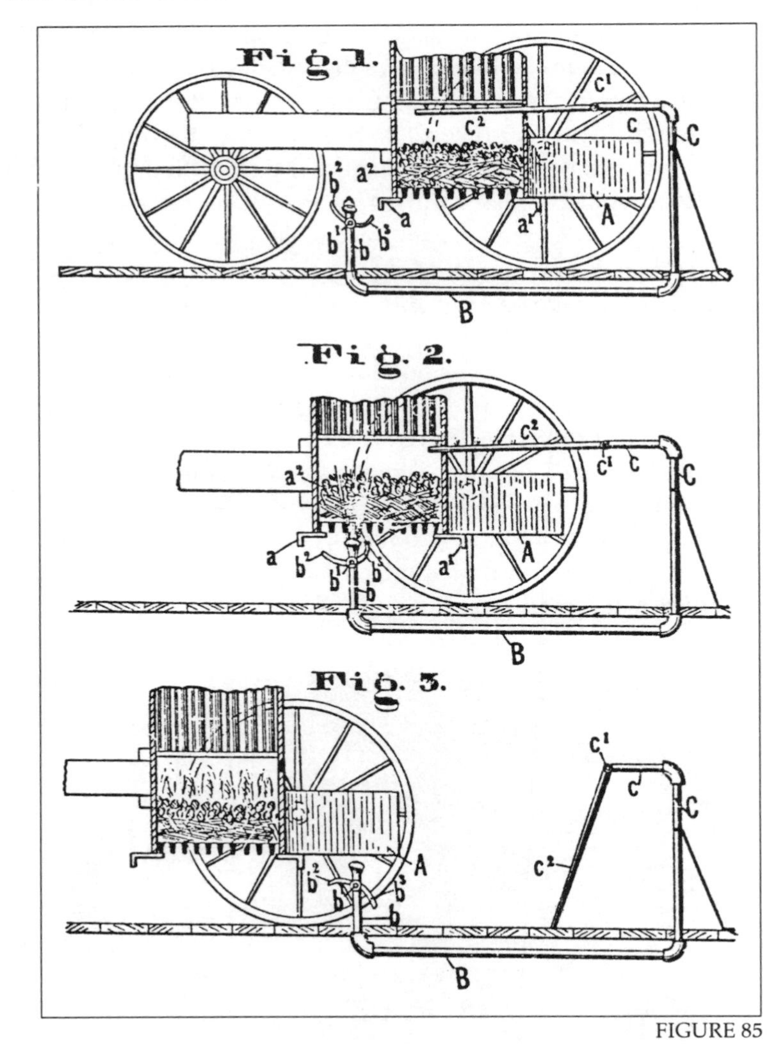

FIGURE 85

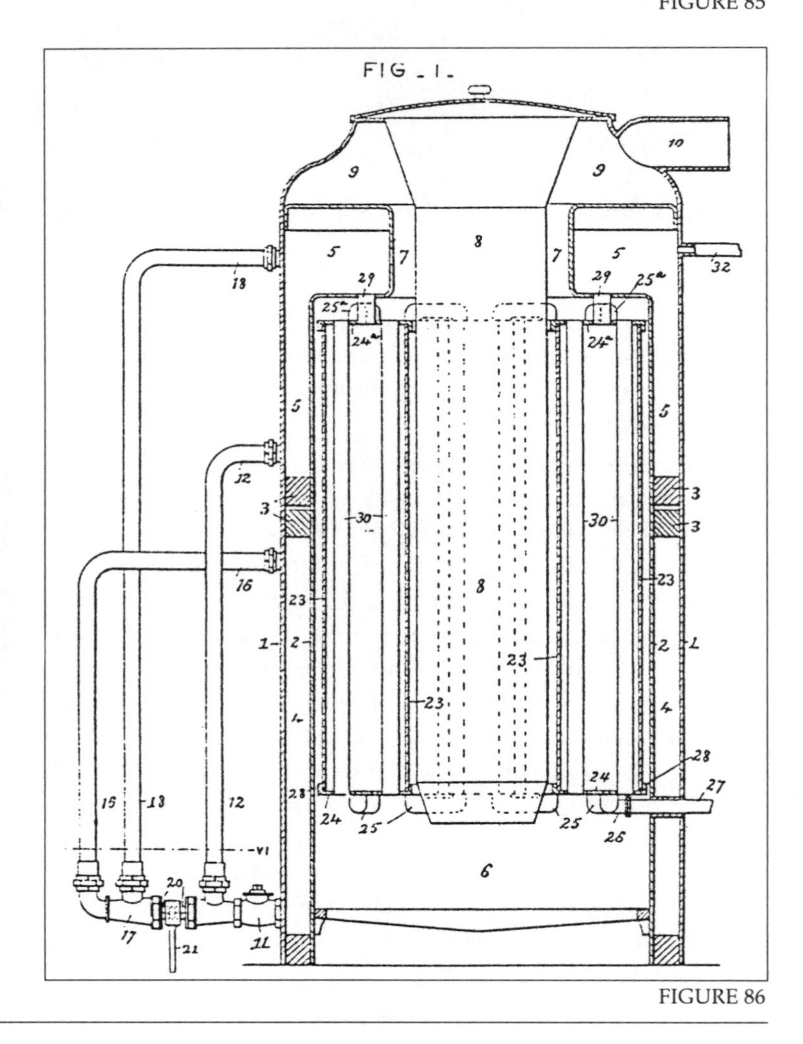

FIGURE 86

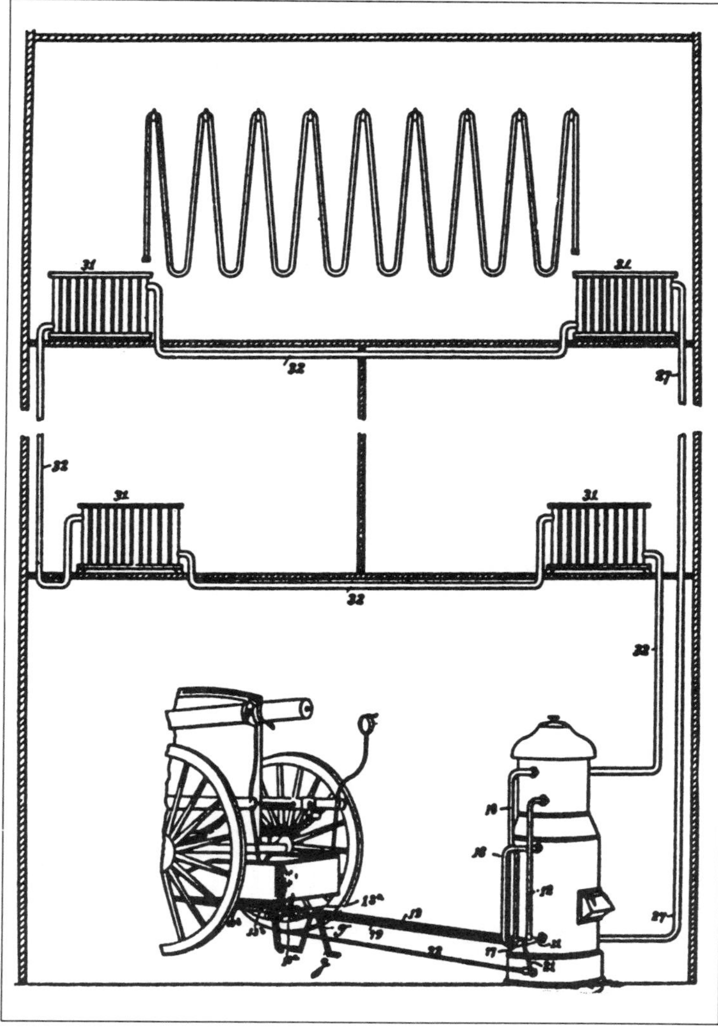

FIGURE 87

in fire engines while working at the Elmira Union Iron Works. So LaFrance and John Vischer together formed a company to build steam fire engines which incorporated some of LaFrance's patented improvements. The LaFrance Manufacturing Company was formed in 1873, was capitalized to build steamers, and it employed Truckson as an engineer in the organization bearing his name. In 1880 the company was reorganized and renamed the LaFrance Steam Engine Company. The purpose of this change was, in part at least, to maximize the profits from the manufacture and sale of LaFrance's latest invention, the rotary, nest-tube boiler (patent number 231,336, FIGURE 88).

Both elegance as well as design simplicity characterized this invention, to be described in some detail. Note in the drawings the cylindrical fire-box, designated by the letter A, set within the boiler, B, leaving a water-filled annular space, C, between. The circular crown sheet of the boiler was D, and E was the steam and water space. Reaching about twelve inches above the crown sheet was a ring, F, an upward continuation of the fire-box wall. The purpose of this ring was to form a sort of reservoir to hold water on top of the crown sheet even if the annular space water level dropped.

Placed high up on the inner surface of the boiler was a circumferential shelf, c, with a raised inner edge, d, which LaFrance termed the "mud-drum," G. He expected that most of the sediments and scale from the water would be deposited here.

Located inside the fire-box were rings of tubes, H, and nests of tubes, H', arranged as circular segments. Each tube ring consisted of a lower, annular segment from which was projected upward four straight tubes, all terminating within a hollow cap. A short tube, i, reached upward from the cap and through the crown sheet. Each of these tubes had an opening in the side, near to the sealed top. A curved pipe, I, connected each ring of tubes to the bottom of the annular water space. Tube nests were formed from curved, horizontal tubes from which small tubes, m, reached upward to another horizontal pipe, n, from which, in turn, closed-top tubes, o, with openings, p, penetrated the crown sheet.

The reservoir above the crown sheet was tapped by a pipe, K, which then extended downward along the outer face of the boiler. At the bottom of the boiler it jointed with the tube from the tube nest. Flow through pipe K was controlled by a valve, L. Flues, M, passed upward through the crown sheet, facilitating the escape of smoke and combustion products from the fire-box.

One of the major advances of this total design was that the crown sheet of the boiler could not become uncovered, i.e., lose its water cover, if the force pumps functioned improperly. The fire would, of course, damage the crown sheet if it was not covered with water. According to inventor LaFrance the water on the crown sheet was enough not only for its own protection, but also by flowing downward over the fire-box walls it shielded them as well, until either the force pump functioned again or the fire was extinguished.

"A further advantage possessed by this device," said LaFrance, "is that when a quick head of steam is required it can be obtained by closing valve L in pipe K, and thus holding the water on the crown sheet and preventing its circulation. Then by opening a blow-off valve and thereby drawing off water from the tube rings and nests the amount is reduced to a suitable quantity for rapid steam generation."

Clearly, inventing, designing, maintaining and operating steam fire pumpers were not jobs for the intellectually deficient!

Another interesting steam fire engine was the one designed by John Blake Tarr of New Bedford, Massachusetts, for which he received patent number 568,379 in September, 1896. This was a light-weight, dual-steam cylinder, dual-vertical piston pump outfit. Notice in FIGURE 89 how the front wheels carried all of the weight of the pumps and their associated mechanisms.

Charles Hurst Fox

U.S. Patent 569,966 dated October 20, 1896 was another prestigious invention by a most prestigious gentleman, Charles H. Fox of Cincinnati, Ohio. The patent was assigned to the American Fire Engine Company of Seneca Falls, New York, and is illustrated in FIGURE 90. According to Ed Hass, author of "The Dean of Steam Fire Engine Builders," this pump became known as the "Metropolitan", and it was used on the American, Columbian, and Metropolitan lines of American Fire Engine Company steamers from 1893 to 1904, and after-

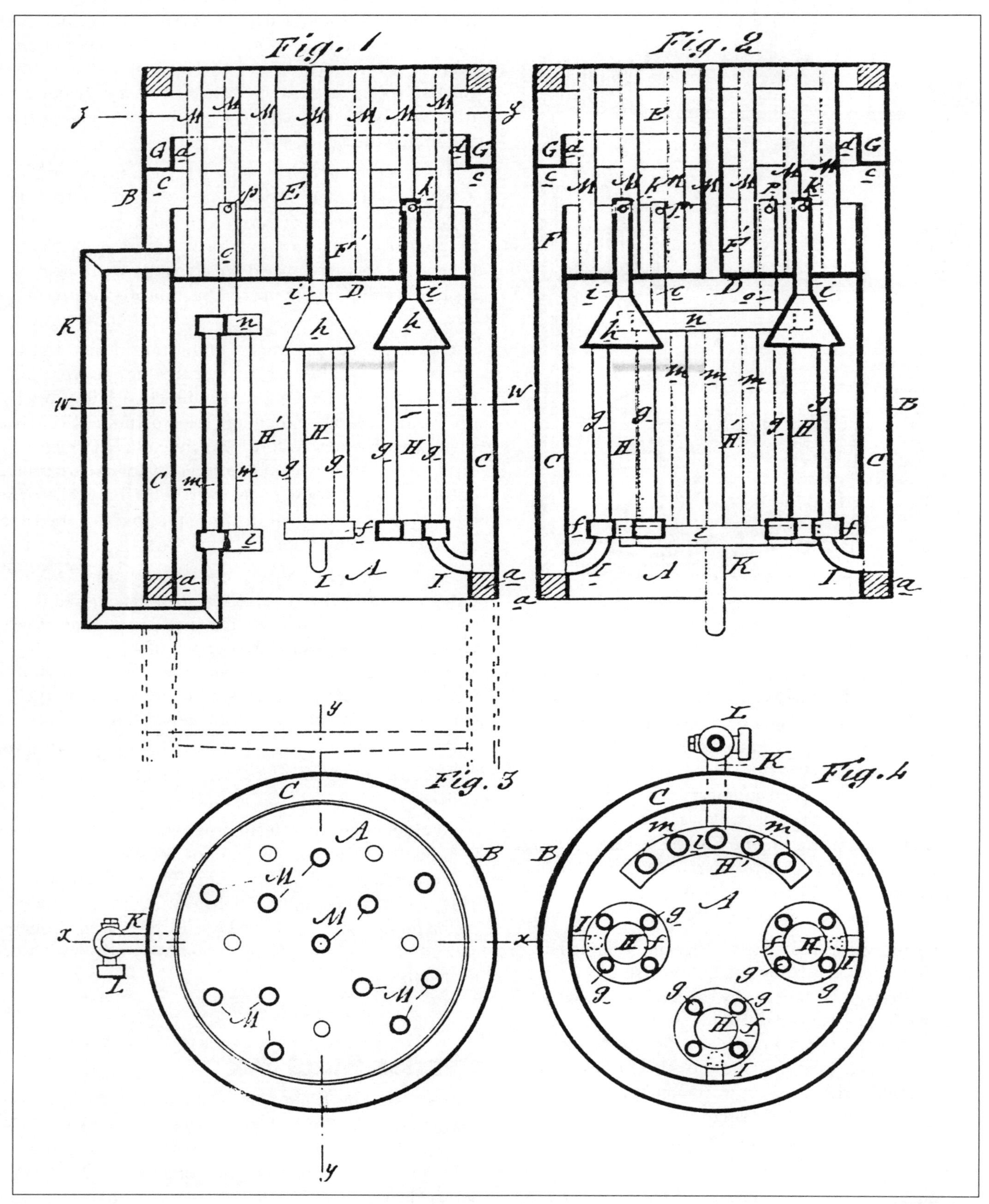

FIGURE 88

ward by American LaFrance from 1904 to 1917. Fox, who was the husband of Chris Ahrens' daughter Lillie, described this invention with the following words:

"The pumps are united in a gunmetal casting which forms a single body, embracing the suction chamber or reservoir, the suction mouths, and reciprocal suction and force chambers of both pumps, and permits the pumps to be placed much closer as to centers than could otherwise be done. Uniting the two pumps in a single structure also obviates the usual weakness in construction attending a connection between separate pumps for the reason that the combined form furnishes a rigid basis for the entire structure, simplifies the driving mechanism, and enables it to endure extraordinary strains without vibration."

And no manuscript such as this would be complete without mentioning the name of Fox's co-worker and colleague William H. Shafer. His patent number 821,6574 is illustrated in FIGURE 91. Shafer was another one of those incisive thinkers who was able to recognize an invention when he had invented it. In this particular steam pump the bed-plate was to have been a single casting, as this was another avenue to promoting strength and decreasing noise, wear and tear due to the vibrations of operation.

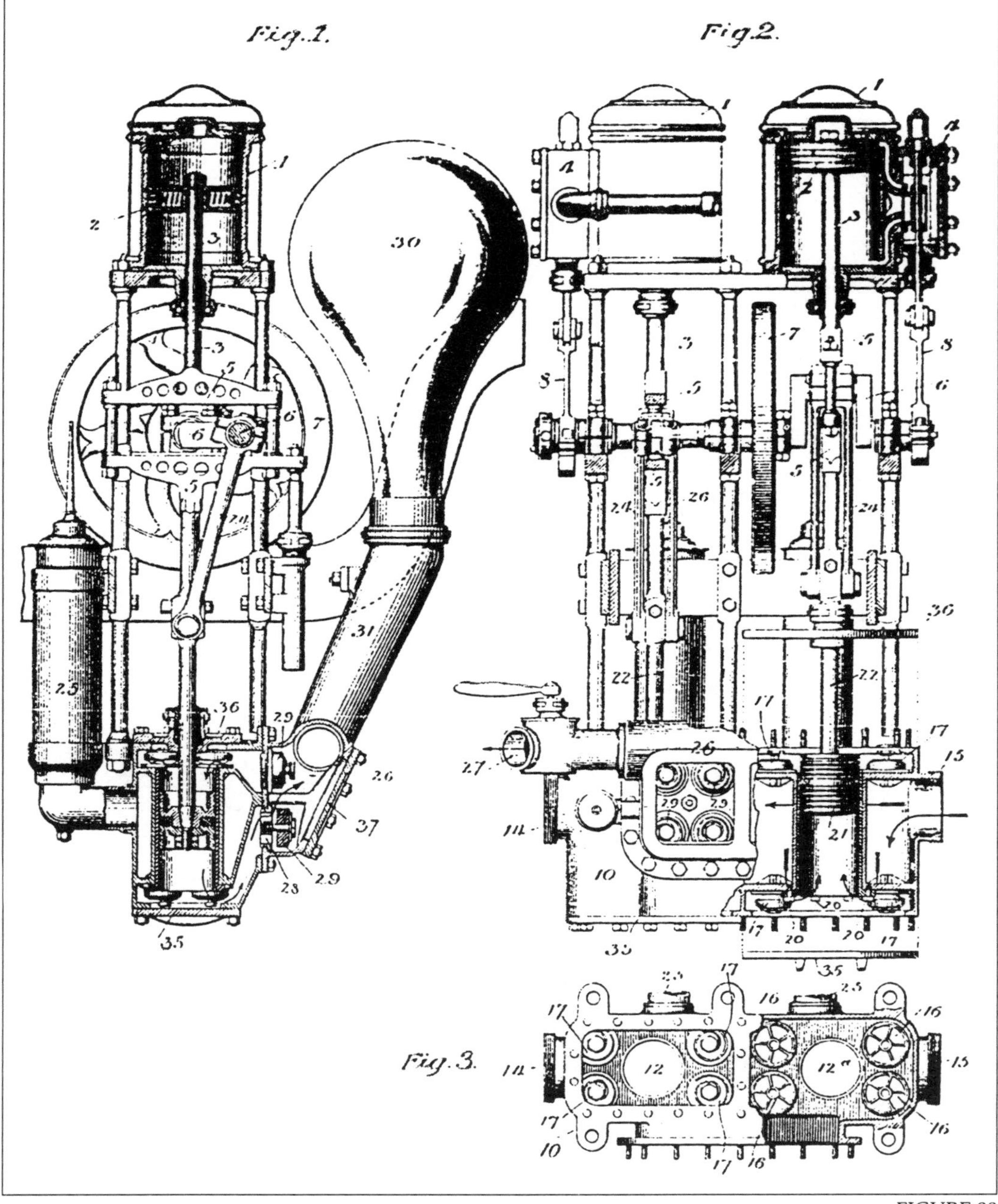

FIGURE 90

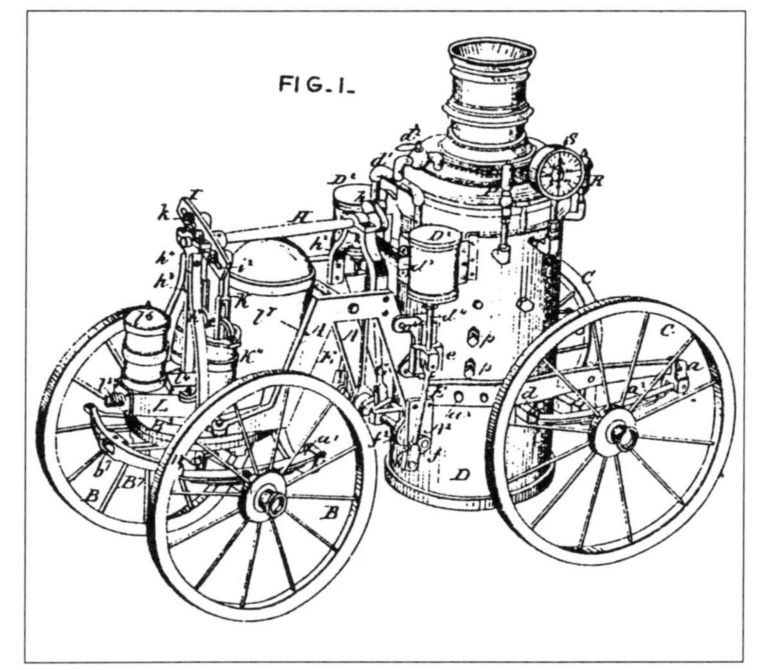

FIGURE 89

Birdsill Holly

One of the neat devices which came into prominence for the first time by reason of its use with steam power was the rotary pump, and an inventor whose name was practically synonymous with these pumps was Birdsill Holly, of Seneca Falls. Pumps built in accordance with his patent were used for many years by Silsby, Mynderse and Company and its successor, Silsby Manufacturing Company. And of course even to the present day rotary pumps are used as priming pumps because of their self-priming characteristic. The drawing for Holly's patent, number 12,350 dated February 6, 1855 is reproduced as FIGURE 92, and the text is shown as FIGURE 93. Notice that Holly used the word "piston" in writing his description of the inner working parts of the pump, most likely because he had no alternate word to select, and he preferred not to coin a new term.

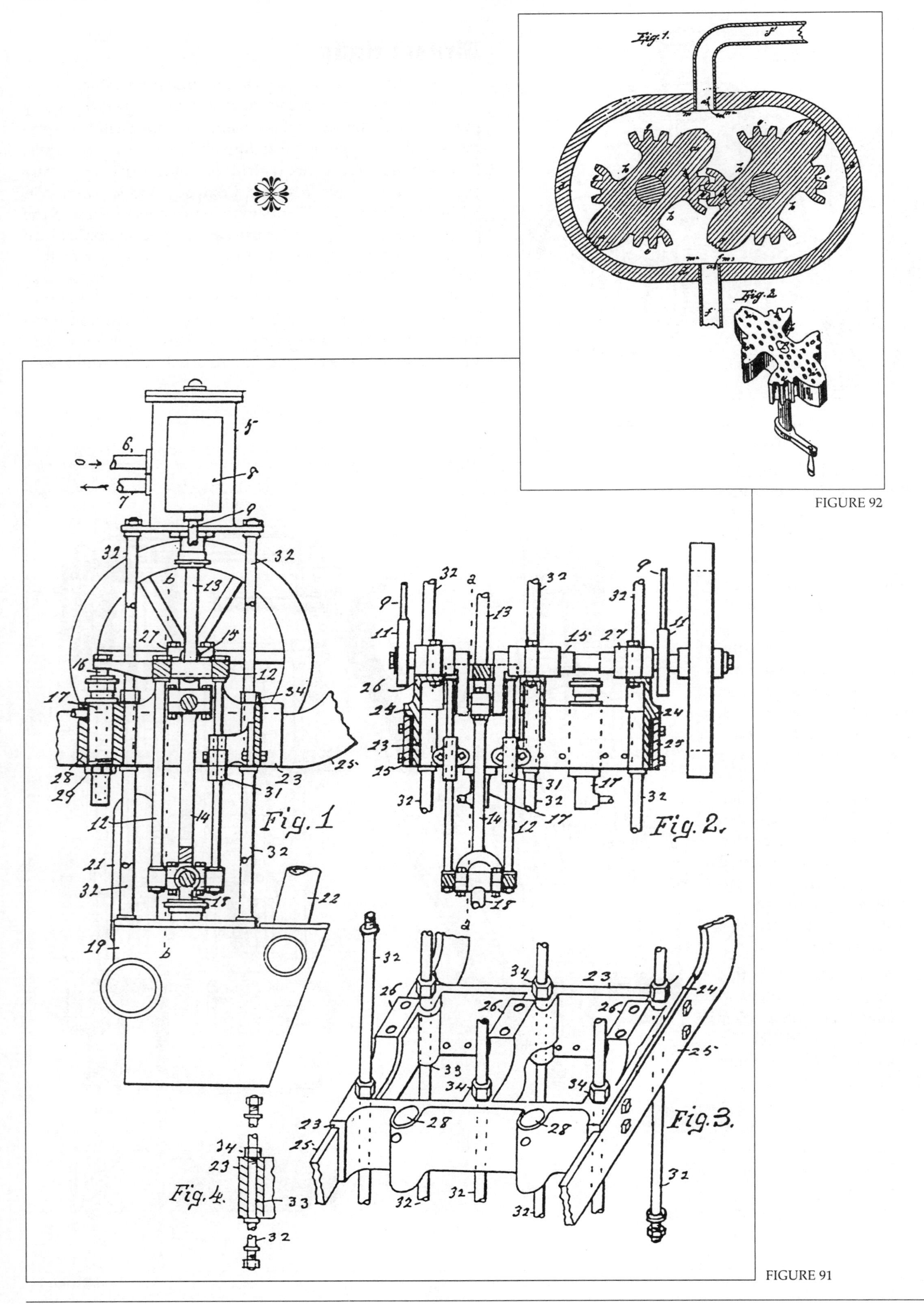

FIGURE 92

FIGURE 91

UNITED STATES PATENT OFFICE.

BIRDSILL HOLLY, OF SENECA FALLS, NEW YORK.

ELLIPTICAL OR ROTARY PUMP.

Specification of Letters Patent No. 12,350, dated February 6, 1855.

To all whom it may concern:

Be it known that I, BIRDSILL HOLLY, of Seneca Falls, in the county of Seneca and State of New York, have invented an Improvement in Rotary or Elliptical Pumps, and that the following is a full, clear, and exact description of the principle or character which distinguishes it from all other things before known and of the usual manner of making, modifying, and using the same, reference being had to the annexed drawings, of which—

Figure 1 represents a section of the pump showing the interior; Fig. 2, a perspective of one of the elliptical pistons.

My invention consists in certain improvements in the class of pumps, known as the rotary or elliptical pumps, shortly to be described, by which I obtain a more efficient and durable pump, than by any other plan with which I am acquainted.

Very many forms have been essayed for rotary pumps but they have gone into disuse, or are but little used, although a rotary pump is desirable, if it can be made serviceable.

By reference to Fig. 1 it will be seen that the elliptical pistons P P′ are provided at their transverse ends, with large cogs *a*, *a*, which are to fit, as the pistons revolve, into corresponding recesses *b*, *b*, in the line of their conjugate diameters. There are also in each piston between the large cogs *a*, *a*, and the recesses *b*, *b*, two sets of smaller cogs *e*, *e*, one set on each side of each piston, and it will be seen that by revolving the piston P on its axis the piston P′ will be carried around by it. These pistons revolving in the air tight (*d*) of elliptical form, are continually producing variable chambers between their surfaces and the walls of the case (*d*). The induction chamber *d′* and eduction chamber d^2 are at their minimum of capacity, when the transverse diameters of the pistons are at right angles to each other, and a glance at the drawings, Fig. 1, will show how the increase and diminution of these chambers will cause the water to enter the induction pipe *f*, and be forced out of the discharge pipe *f′*.

In rotary pumps of this general character a great difficulty has been, the packing of the piston or that part which rubs against or passes near to the inner wall of the case or shell. I meet this difficulty by means of a water packing by making grooves on the periphery of the cogs (*a*) (*a*) and as these become filled with water the centrifugal action of this water becomes opposed to the resistance of the water to be forced, and it is found to be equivalent, in practice to a tight packing. The friction of packing in such pumps is enormous, especially if any degree of tightness is aimed at, and the expense of tight fitting would prevent their common use. The water packing more than compensates for the want of tightness or nice fit by the diminution of friction. The cogs (*a*) (*a*) come into bearing upon the inner wall of case (*d*) at all points between the abutments *m*, *m′*, and m^2 and m^3.

It is obvious that such a pump is convertible to a rotary engine for water, steam, etc., only that in the case of steam or gaseous fluids the packing would require modification.

What I claim as my invention is:

The corrugated or grooved pistons or cogs in the manner and for the purposes specified.

BIRDSILL HOLLY.

Witnesses:
GEO. W. MEAD,
GEO. O. DANIELS.

FIGURE 93

Railroad Steam Engines

Finally, a few words about railroad locomotives of the old kind, those which were most assuredly steam engines. It ought not to be a surprise to find that inventors had devised ways for employing these as steam fire engines. One such inventor was Dyer Williams, Master Mechanic of the Middle Division of the New York Central Railroad. His patent, shown in FIGURES 94 and 95 was number 35,053, received in 1862. Williams' design called for a pair of Holly rotary pumps to be mounted on the top of a locomotive. Steam from the boiler was supposed to turn one pump, and the axle of that pump was to revolve the other. Water was supplied from the large tank carried by the coal tender. This apparatus was fitted to a locomotive which operated out of the Syracuse, New York station. Williams noted that, "The tank holds about 2,500 gallons of water, and it requires about eight minutes to be pumped to empty, through an inch and a quarter nozzle, running the engine at as great a speed as the hose will bear."

Twenty years later, in 1882, patent number 259,463 was received by Robert C. Blackall of Albany, New York (FIGURE 96) for a method of extinguishing fires within railway cars. Again, water was to have been pumped from the tender. During their peak years of steam locomotive utilization the old Pennsylvania Railroad had no less than 612 engines equipped for firefighting. As well as the pump, each of these locomotives was also outfitted with 150 feet of two and one-half inch linen hose and a 15-inch cast iron nozzle with a five-eighth inch discharge opening.

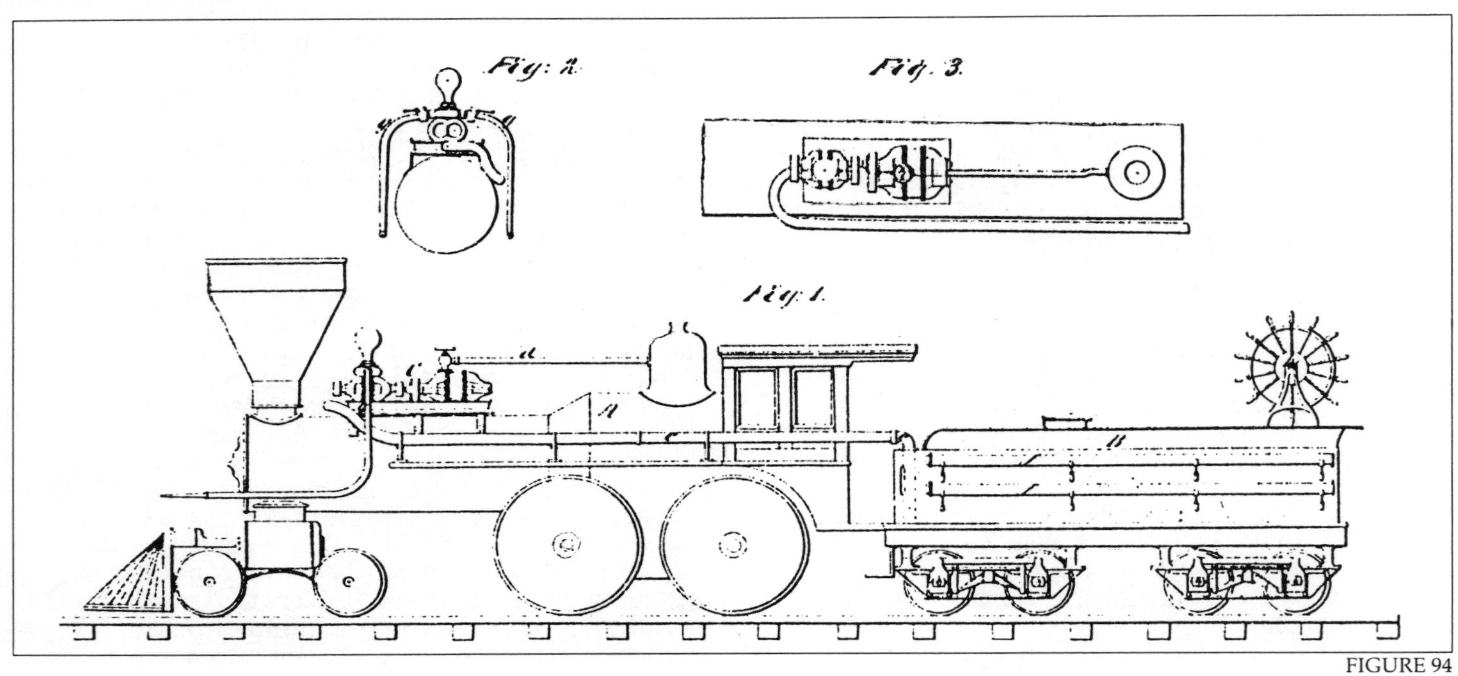

FIGURE 94

FIGURE 95

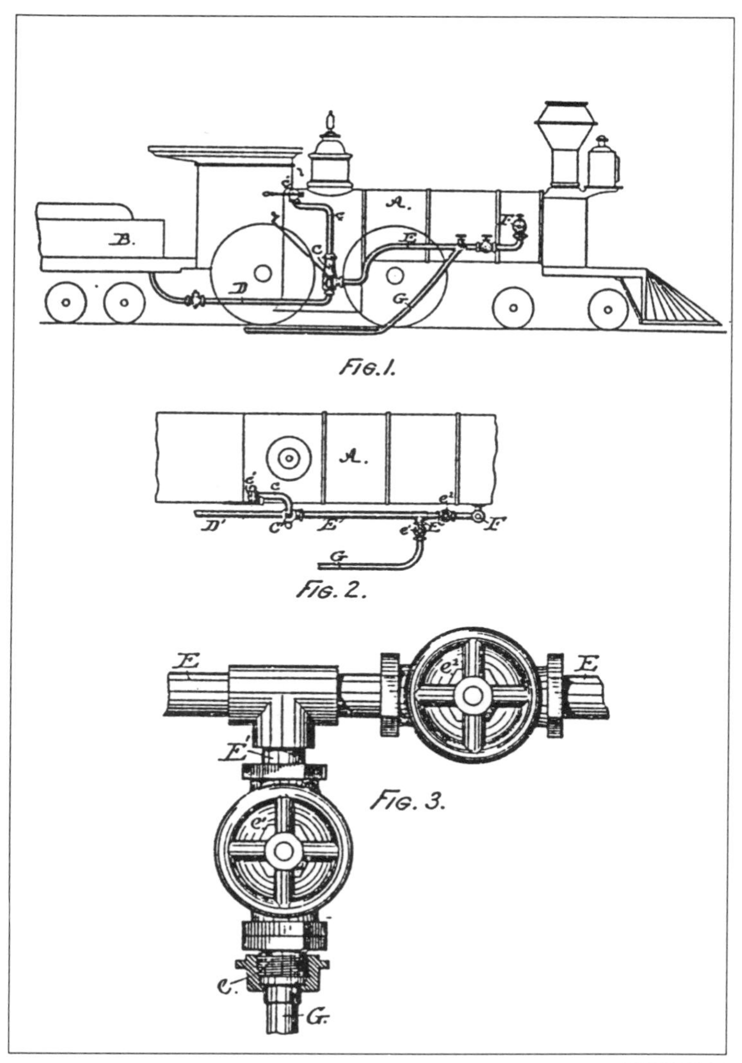

FIGURE 96

FIGURE 97

FIGURE 98

FIGURE 99

FIGURE 100

A Steam Fire Engine Album

A group of steam fire engine illustrations, presented in alphabetical order in accord with their builders' names, comprises the following album. These drawings and photographs have been selected from a pair of sources. The Jerome DeFreest Scrapbook of American and Foreign Fire-Fighting Equipment, which is in the collection of the Rensselaer County Historical Society of Troy, New York, and William T. King's History of the American Steam Fire Engine, published in 1896, were the sources consulted.

The Ahrens Manufacturing Company was the successor outfit to A.B. & E. Latta, who sold out to Lane & Bodley in 1863, and who, in turn, was taken over in 1868 by Ahrens. In FIGURE 97 there is illustrated a single-pump machine, and in FIGURE 98 a double-pump rig, both products of the Ahrens Manufacturing Company. The "Columbian" and the "American" steam fire engines, FIGURES 99 and 100, were products of the American Fire Engine Company. Incorporated on December 12, 1891 in New York state, and capitalized at $600,000, it was a consolidation of the Silsby, Ahrens, Clapp & Jones and Button Companies. Both fire engines employed Charles H. Fox's pump invention, patent 569,966.

FIGURE 101

FIGURE 102

FIGURE 103

A straight-frame, single-pump, U-Tank Amoskeag steamer is seen in FIGURE 101. The following illustration, FIGURE 102, also shows a straight-frame vehicle, this one fitted with double pumps. Later Amoskeag models, of course, were built with crane necks, such vehicles being much easier to turn about. The Button Fire Engine Works was located at Waterford in Saratoga County, New York. Hand-powered fire engines were offered by this concern beginning in 1834, and steamers after 1862. A Button & Blake machine is the subject of FIGURE 103. Clapp & Jones of Hudson, New York, also commenced building steam fire engines in 1862, and, along with Button & Blake, and others, went on to form the American Fire Engine Company. A double-pump, crane-neck Clapp & Jones engine is shown in FIGURE 104.

From Pawtucket, Rhode Island, the Cole Brothers were active in building and selling their fire engines between the years 1867 and 1880. FIGURE 105 is a steam fire engine which the Coles marketed for manpower or horsepower traction (also see patent 108,452).

The gentleman who has been credited with the crane-neck frame style invention was John N. Dennisson, for many years an employee of R.J. Gould. Gould, beginning in 1865, built steam fire engines in Newark, New Jersey, selling out in 1875 to B.S. Nichols & Company of Burlington, Vermont. A Gould steamer is seen in FIGURE 106.

LaFrance steam fire engines are shown in FIGURES 107 and 108, the first illustration being a rotary-pump vehicle while the second shows a piston pumper. As is well known the LaFrance Company went on to become one of the most successful of all builders of fire engines.

The "John F. Torrence," an early steam fire engine built by Lane & Bodley is the subject of FIGURE 109. Notice the tricycle-style configuration of the wheels of this pioneer self-propelling machine. A.B. & E. Latta produced the steamer shown in FIGURE 110, in 1861, for the fire department of Memphis, Tennessee. Most Latta machines were also three-wheelers, with this crane-neck engine being one of the exceptions.

The Lee & Larned Company of New York City built a small number of self-propelling steam fire engines, such as the one illustrated in FIGURE 111, but their

FIGURE 104

FIGURE 105

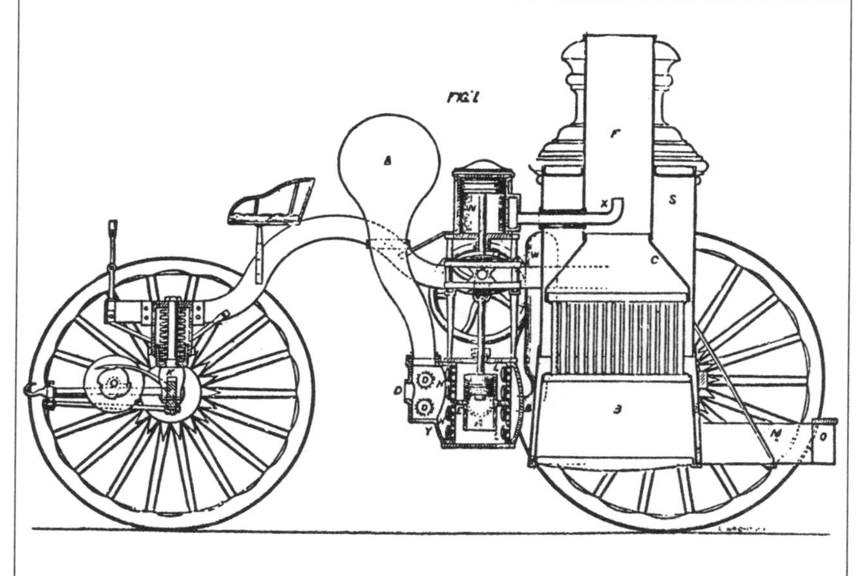

FIGURE 106

FIGURE 107

FIGURE 108

FIGURE 109

FIGURE 112 is an example of the work of Thomas Manning, Jr., and Company, located in Cleveland, Ohio, and founded in 1886. Manning was a successful steam fire engine builder, producing four sizes of these machines. Eventually Manning and others joined with the American Fire Engine Company to form the unsuccessful conglomerate International Fire Engine Company.

W.S. Nott Company, Minneapolis, Minnesota built a number of steamers beginning around the turn of the century. FIGURE 113 is a Nott fire engine built in January, 1901. On March 20, 1911 the New York Fire Department began operating their first motor-powered steam fire engine, a Nott. Engine Company Number 58 used this vehicle which was a 700 gallon per minute pumper with a top speed of 30 miles per hour. It cost the City of New York, $9,722. The Nott Company, however, did not make a successful transition into the modern world, and it disappeared about 1915.

Abel Shawk was a well-known builder of hand-powered fire pumpers who fabricated his first, out of a total of only five, steam fire engine in 1855. FIGURE 114 is a drawing of his machine named the "Missouri," said to have been the first steamer ever used on the west bank of the Mississippi River.

The Silsby Manufacturing Company of Seneca Falls, New York, was established as a manufacturer of hand-powered fire pumpers in 1845. In 1856, with the firm now named Silsby, Mynderse & Company, their first steamer was built. This experimental prototype, called "The Neptune" was a 9,500 pound monstrosity (FIGURE 115). The Silsby firm was perhaps the most successful of all of the steam fire engine manufacturers, producing more than 1,000 such vehicles. Silsby machines were quite refined in many respects, but it was Birdsill Holly's rotary pump which brought them fame. Additional Silsby steam fire engines are shown in FIGURE 116.

Finally, from Fire and Water, issue of July 25, 1896, FIGURE 117, information concerning the Waterous Engine Works of St. Paul, Minnesota is presented.

great weight precluded their acceptability by the fire services. Their rotary pump-steam piston "mongrel" engines, however, were well received. One of these, named, "Manhattan No. 8," together with a crew of firemen traveled to England in 1863 in response to an invitation from the London Fire Brigade. The American machine was to have been demonstrated during a public exhibition at the famed Crystal Palace. As the story goes, the host Londoners caused the Manhattan No. 8 to be damaged in an accident, and an unsatisfactory pumping performance was the result. Thus the London Times newspaper reported that, "It must be understood that the American steam fire engines are as much behind the steam fire engines of other countries, as that most pretentious political association called the 'New York Fire Brigade' is behind any fire brigade in Europe in real usefulness."

We next move on to contemplate the subject of electric fire vehicles: fire pumpers and other apparatus powered by electricity. Perhaps nowhere else in the entire history of civilization is there to be found another event with such special significance for the firefighting fraternity as the advent of electricity. First, it was the introduction of electric power, culminating in the invention of the incandescent light bulb which finally terminated the era of night-time illumination by the flame. And second, the ever-expanding networks of wires and electric devices, especially as found inside of buildings, has inaugurated a new spectrum of hazards. And electric motive power for vehicles and for pumps was first devised long before the final runs of horse-drawn rigs, has persisted for many years, and continues as an object of active study.

FIGURE 110

FIGURE 111

FIGURE 112

FIGURE 113

FIGURE 114

FIGURE 115

FIGURE 116

WATEROUS STEAM FIRE ENGINE.

IN this number of FIRE AND WATER are given a cut and description of a new type of steam fire engine manufactured by the Waterous Engine Works Company, of St. Paul, Minnesota. The aim of the designers of this machine has been to reduce and simplify the working parts of the entire engine, yet at the same time to accomplish in the way of a positive motion what has hitherto been capable of being done only by the ordinary double crank engine. That they have undoubtedly succeeded in this is

FIG. 1

WATEROUS STEAM FIRE ENGINE, FIGURE I.

beyond dispute, as an examination of Fig. 1 will show. One crank and one eccentric do the same work as that of double cranks and double eccentrics, thus reducing by one-half this amount of extra machinery, with its consequent additional friction, both pistons working in the same relative position with each other as if employing two cranks and two eccentrics. (Patents are now pending for this method of construction as applied to steam fire engines). A still greater advantage, however, to be had from this method of construction is the fact that the pumps are given a long stroke, while at the same time the pumps and steam-cylinders are brought closer together.

Mechanics will quickly perceive the immense advantage of using a long stroke on piston-pumps, as it reduces materially the work and wear and tear of the pump-valves, and also reduces in a like degree the slippage or loss of water consequent upon the opening and closing of the valves. The valves in the pumps of this engine do not open or close within 30 per cent. as often as those in any other type of steam fire engine pump of equal capacity.

The materials and workmanship throughout are of the very highest standard. The pumps themselves are made entirely of pure phosphor bronze; no other metal whatever entering into their construction, and are so arranged that all the valves are exceedingly easy of access. By simply removing the outer plate or case, which requires only the unscrewing of fourteen nuts, the valves are exposed, making it possible to get at the valves quickly and easily when occasion requires. The area of all valve-openings and water-ways are in excess of the area of piston-displacement, thus insuring an easy flow of water through the pumps and doing away with all pounding of water so often found in piston-pumps.

The boiler used is the "Waterous Patent," having the double upper head and water-chamber, whereby the tubes are constantly submerged in water, which prevents the possibility of their becoming overheated and leaking. The large amount of heating surface obtained in the construction of this boiler enables it to steam easy and fast without undue forcing; thus adding greatly to the durability and life of the boiler.

The running gear is unusually strong and very handsomely arranged. The front of the engine rests on full platform springs and the rear of the engine on a pair of long half elliptic springs, hung below the frame with togle-joint arrangement.

At a recent test of one of these engines (third-class in size), a 1 1-2-inch stream was thrown a distance of 245 feet. Although the test lasted several hours and was of unusual severity, yet nowhere was there perceptible the slighest weakness or defect; the machinery ran so smoothly and easily that the engine stood apparently as steady as when not in operation; there being an entire absence of the jumping and springing so often found in steam fire engines where double cranks are used.

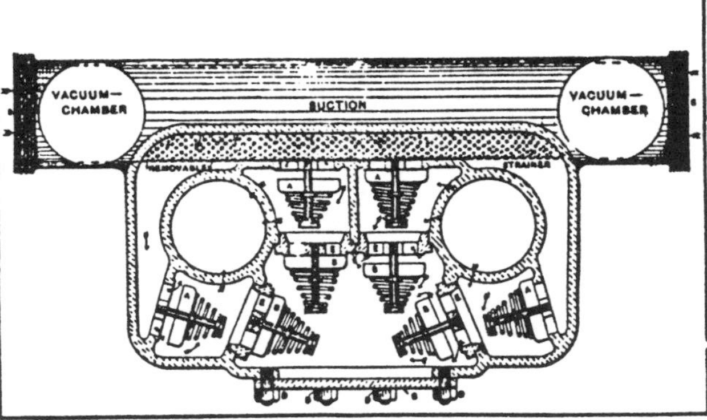

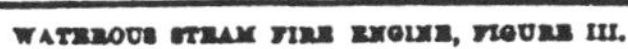
WATEROUS STEAM FIRE ENGINE, FIGURE III.

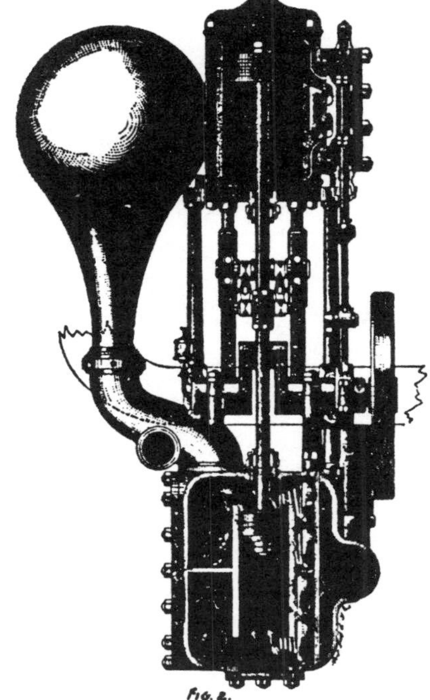

WATEROUS STEAM FIRE ENGINE, FIGURE II.

FIGURE 117

Fire Engines Powered By Electricity

Introduction

While the concept of an electric fire engine seems to date from the 1870s, it is necessary to look back just a little farther in history in order to understand the evolution of this idea. So we begin with a man named Abraham Brower, and the development of public transit in the city of New York.

In 1826 the two hundred and some thousand New Yorkers, together with Brooklynites and assorted other folk from the suburbs had no running water, no street lights, no professional police or fire protection, and no public transportation. Unless you were the owner of a horse or a mule and were able to ride same, or unless you could afford to hire a carriage, you walked.

But in 1827 Brower contracted with the firm of Wade and Leverich, coach builders, for the construction of a twelve-passenger stage coach-like vehicle. It had open sides and four benches which alternately faced front and rear, and it was called the "Accommodation" as it operated along Broadway. A couple of years later Brower added a second vehicle to this route. And in 1831 he introduced his first "omnibus" to the traffic on Broadway. This bus was without doubt based on a concept which he purloined from its English originator, George Shillibeer. In an omnibus the seating was arranged lengthwise, the driver was placed high up upon the front, and small boys were stationed at the rear, where passengers entered and alighted, to collect fares. A ride from the Battery to Bond Street cost twelve and one-half cents.

By 1836 the number of such vehicles using the streets of New York was well over 100. And twenty years later there were at least 600 omnibuses in operation. But years earlier, in 1832 to be precise, a horse-drawn street railway service had been started by the New York and Harlem Railroad, and this operation was the direct ancestor of the subsequent electric trolley and the subsurface electric car. Many of the major metropolitan street railway lines eventually owned more than one thousand horses each. Horse car enterprises of this magnitude actually pursued two simultaneous business endeavors. Their function was to carry paying passengers, of course, but they were also engaged in peddling their horse manure. By 1885 there were some 105,000 horses and mules which were employed in pulling 19,000 street cars over the 3,000 miles of tracks within our cities. It was not until 1917 that the last horse-car line finally disappeared from the streets of the city of New York.

Surviving today only in San Francisco, there were by 1890 about 500 miles of cable railway with about 5,000 cable cars in the country. This particular fad was started in the mid 1870s and had all but disappeared 30 years later.

The first commercial electric-powered street railway system was carrying passengers in Montgomery, Alabama in 1886. Following this introduction the story of the street railway car was that of its rise, its plateau, and finally its collapse in favor of the gasoline-powered bus. But during many of those active years the overhead electrical power line was there, as shall be related, to tempt and excite the inventors of electric fire engines.

The electrical industry began in 1869 with the Gray and Barton Company, the antecedent of Western Electric. Products such as bells, telegraph equipment, and fire and burglar alarms were their first manufactured offerings. But it was the 1876 Centennial Exposition which clearly marked out the birth of this new industry of electrical manufacturing. Alexander Graham Bell brought his new telephone device (considered to be the single most valuable invention ever patented) to the Exposition, and there were displays of direct-current generators and dynamos.

By 1880 Thomas Alva Edison was making DC generators and lamps, and there were several arc light systems on the market. September 4, 1882 marked the first full day of operations at Edison's Pearl Street generating station in New York. He and his engineers had installed six direct current generators driven directly by steam engines of 900 horsepower overall. And by 1887 at least fifteen competing companies were producing small motors. Fire apparatus inventors were not inattentive to these developments.

L.G. Woolley

Leonidas G. Woolley lived in Mendon, Michigan, a very small town near the southwest corner of that state and, curiously, just a few miles along State Road 60 from a town named Leonidas. Woolley was the recipient of U.S. Patent 207,377 dated August 17, 1878 for his invention which he called an "Improvement in Electromagnetic Fire Engine and Alarm." As this seems to be the very first ever patent granted for an electric fire engine, the word "improvement" does impress one as gratuitous, but such was and still is the type of terminology beloved to patent lawyers.

What Mr. Woolley had designed was, in fact, a complete system composed of a fire alarm, fire engine response, and fire suppression. Whether or not it would have worked is open to serious question. Thankfully no record exists that it was ever built and tested as the inventor desired. The concept was neat and clean, and some of the inventor's ideas were well in advance of the technology of his time. Woolley wanted to operate a rotary pump with an on-board electromagnetic motor, but he wanted to do so without regard to the distance from his direct-current battery power supply.

FIGURES 118, 119, and 120 are the drawings for Woolley's patent. As can be seen he designed a side-by-side siting of the motor and the pump and he placed them both directly over the rear axle, characteristics which would have tended to unbalance the vehicle and make it objectionably wide. It was the inventor's intent that his fire engine be hand pulled to the fire location and then be attached–plugged in, so to speak–to the electrical source which was to energize its motor and thereby power its pump. The on-board reel of electrical wire designated by the number 4 in the drawings, was to serve as a sort of extension cord so that the engine could be moved closer to the fire after it had been plugged in.

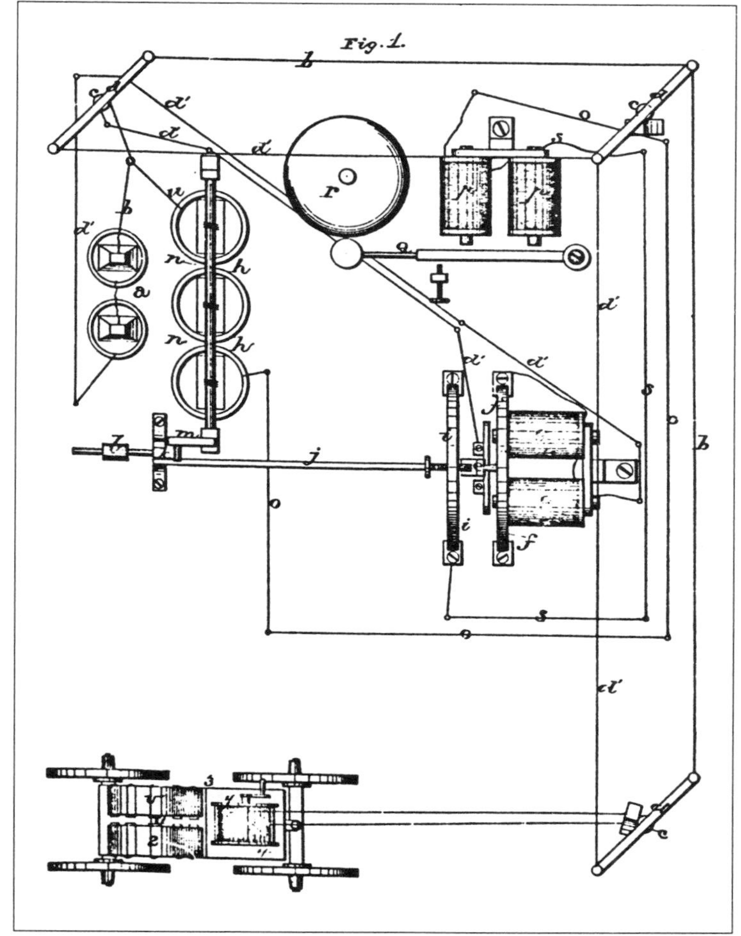

FIGURE 118

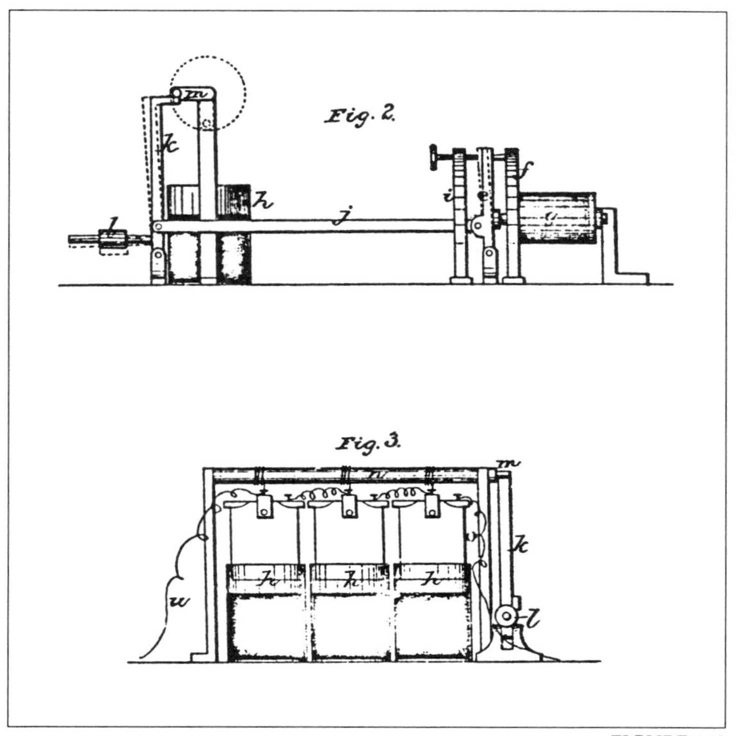

FIGURE 119

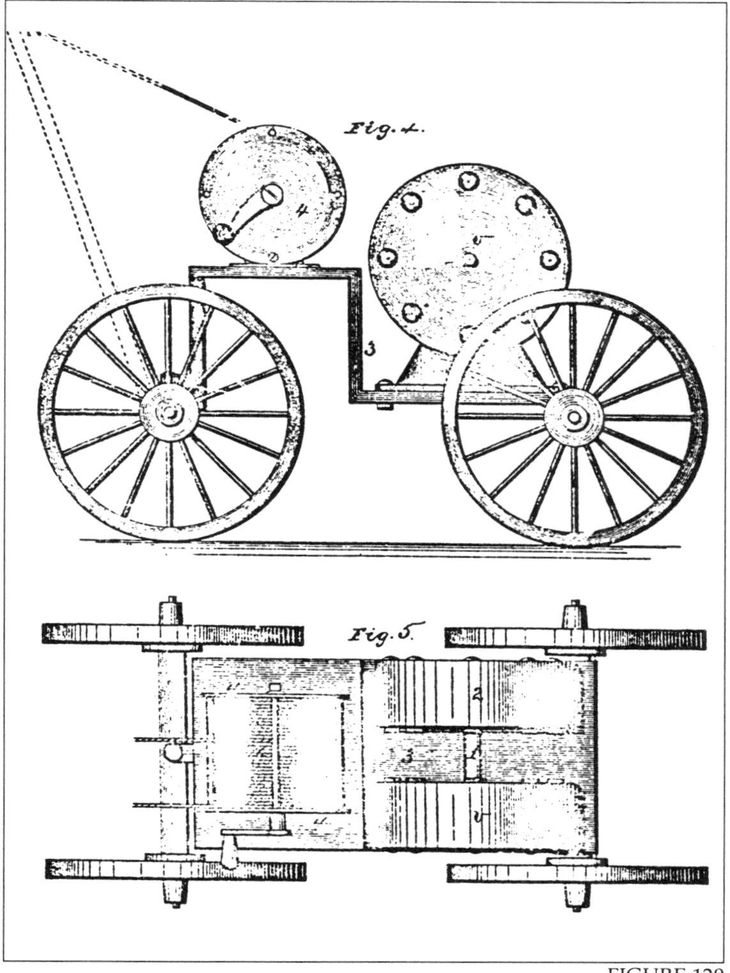

FIGURE 120

"Should more than one engine be required," said inventor Woolley, "the current will be passed through the first engine into the second one, and so on to any desired number of engines which may be needed. Where the conducting cords must cross a street the tongue of the engine will be made longer than usual and will be elevated so as to raise the wires sufficiently high to be out of the way of horses or vehicles passing along."

"My combined electro-magnetic engine and fire alarm is especially adapted for use in villages and towns, as the pump and engine together will weigh but two or three hundred pounds, and hence can be drawn by one or two men. There being no parts to get out of order and nothing to superintend no engineer is required to run or operate the engine while in use. It is always ready at a moment's notice, and being so light it can be taken into places where the ordinary steam engine could not be drawn. As there is no necessity for any men to work upon the engine, as is the case with the usual village fire engine, the efforts of the men can be directed toward removing furniture and other articles to places of safety. By the use of electricity in operating the fire engine it will be readily seen that the first great cost of a steamer is entirely avoided; that the engine has but to be taken to the place of the fire and the pump will operate of itself with no further supervision. The engine as constructed contains no water nor steam pipes and no valves, and needs no regulating nor adjusting while in operation."

Well, the most charitable reaction to the foregoing is to characterize it as hopelessly utopian. That Leonidas Woolley was a thinker too far in advance of his own time seems evident, given that the day of the electric vehicle has yet to dawn. One of the problems with these vehicles, one problem among many, has been and still is the battery system. A lead-acid battery of the traditional kind, even if it was 30 to 35 times larger than the average gasoline tank, could power only a very light vehicle, for only a very short distance, and at only a very slow speed. A battery system using sodium and sulfur electrodes produces much more power per unit of its weight, but it is markedly exothermic and markedly risky: metallic sodium burns and explodes when contacting water, such as could occur in a fire vehicle accident. The quantity of power which can be extracted from a nickel-cadium battery lies between these other two kinds, but NiCad systems are very, very costly. Small wonder, then, that electrical vehicles are still highway curiosities, whether they have been Anderson Electric cars of 1920, Ford Comutas of 1967, English GM Griffons, GM 60-mile range G-Vans, nickel-iron battery-powered TEVans, or 1992 GM Impacts.

S.S. Wheeler

Another electric fire engine invention, and one which was in fact built and given a trial, was described in patent 312,939 dated February 24, 1885. This vehicle was configured to carry its motor and its pump in a front-to-rear axis, and it was mounted on a horse-drawn carriage. The inventor was Schuyler Skaats Wheeler, a former President of the American Institute of Electrical Engineers, and the founder of the Crocker-Wheeler Electric Company of Ampere, New Jersey. FIGURES 121 and 122 are the patent drawings. Notice the reference to "Attorneys Curtis and Crocker." Francis B. Crocker was Wheeler's partner. The photograph of this fire engine, FIGURE 123, is from an original in the collection of the Rensselear County Historical Society. FIGURE 124 first appeared in the August 13, 1910 issue of Scientific American. The carriage was, clearly, in the Silsby style.

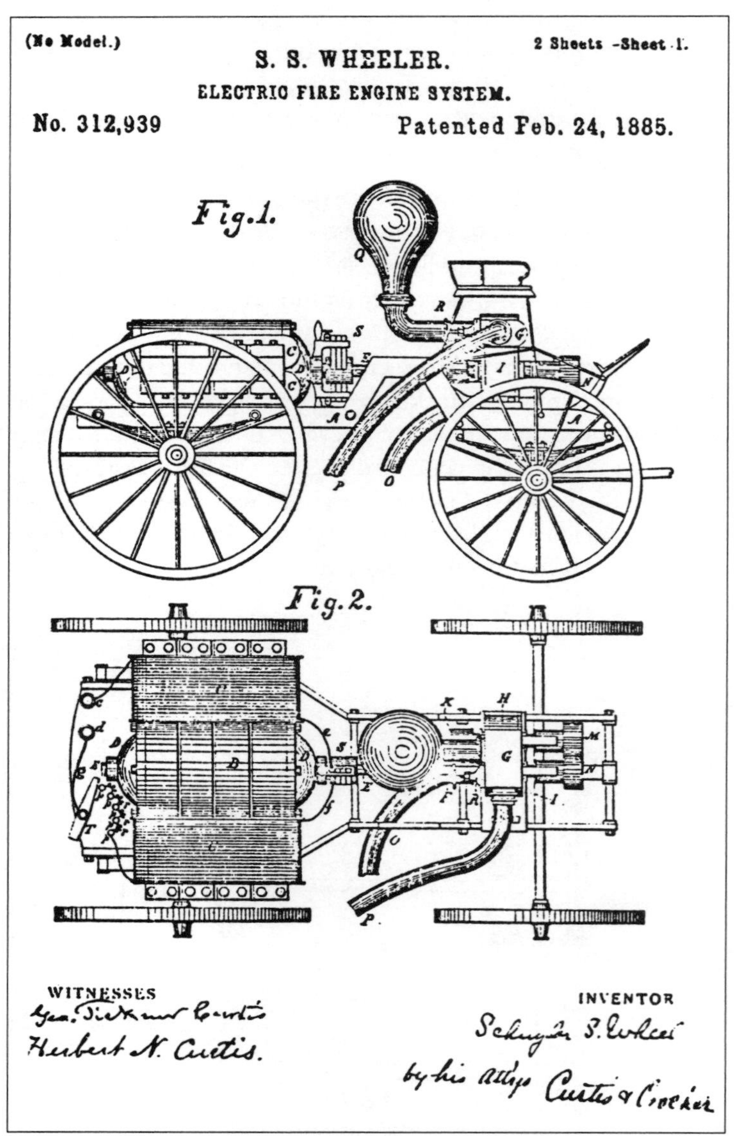

FIGURE 121

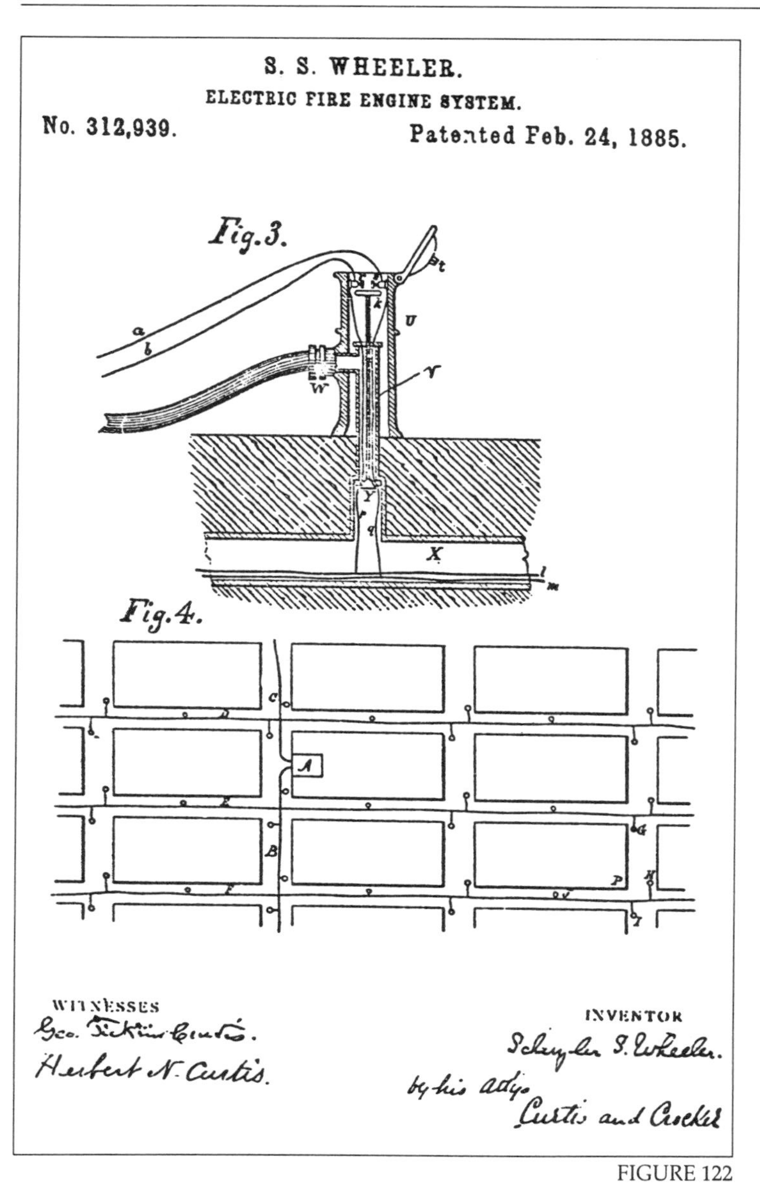

FIGURE 122

S.S. Wheeler's idea was not to use battery power at all, but rather to "plug in" to power lines which paralleled the water mains and hydrants. Among the essential components seen in the patent drawings were the following: the fire engine frame, designated A; the motor, B; the magnets, C; the armature, E, which was connected to the shaft, F, of the pump, G. Water was to enter from the hydrant connection by a feed pipe, O, and was to have been pumped out through a fire hose, P, with uniformity of flow from the piston pump being maintained by the air chamber, Q. With respect to the hydrant, letter U designated a cast iron case for the vertical pipe, V, which was to connect the water main, X, and the nozzle, W, with flow being controlled by a valve, Y. The electrical cables were to have been carried inside of the water mains. The central electric generating station was supposed to be able to develop about 375 kilowatts, enough, "to actuate the maximum number of engines in operation at any one time during a fire."

There was a brief note about this apparatus in the November 10, 1888 issue of Scientific American: "A recent invention is Professor S.S. Wheeler's electrical fire engine. It is intended to be worked by the current of an electric light wire, which can be tapped for temporary service anywhere that it is wanted. Each engine will carry on a reel some 500 feet of insulated fine copper wires, bound together, cable fashion, so as to equal a number 3 wire, for transmission of the current. As it is a good deal easier to squirt electricity than to squirt water, the engine, it is intended, shall be placed near the fire and the electric connection made as is convenient. The powerful current of an arc light wire will not be required, that of the ordinary incandescent light circuit, which is much lower in intensity, being amply sufficient to run the motor of the engine. The great advantages claimed for the electric fire engine are that it can be instantaneously started up at full speed; that it is much lighter than a

FIGURE 123

steam fire engine of equal power; that it costs one-third less; that it is safer and easier of control; that it is noiseless in its operations; and that it is economical. Where there are no electric light wires in the street to be tapped, it will not be impractical to run it by means of storage batteries charged from a dynamo at the engine house or at any other convenient established point."

AN EARLY ELECTRIC FIRE PUMP.

The application of electricity to the elimination of the fire hazard in large cities is not so recent as many people imagine. This application is now accomplished in big cities such as New York by locating small electric pumping stations in various parts of the city and connecting them hydraulically to the city water supply and electrically to some source of power, usually the lighting company's generating station. A system of pipes leading out from the pumping stations gives them control over a large area. In the event of a fire, the pressure at any part of the system can be raised in a few seconds to the necessary tension.

In the first development of this idea the pumping station was mounted on four wheels and carried directly to the scene of action. The accompanying illustration shows its construction. This practice had several advantages over the usual steam boiler and engine outfit. It was much lighter and would develop the necessary pressure in less time. By its use the usual accompaniment of smoke was entirely avoided.

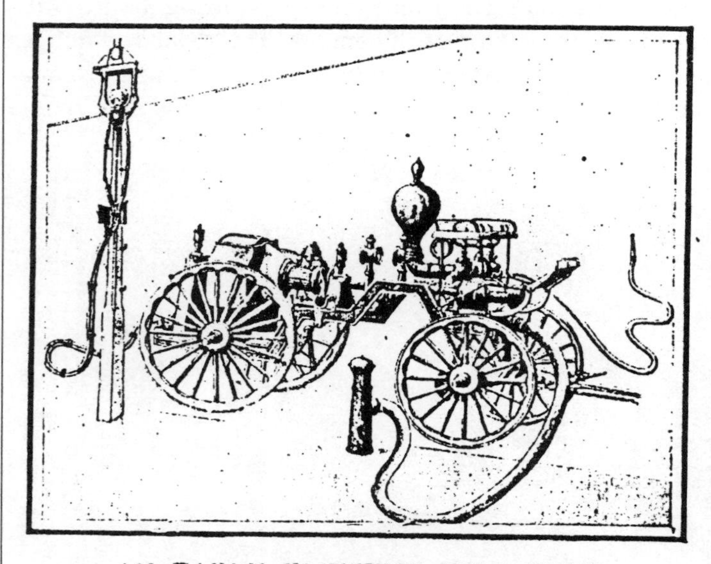

AN EARLY ELECTRIC FIRE PUMP.

It was far less expensive than the present electrical method, for its hydraulic system consisted merely of a few feet of hose and its electric system of a few feet of wire. The apparatus was patented by Dr. Schuyler Skaats Wheeler, past president of the American Institute of Electrical Engineers, in the early years of the Crocker-Wheeler Company, which he founded.

FIGURE 124

M.W. Dewey

Each of these next three electric fire vehicle patents differs from those already described because their inventor incorporated electrically-powered self propulsion as an important, coherent design component. This inventor was Mark W. Dewey of Syracuse, New York, and he assigned his patent rights to his own company, the Dewey Corporation, also of Syracuse.

Dewey declared that, "The object of my invention is to electrically propell an electric fire engine to and from the fire, so that it will not be necessary to employ horses for this purpose, and to save the time heretofore lost in hitching them to the engine, as the electrically-propelled engine is always ready to start at the instant the alarm is sounded and does not consume any energy while not in use. The object of my invention, also, is to electrically steer an electrically-propelled vehicle so that it may be handled easily under the perfect control of the steersman."

These were U.S. Patents number 446,703 dated February 17, 1891 (FIGURES 125, 126, and 127), number 464,244 dated December 1, 1891 (FIGURES 128 and 129), and number 464,245, also dated on the first day of December, 1891 (FIGURE 130). The electric motor was designated as letter A in the first patent drawings, and it was supported upon a framework, B, along with the battery case, D, and the piston pump, E. An additional motor, J, was to have been used for turning the steering mechanism. Dewey made no comments at all respecting the capacity, the construction, or the characteristics of his massive main battery. He replaced the piston pump with a Silsby rotary pump for his second patent, and also designed for it a compensating gear to govern rear wheel angular velocity during cornering. The final one of these three patents was for a battery-powered, electrically self propelled hose wagon which mounted an electrically-operated hose reel.

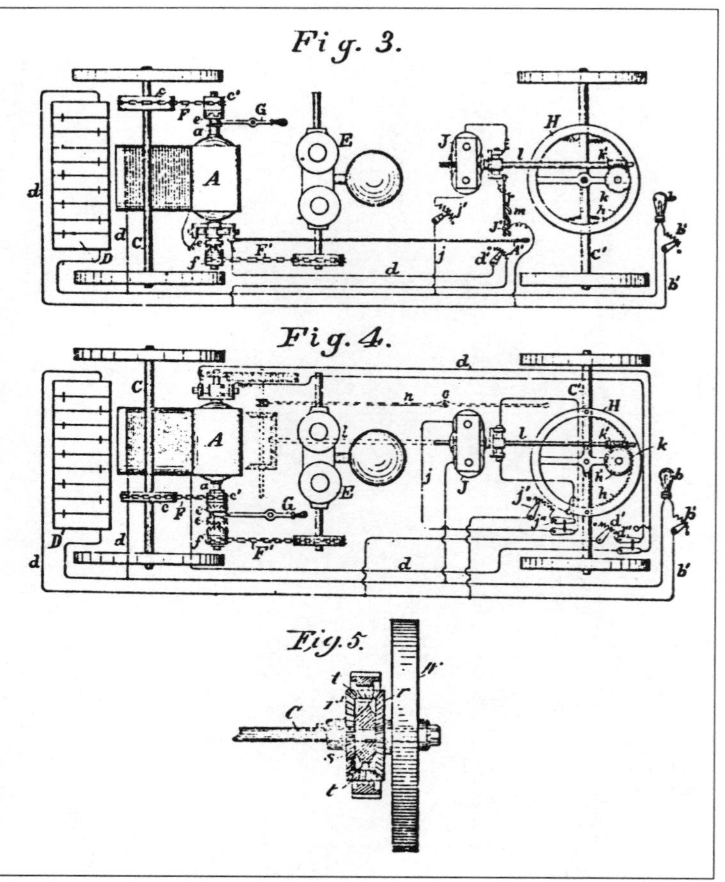

FIGURE 127

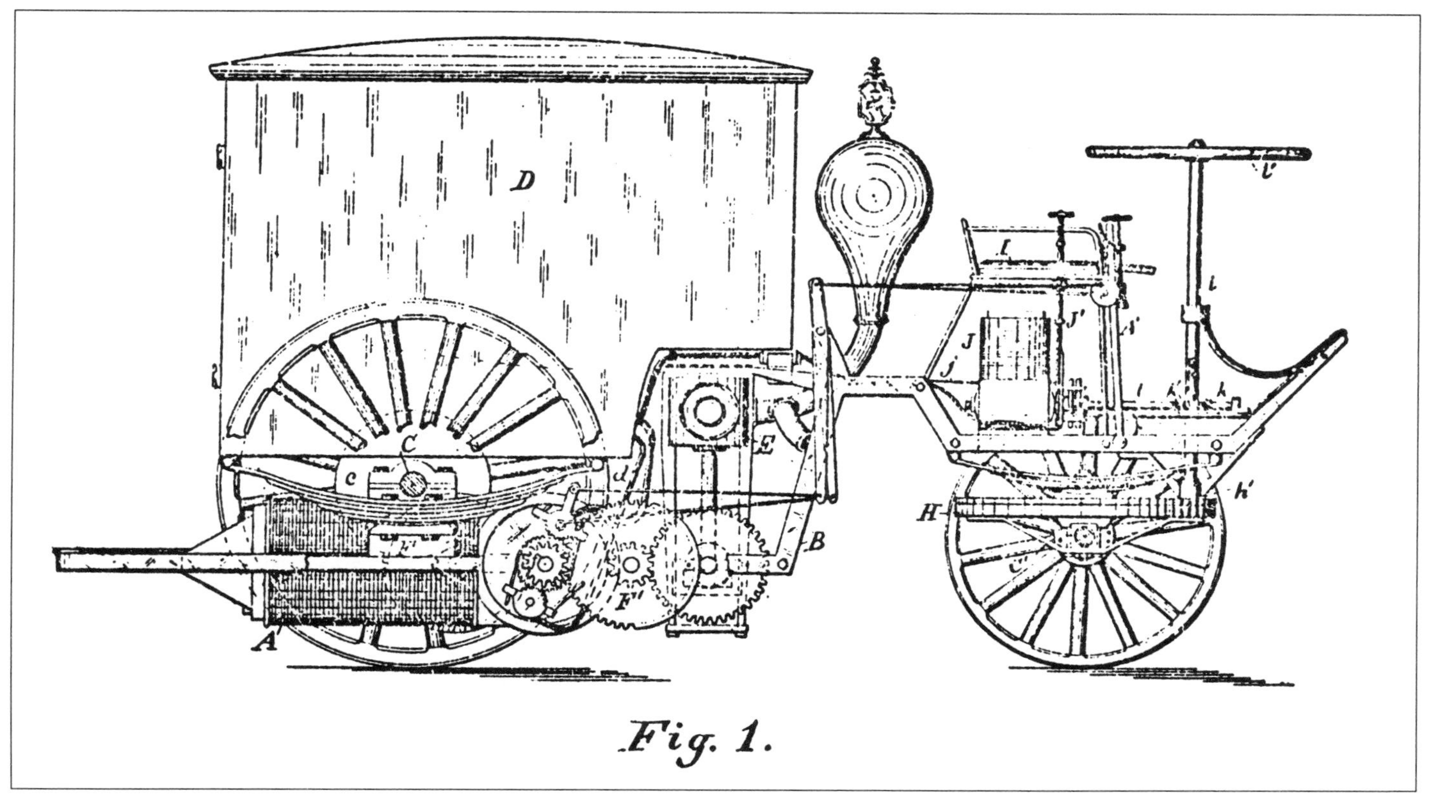

FIGURE 125

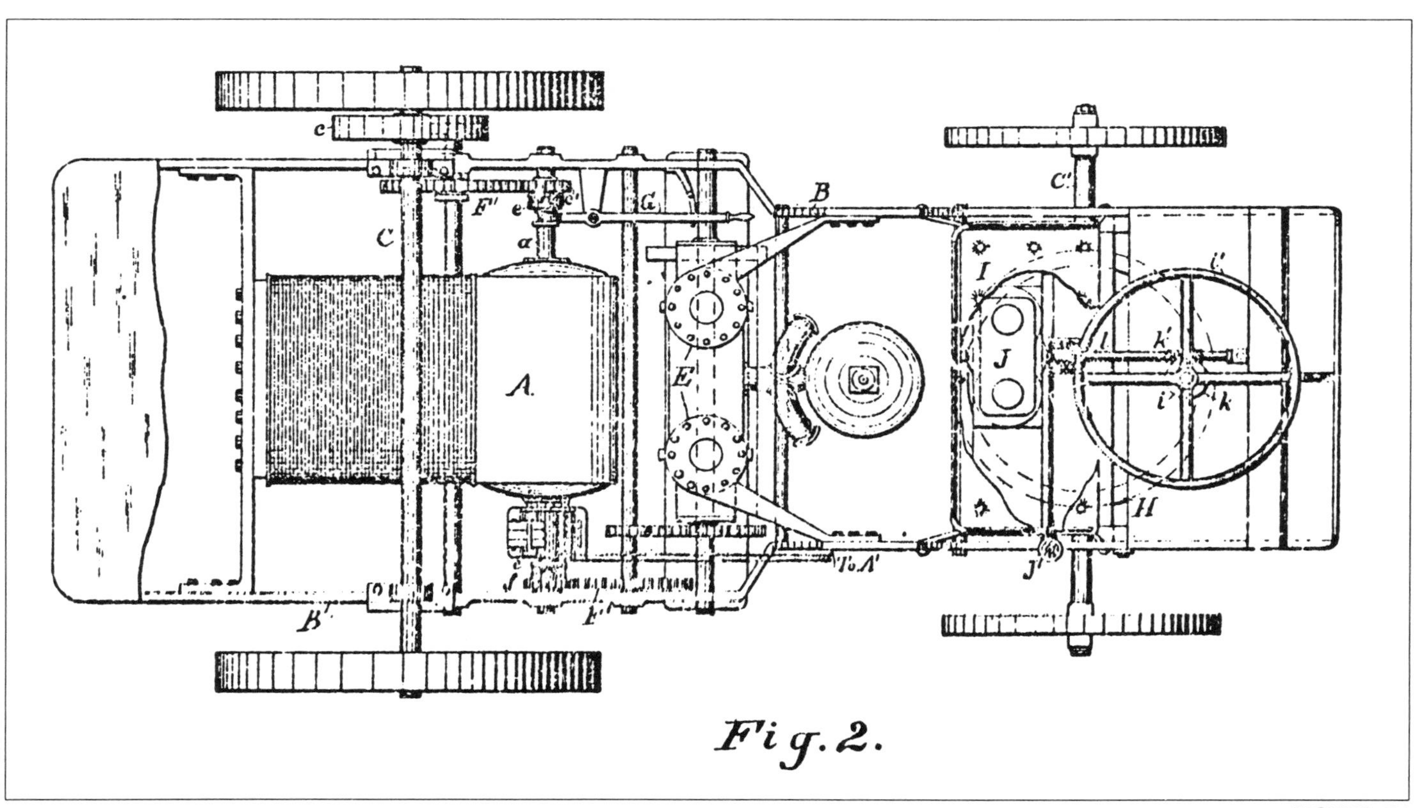

FIGURE 126

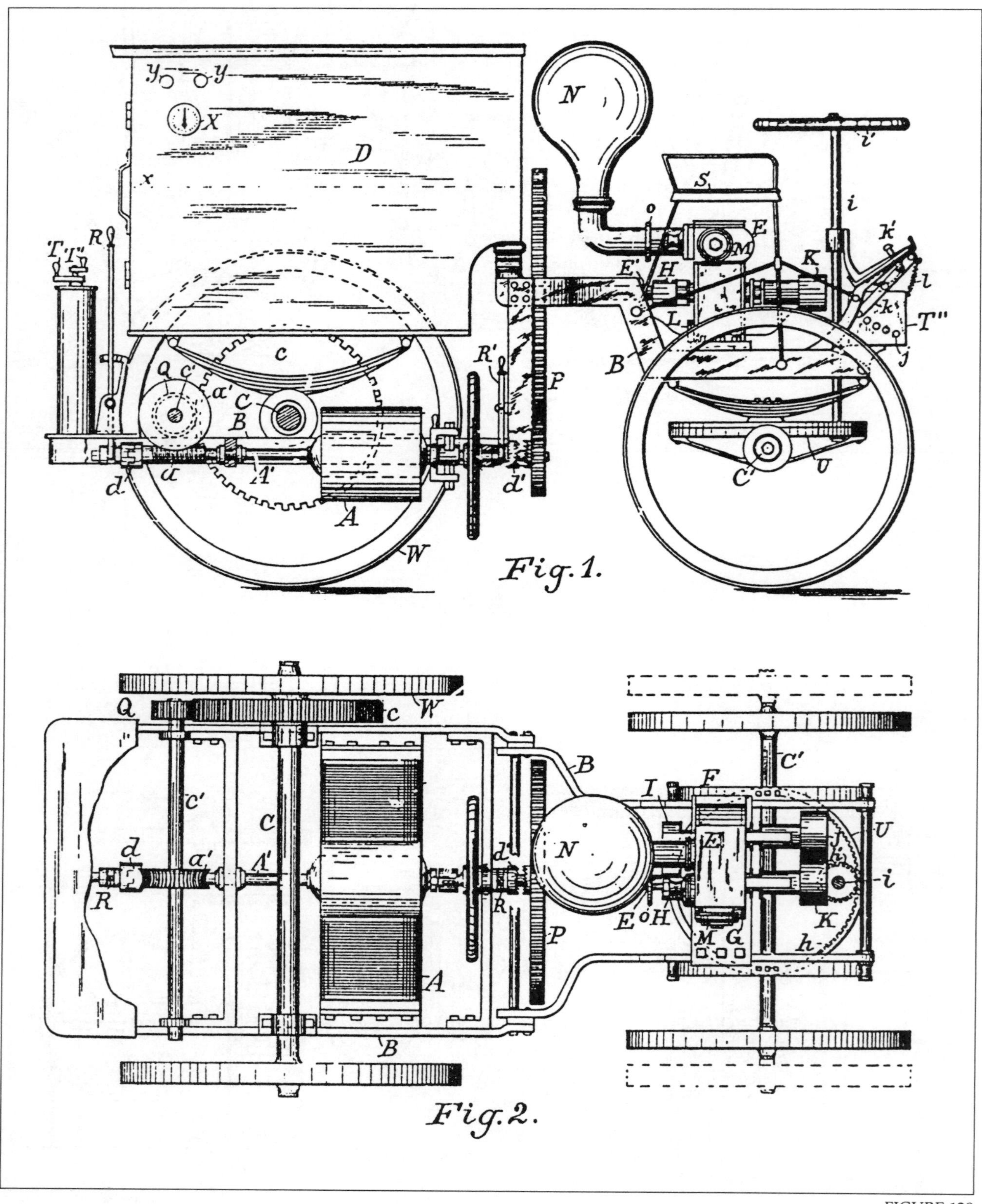

FIGURE 128

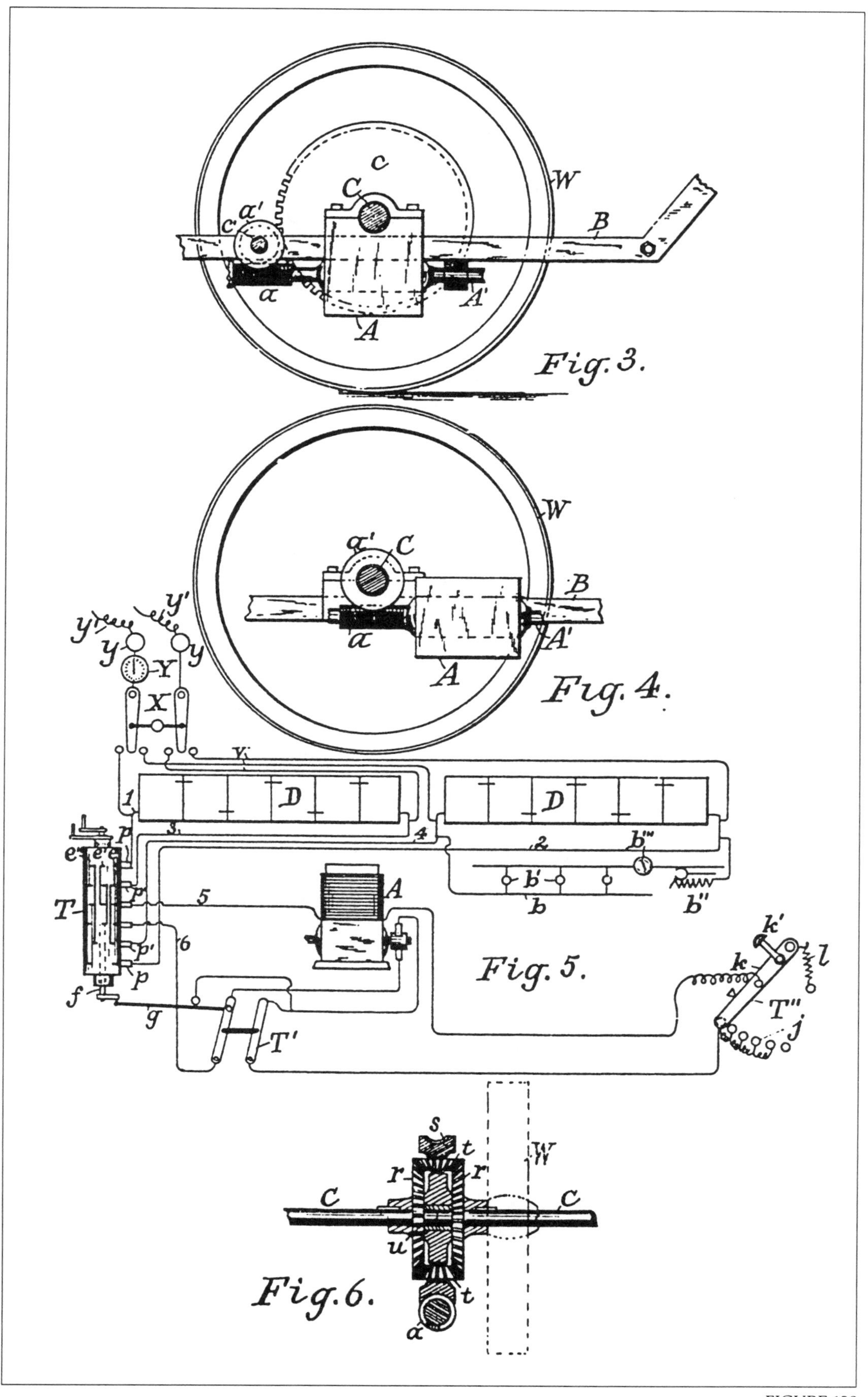

FIGURE 129

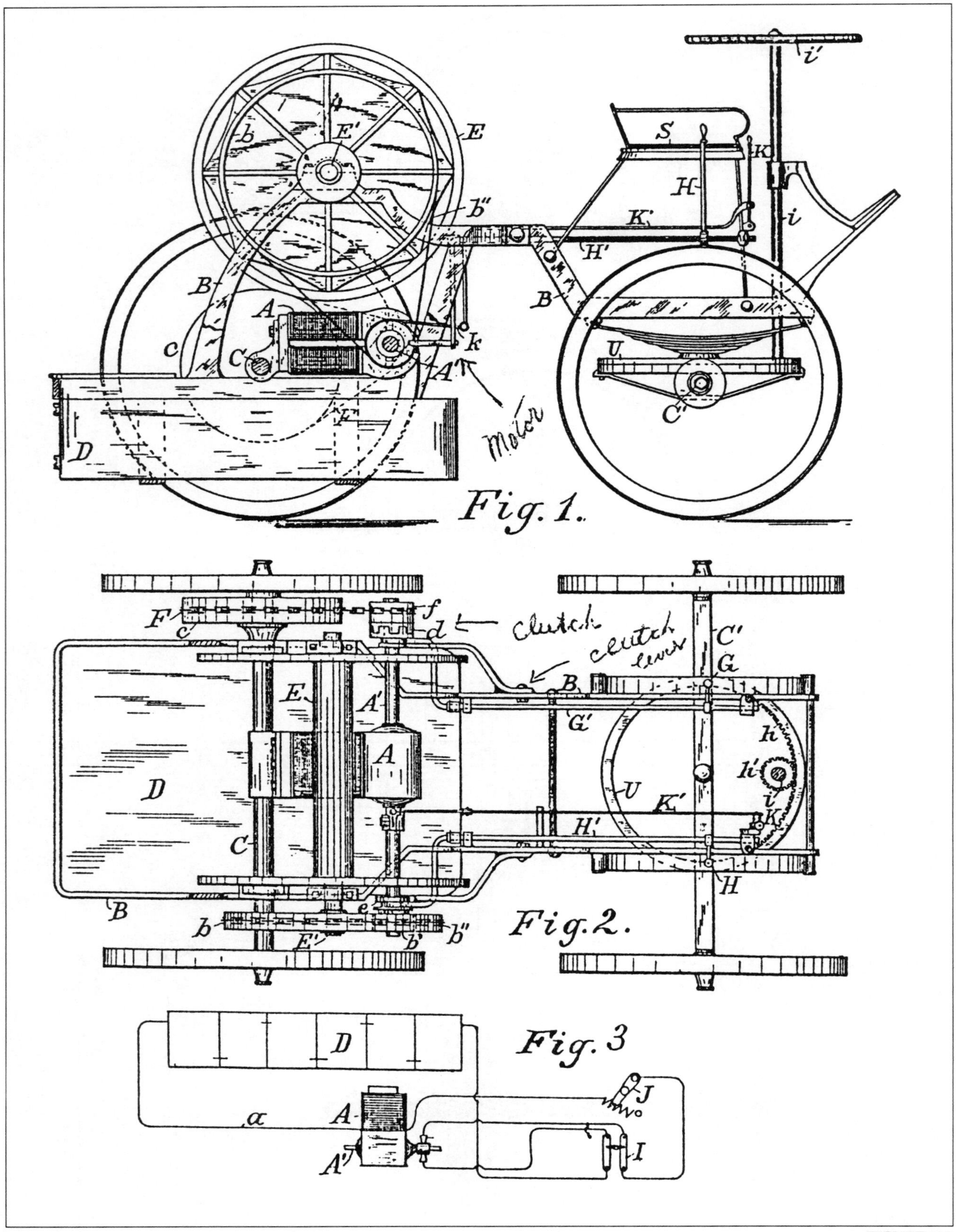

FIGURE 130

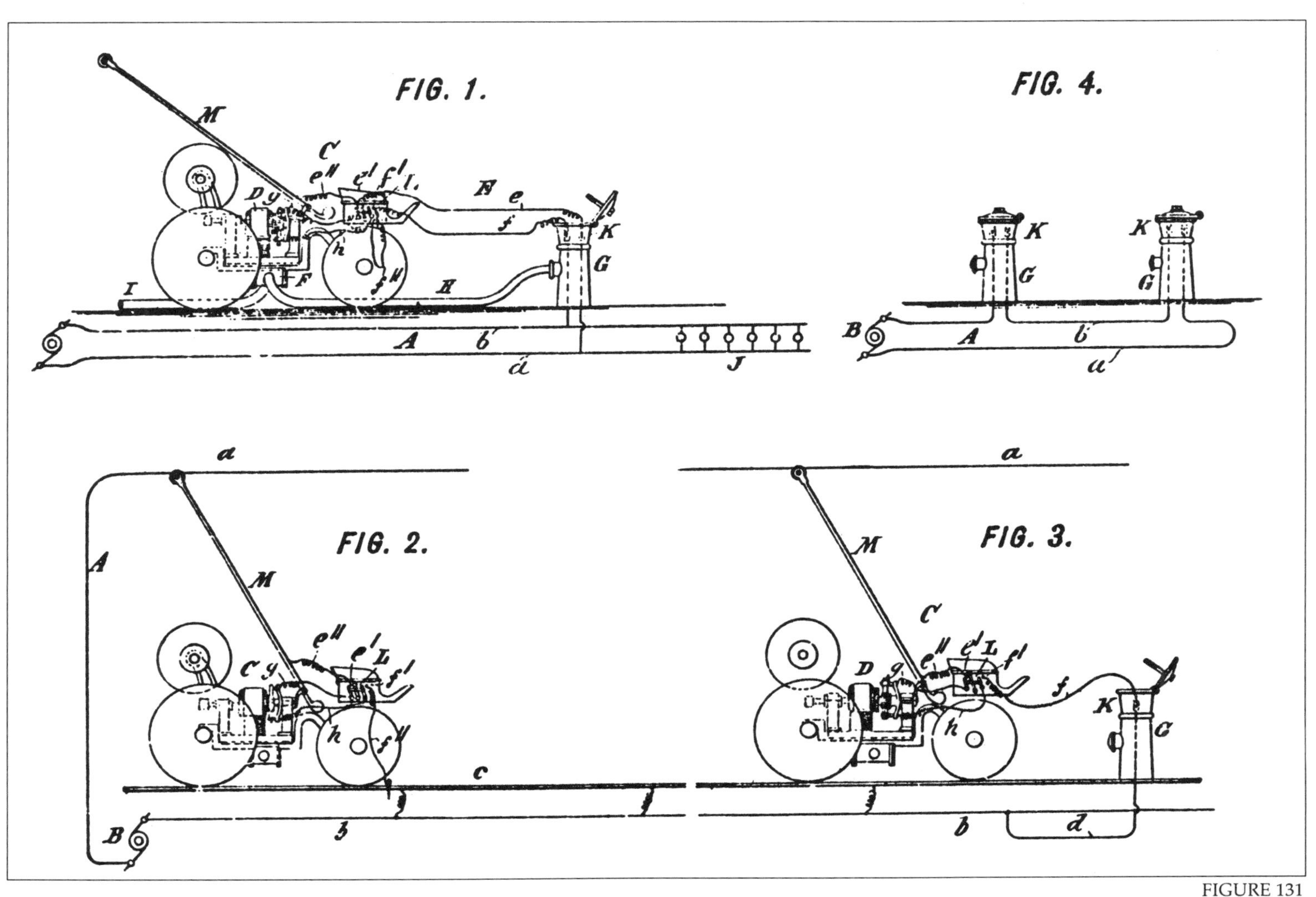

FIGURE 131

Trolley Wire Power

Overhead trolley wires, as well as either overhead or underground lines for incandescent lights were all suggested by inventors John Birkett and William McElroy as potential power sources for their electric fire engine. FIGURE 131 was one of the illustrations for their patent number 540,266, received on June 4, 1895. Although there exists no evidence that this distinctive system, designed for a horse-drawn vehicle, was ever built, we shall presently see that such apparatus was for a time employed in at least one European city.

The intriguing fire apparatus shown in FIGURES 132 and 133 was invented by Stuart, Iowa resident George W. Cox. Stuart, by the way, remains to this day a very small town, just about at Exit 93 on Interstate Route 80. Cox's patent number 588,399 was issued to him during 1897. This was a battery-powered, self-propelled vehicle which carried power-line connectors and a water tank, as well as the electric motor and pump, and which pulled along behind it a hose reel.

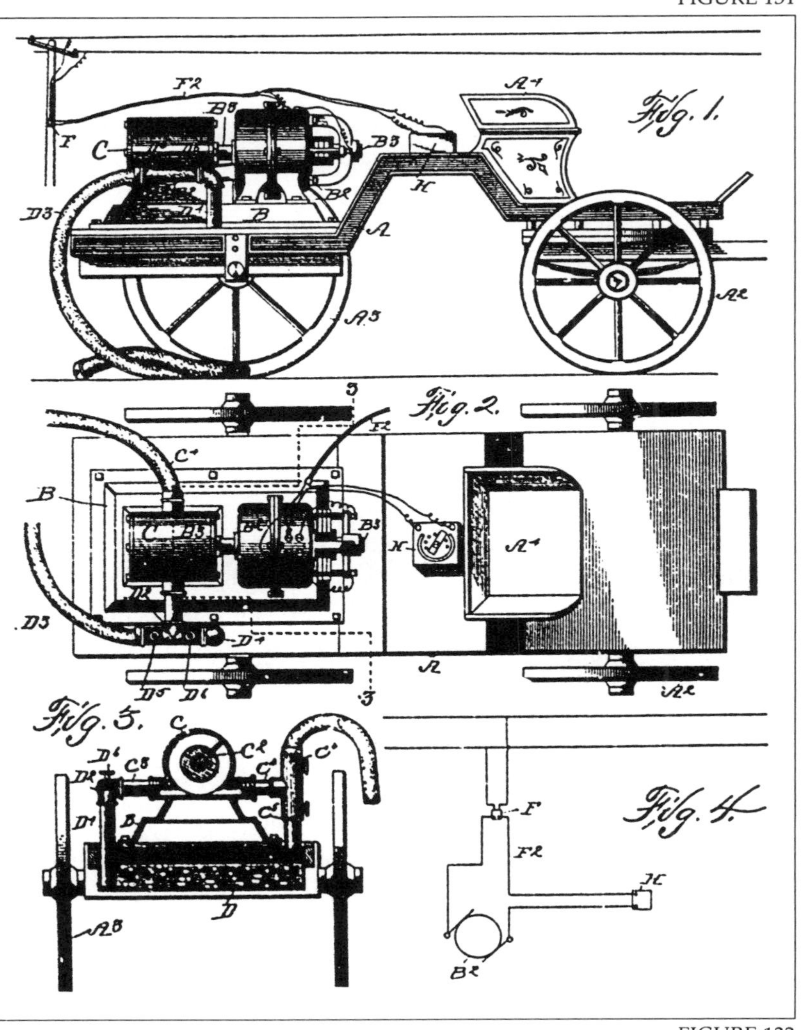

FIGURE 132

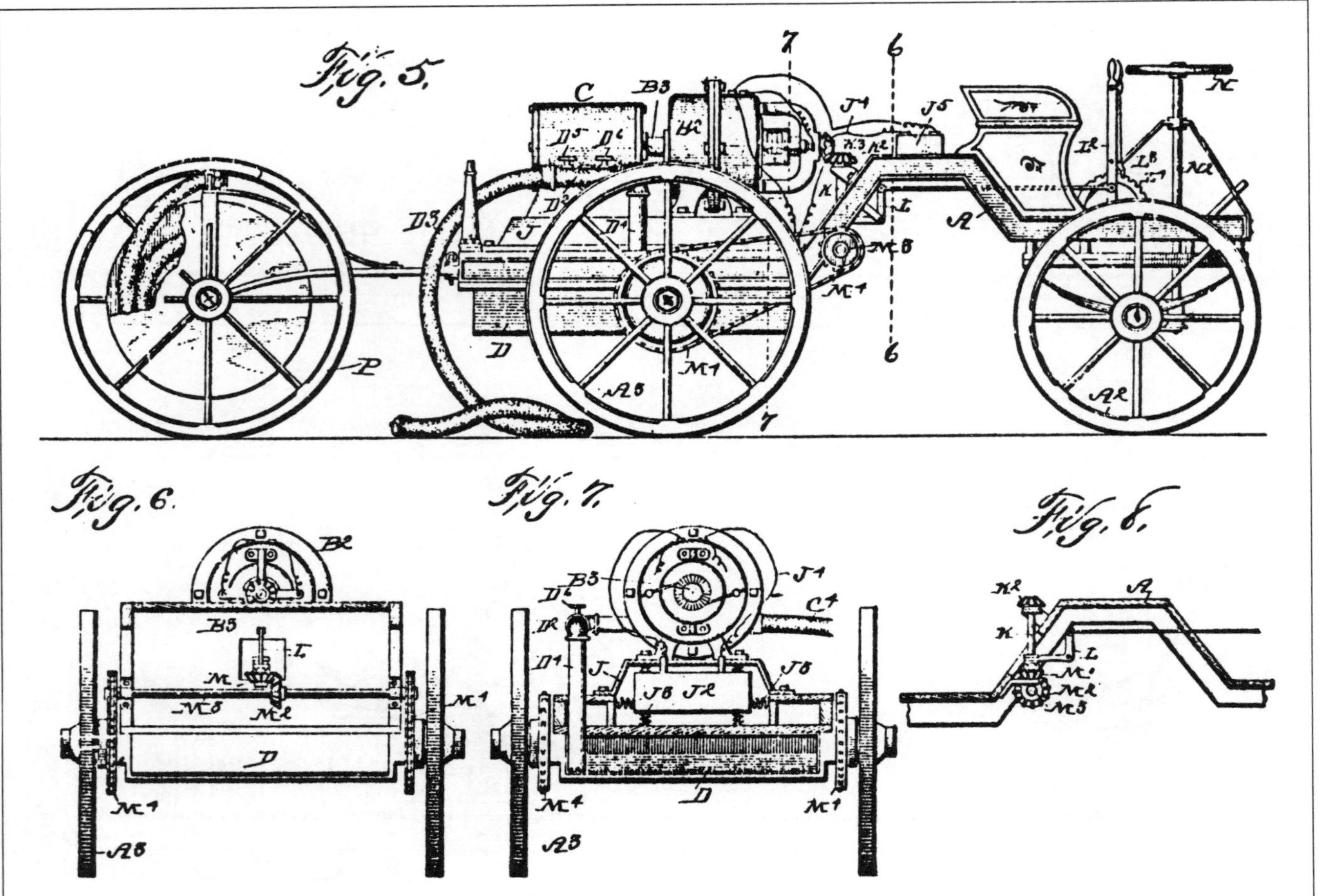

FIGURE 133

The Electric Engine Truck

Another development during 1897 was the introduction of the "electric engine truck" by the Springfield, Massachusetts-based Wason Manufacturing Company. This was an eight-wheeled track-riding car which could be used on streetcar tracks and interurban train lines. Platforms for carrying hose and tools as well as firemen were located over the wheels, both fore and aft. The 14,000 pound car was just under 31 feet in length and just about eight feet wide (FIGURE 134).

In order to load a steam fire engine the platform deck was first lowered to the rails. Next, the front wheels were disconnected and moved a short distance away, down the tracks. Following this the steam fire engine was backed into position and then drawn into place using a rear platform-mounted winch. The front parts were returned to their proper position, and the platform deck was raised above the rails with chain hoists, and then, as soon as the side girders were locked in place the vehicle was ready to roll. According to reports it was possible for a skilled crew to load and move within three minutes. During unloading it was possible to detach the front platform and wheels in 45 seconds, and 75 seconds later the horses had been attached and the rig was set to go.

The inventors felt that, "The great extension of the trolley and railroads in connecting widely-separated towns will enable such fire trucks to be used with great effect, as, in the case of a serious fire, engines from a number of towns may be brought into action in a very few minutes and will enable small hamlets and isolated houses to receive fire protection which they could not get in any other way."

AN ELECTRIC FIRE ENGINE TRUCK.

LOADING THE TRUCK WITH A FIRE TRUCK.

FIGURE 134

Whether or not this invention functioned in the grand manner as described has been lost to history. But even though questions remain without an answer, one can still admire this attempt at using electric transportation instead of running the horses until near dead.

Whiting, Whitlock, Brunau and Others

As the turn of the century was approached, and passed, more electric fire engine patents were receiving recognition. There was, for example, Beachmont, Massachusetts resident William H.H. Whiting who received patent number 632,665 near the end of 1899. This invention was a horse-drawn electric fire engine equipped with a rotary force pump and operated without an air chamber. It is illustrated in FIGURES 135 and 136. A handy source of electrical power was necessary to energize its motor. Wilbur

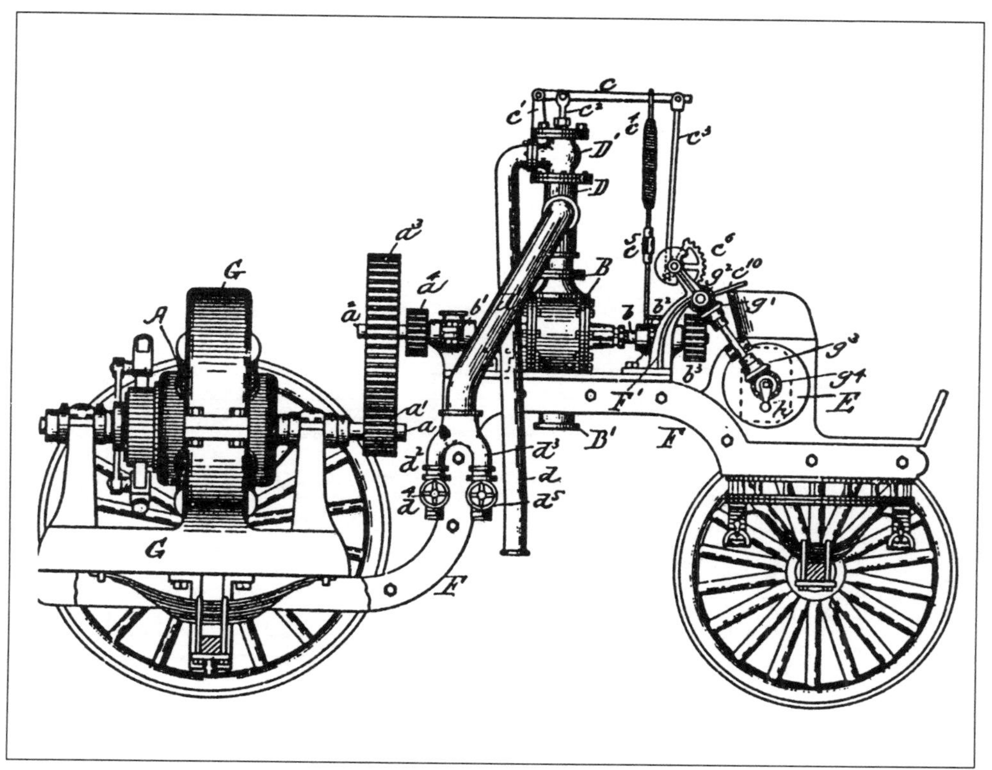

FIGURE 135

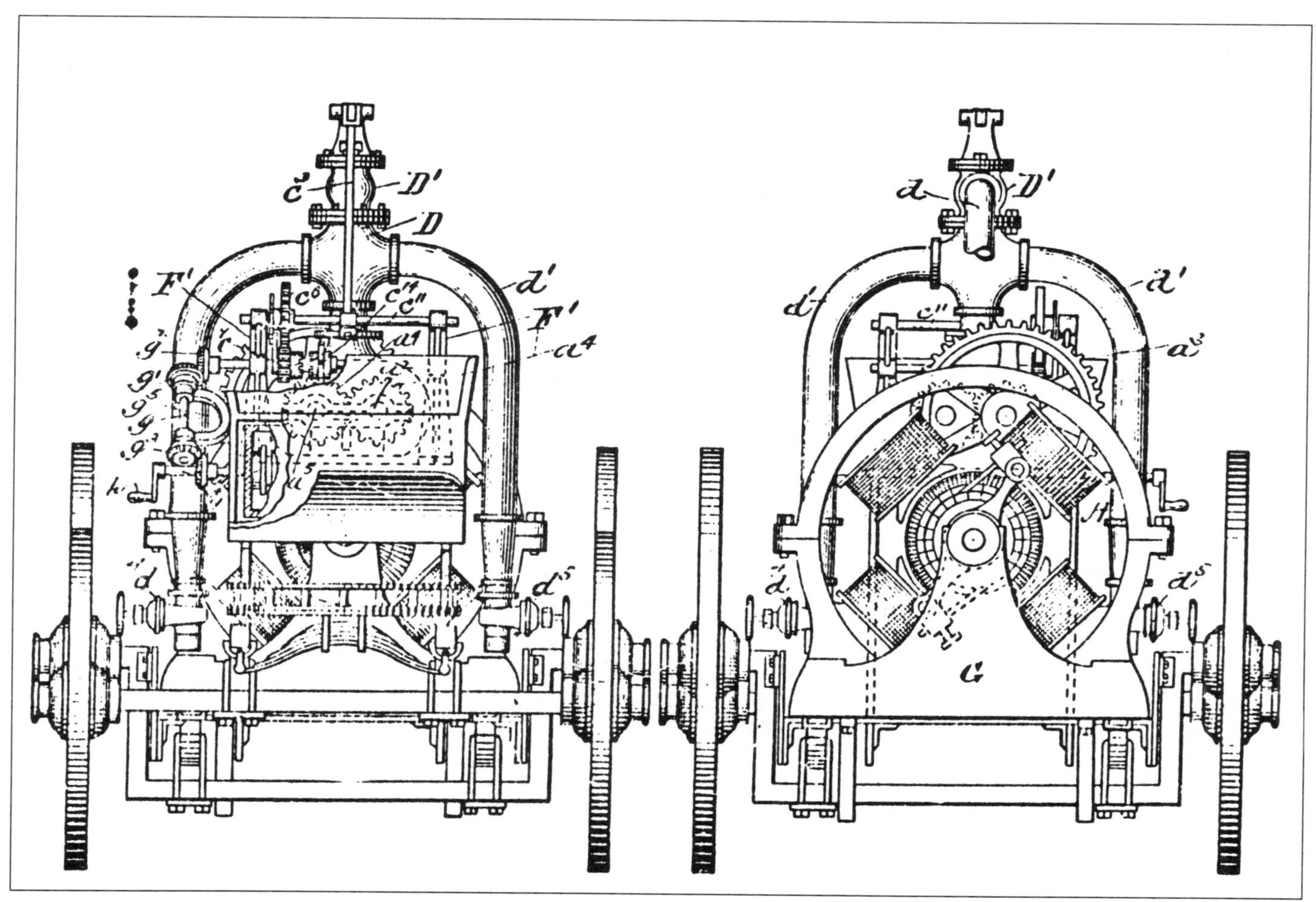

FIGURE 136

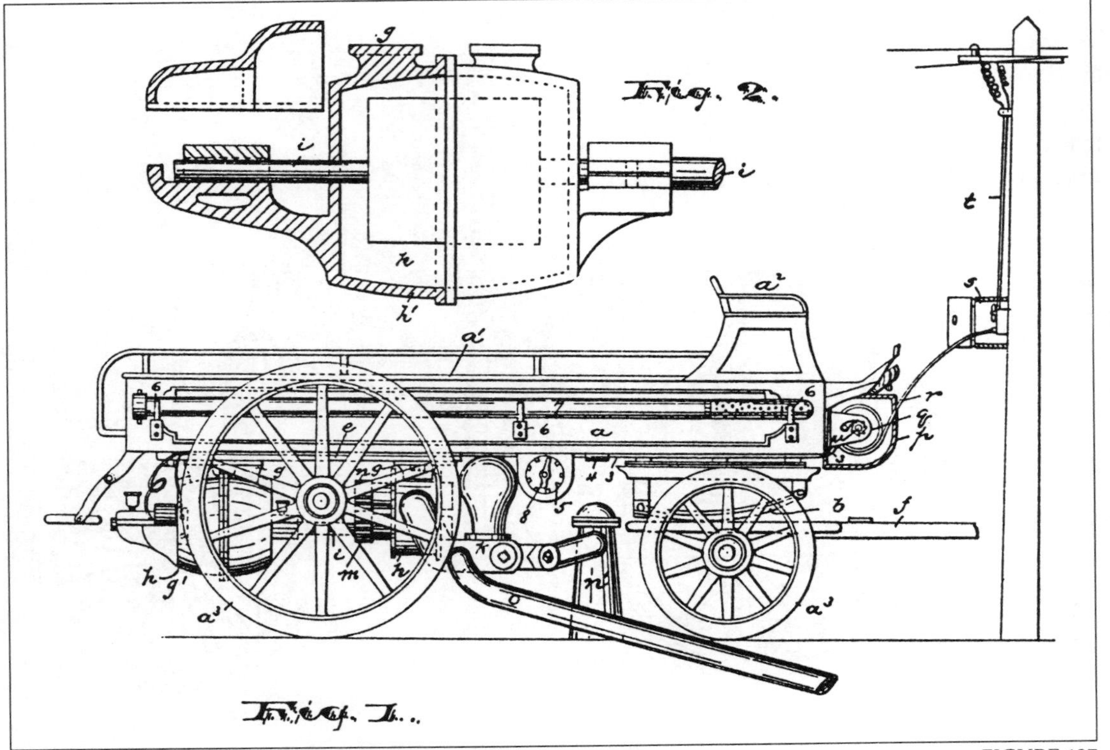

FIGURE 137

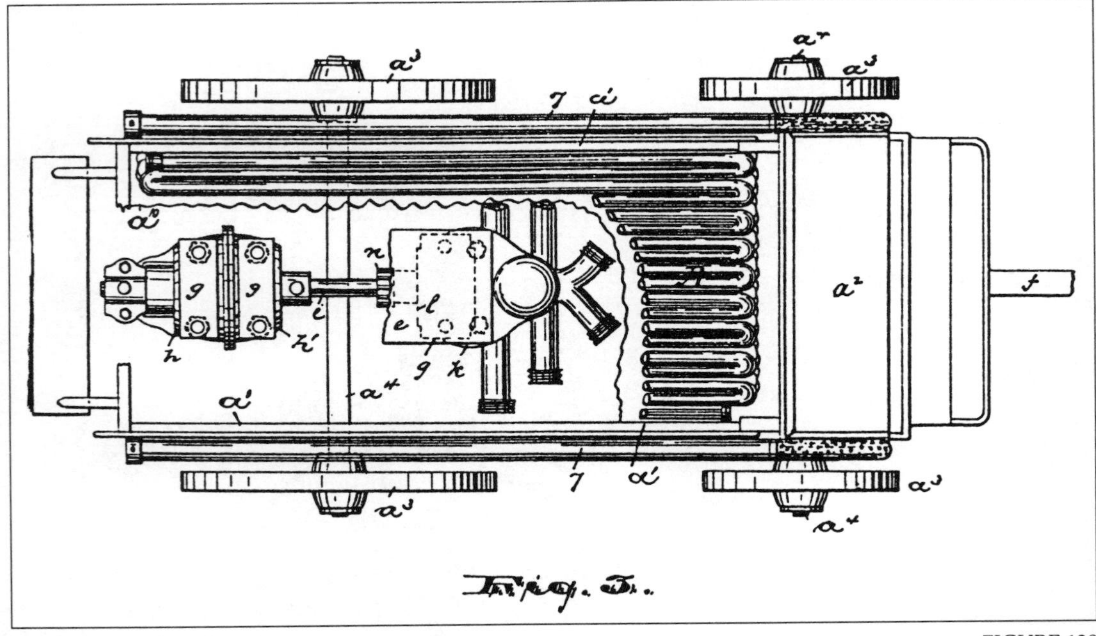

FIGURE 138

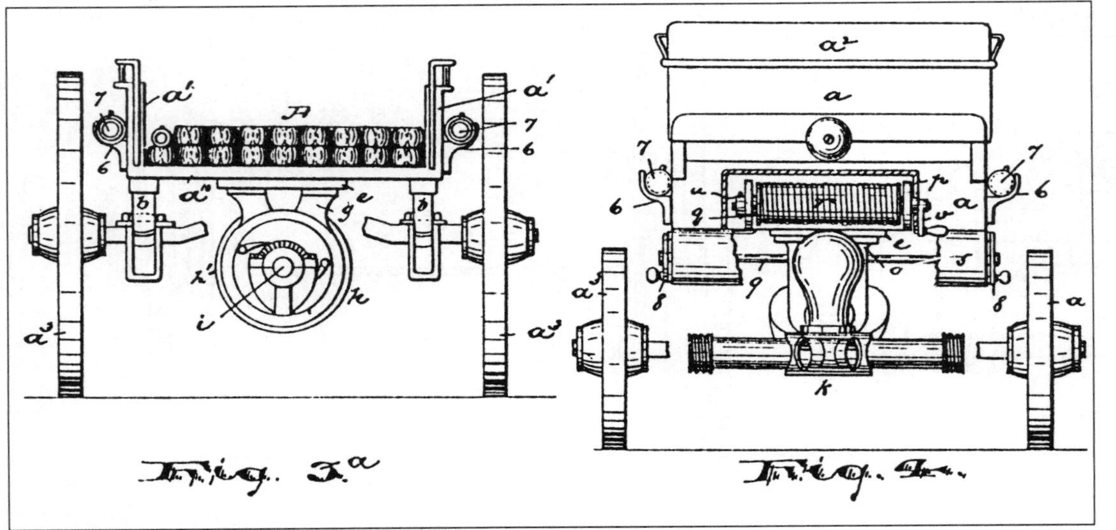

FIGURE 139

M. Whitlock and Edgar P. Harrison, both of Newark, New Jersey, were awarded patent number 670,943 in April, 1901 for their design of a combination fire engine and hose wagon (FIGURES 137, 138 and 139). The most unique feature of this device was the under-chassis suspended location of the motor and the pump.

The electric pumping engine of Sven Brunau is illustrated in FIGURES 140 and 141. This patent, number 701,536 dated June 3, 1902 was assigned to the De Laval Steam Turbine Company by this Swedish inventor from the town of Koping. In describing the potential utility of his apparatus Mr. Brunau wrote that, "While my invention is an electric pumping engine suitable to many uses, I especially design it for employment as a fire engine, and especially in villages or other small settlements where there is no water supply under pressure, but where there is already established an electric lighting plant. The electric mains necessarily go to every house of the settlement, no matter how far distant it may be from others. Every house also necessarily has some source of water supply, as, for example, a well, driven or otherwise. All that would be required, then, in order to make my engine at once available for the protection of the house would be the establishment thereat of a place of electric connection and a suction connection for the well. The persons in charge of the engine on bringing it to the premises would have simply to make these connections–the work of a moment–and the engine could at once deliver the stream. So, also, the engine could be utilized for the irrigation of crops, lawns, and etc., the drainage of ponds and pools, and a variety of other purposes. In cities having regular water and electric supply my engine may not only be used for fire purposes, but for the flushing of streets and sewers–a class of work which now falls upon the steam fire apparatus, but which could

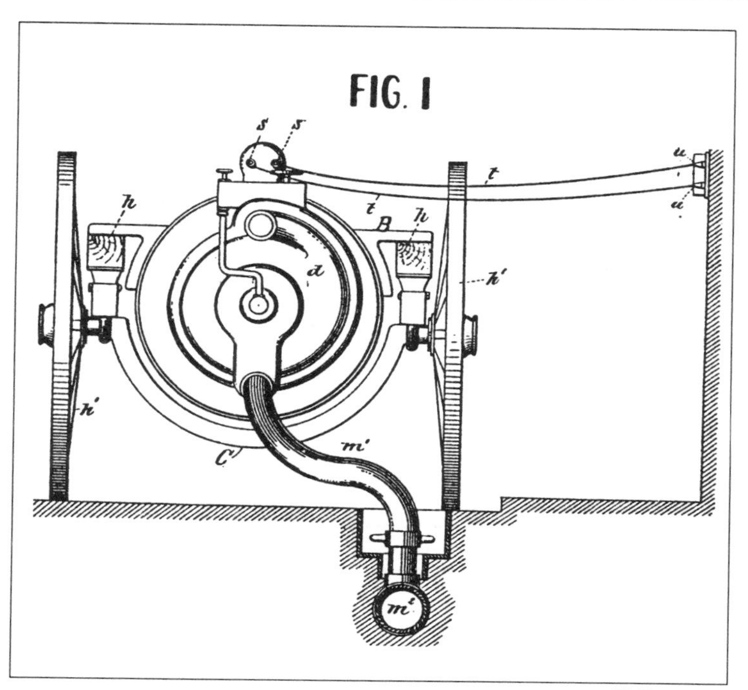

FIGURE 140

be much more conveniently and cheaply done by an auxiliary electric pumping engine taking its current directly from the street electric mains."

U.S. Patent 880,780 reported inventor Gustave A. Drake's idea for combining the electric system of a town with the street hydrants to promote instantaneous current flow upon hooking up a suction hose to the water supply FIGURE 142 is from Drake's patent papers. Years later, of course, there were analogous efforts to build in hard-wire communications channels using hose lines. Fortunately other inventors devised radios!

The General Electric Company's Richard H. Rice of Lynn, Massachusetts received patent number 957,903 on May 17, 1910. The basic novelty of this invention, seen in FIGURE 143 was the truss framework which supported the pump and the motor and held all of the parts together.

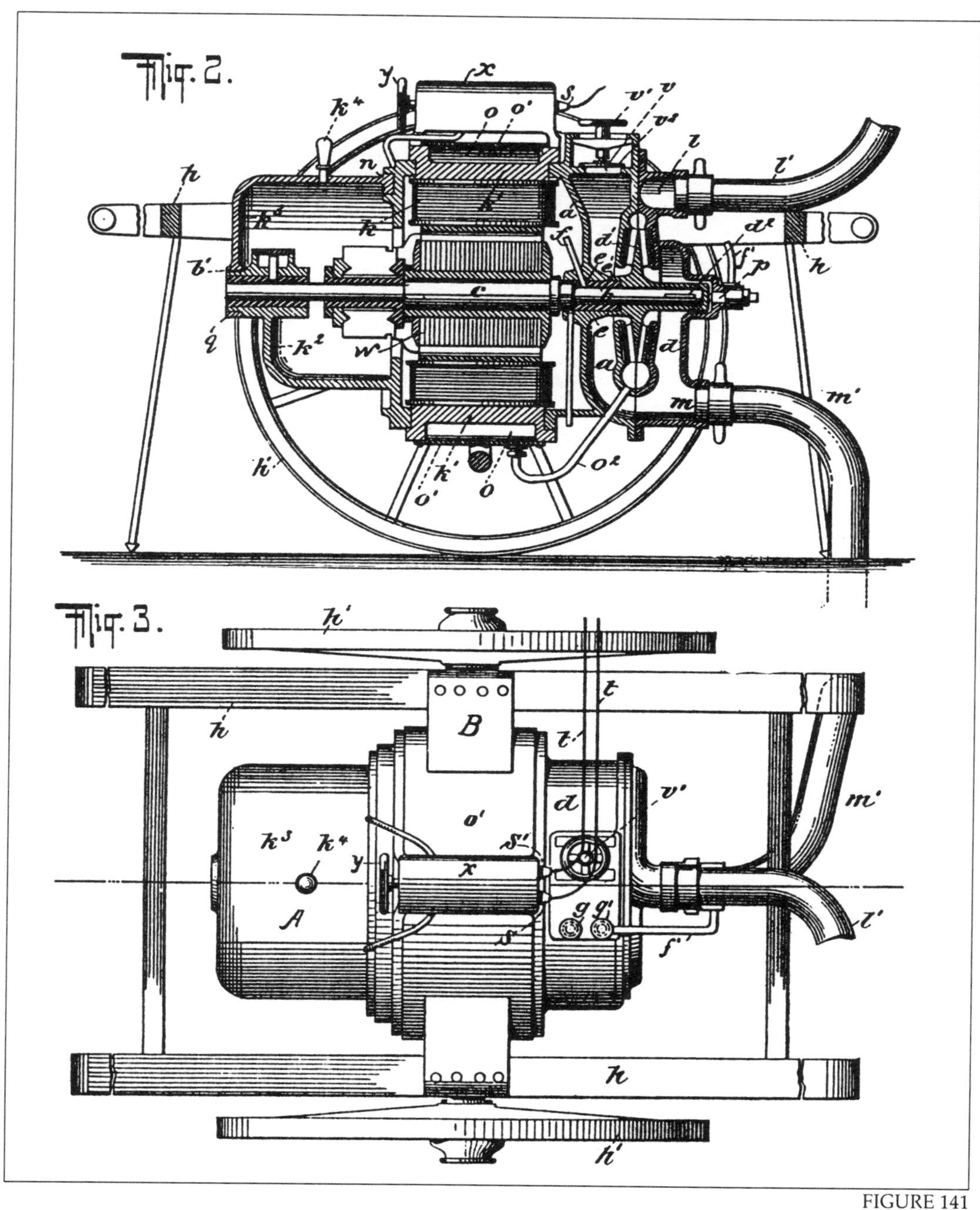

FIGURE 141

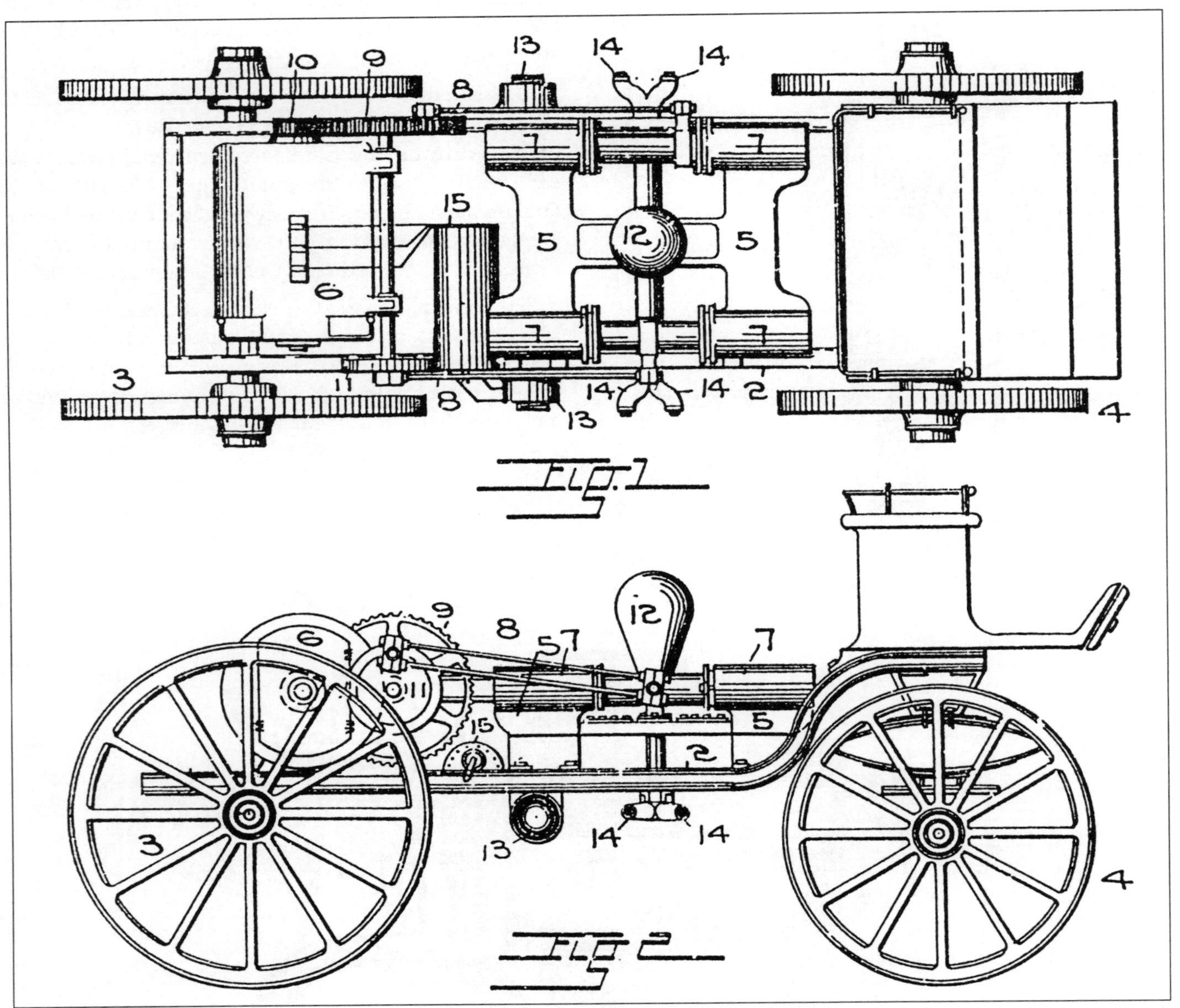

FIGURE 142

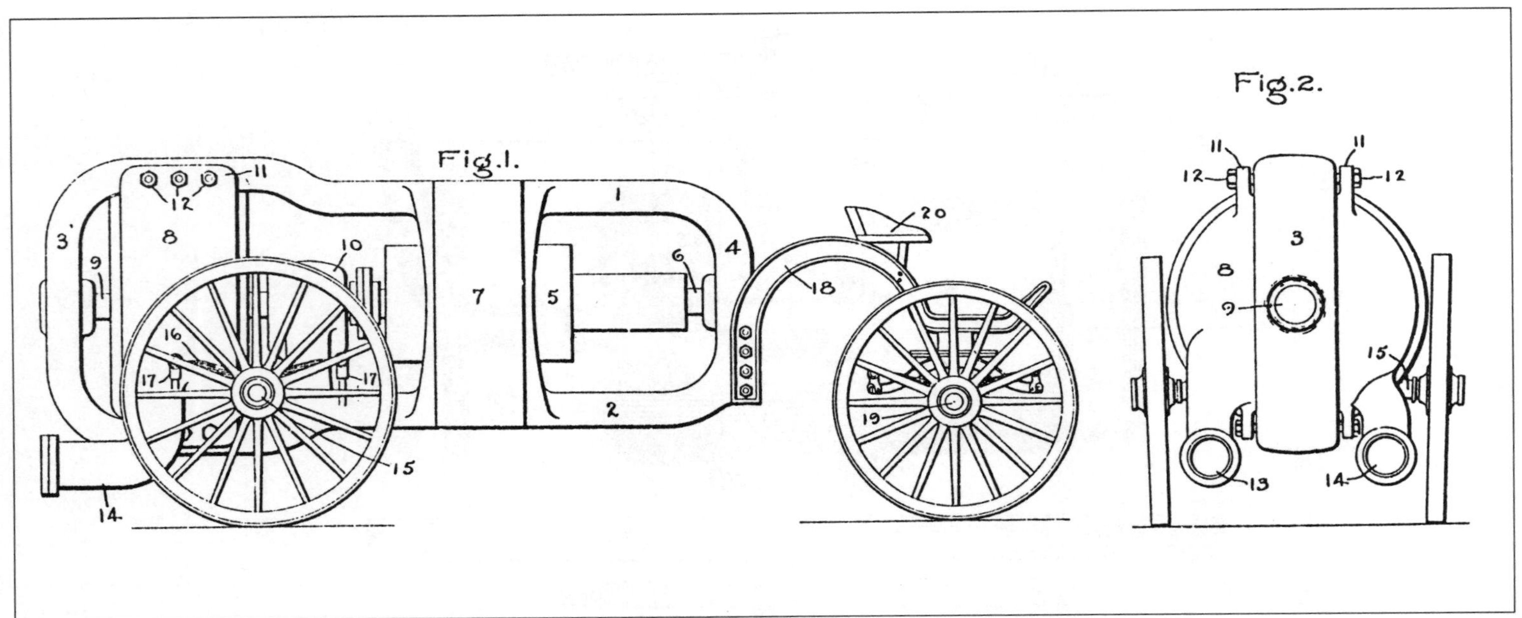

FIGURE 143

A.C. Farrand

A hybrid electric fire engine composed of parts and subassemblies soon to appear on American streets as fire departments began investing in these vehicles was the subject of patent number 1,081,224, dated on December 9, 1913. The inventor was Albert C. Farrand of Atlantic City, New Jersey, and his apparatus is shown in FIGURES 144 and 145. Farrand discussed his design in the following words: "My invention provides a self-propelled apparatus of novel construction, carrying a pump, having means for furnishing light when desired, and carrying a full equipment of hose and chemicals. The advantages of a self-propelled fire fighting apparatus in saving space usually occupied when horse-propelled apparatus are employed is very great, and among the savings may be mentioned two teams of horses, two drivers, one stoker, four sets of harness, stable room, room for hay, feed and bedding, room in the engine house, as two of the self-propelled engines, fully equipped, can be placed in the space now occupied by a steamer and hose wagon, and room taken up for coal storage and manure pits. Another advantage is the obviating of all objectionable odors so common at present in and about engine houses. A still further advantage is the absence of smoke and cinder nuisance while in service at fires, thereby avoiding all danger of setting adjacent buildings on fire, a condition that might readily occur by reason of live coals discharged from the stack of the ordinary fire engine."

Inventor Farrand called for a gasoline engine to power a dynamo which generated the electricity for moving the vehicle and for operating the pump. Power for the vehicular lights, electric brakes and storage battery, as well as the portable flood lights and an air compressor was also supplied by the dynamo. The compressed air thus produced was to have been used for air brakes and for an air whistle which sounded the alarm while on the way to the fire. Among the key components seen in the illustrations were the engine, designated as number 6; the dynamo, 10; the motor, 11; the pump, 12; the search light, 46; water and chemical tanks, 56; a basket for chemical hose, 57; service hose, 59; ladders, 62; a reel of cable for the portable search light, 65; suction hose, 66; hose box, 68; and hand-held fire extinguishers, 74.

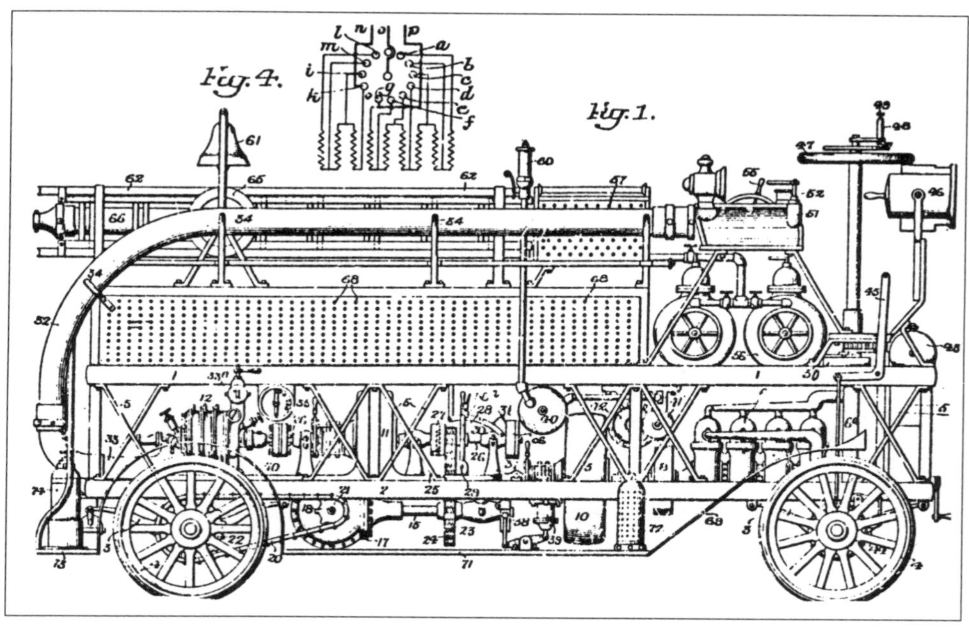

FIGURE 144

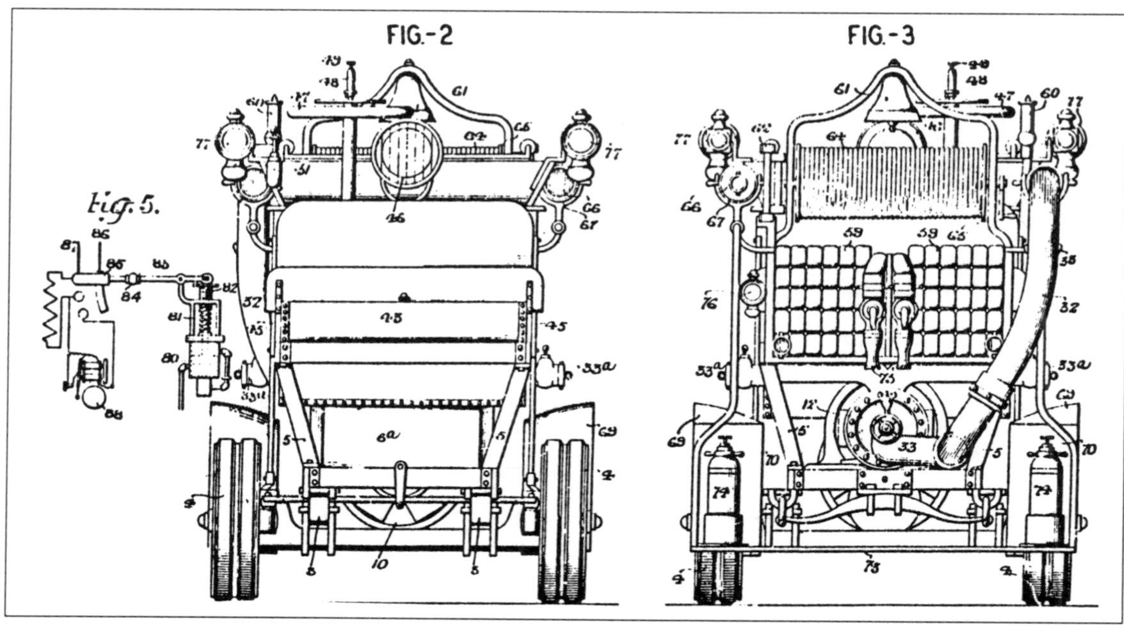

FIGURE 145

C. Loeffler

And finally, and beginning to look a bit like a fire engine as we know it today, was Charles Loeffler's; patent number 1,160,422 of November 16, 1915. Philadelphian Loeffler designed a gasoline-powered vehicle which was to carry an electric motor and, of course, a pump. But the real novelty of this system, and the reason why Loeffler was obviously another utopian, was that the otherwise standard street hydrants were supposed to house electrical connections and controls. The patent is illustrated by FIGURES 146 and 147.

European Electric Fire Engines

Of course during all of these same years, and in particular during 1899 to 1907, any number of European fire departments were also building, testing, and finally using a variety of electric and electric-hybrid types of apparatus. For example in Vienna, Austria the first electric fire engine was deployed in 1899. Its top speed was about fifteen miles an hour, and it carried 600 liters of water which could be expelled from the tank by using compressed carbon dioxide from cylinders.

In 1900, during the summer, the firemen of Paris, France—le Battalion de Sapeurs-Pompiers—hosted their provincial colleagues in order to conduct drills and demonstrate new equipment. Three types of electrical fire vehicles were displayed on the drill ground at Vincennes: a pumper, FIGURE 148, a ladder truck, FIGURE 149, and a hose wagon, FIGURE 150. The wagon was, quite simply, an electric automobile able to seat a crew of six and to carry hose, nozzles, and tools. Also electrically propelled, the ladder truck was a rather heavy rig, weighing about 9,000 pounds. The electric pumper weighed about 6,400 pounds and carried 100 gallons of water. It used the same electric motor for both pumping and propulsion. The three-chambered pump was powerful enough to draw water through 23 feet of suction hose. The vehicle had a reel of fire hose, and this was routinely kept full of water. At the standard pumping rate a flow of about twenty gallons a minute could be sustained with a quarter-inch orifice hose nozzle. Suspended underneath the rig's chassis, enclosed within a case, were the batteries. These were powerful enough to fuel the apparatus for 36 miles at an average twelve miles an hour speed.

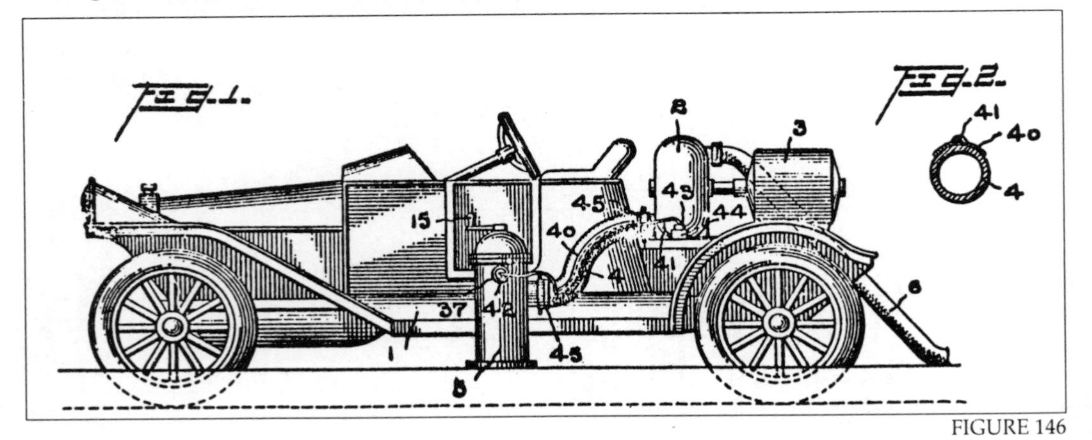

FIGURE 146

The French city of Rouen also received their first electric fire engine in 1900. This rig mounted an eight horse-power motor which was wound for 525 direct current volts and was capable of 2,000 revolutions per minute. Current was supplied from overhead trolley wires by a contact fastened to the top of a bamboo pole. This feed wire was rolled on a reel above the motor, and the circuit was completed when its free end contacted the tramway tracks via a block of cast iron (FIGURE 151).

In 1907 the fire brigade of Hanover, Germany was using an electrically-propelled apparatus which carried 100 gallons of water in a rear-mounted tank along with two cylinders of compressed carbon dioxide for expelling it with gas pressure. Storage batteries were located beneath the front seat and there were two electric motors, one mounted next to each of the rear wheels, powering them independently. The batteries were contained with in four wood cases, each of which held eleven cells. Overall the battery weight was 2,400 pounds. Each of the enclosed, four-pole rear wheel electric motors used small steel pinions to engage large bronze gears on the wheels, a system which provided an 8.5 to 1 speed reduction. The motors made 410 revolutions per minute, and they were run on 88 volts direct current. With a fresh battery charge the 16-foot long vehicle, which weighed about 10,000 pounds including an embarked crew of five, could travel for 15 miles at a speed of 10 miles an hour.

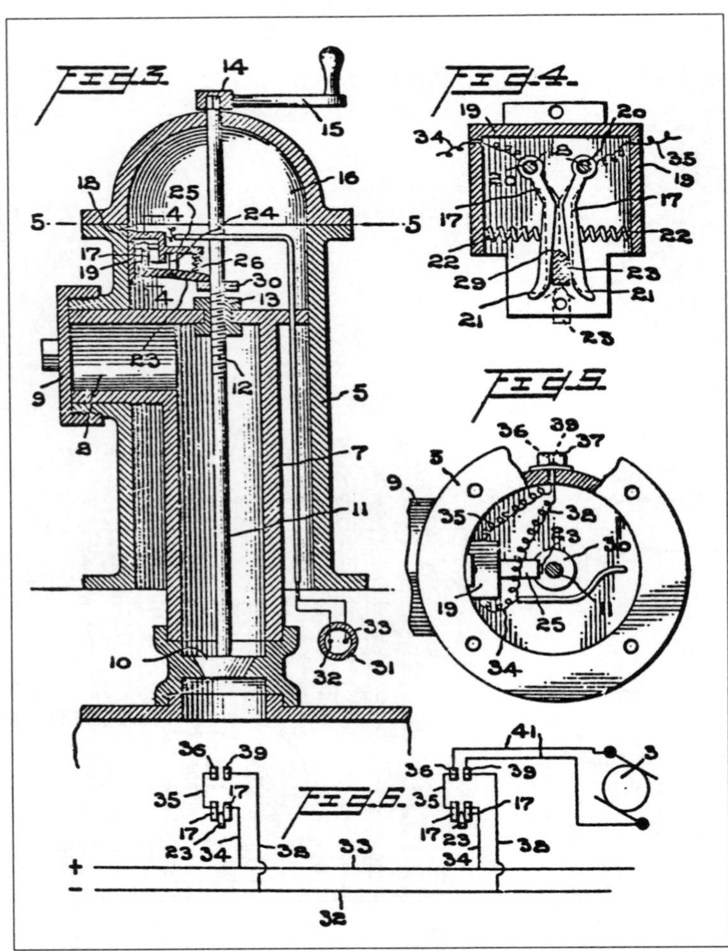

FIGURE 147

FIGURE 148

FIGURE 150

Fig. 1.—THE ELECTRIC HOOK AND LADDER.

FIGURE 149

FIGURE 151

The Fire-Fighting Trolley

Surely the most charming of all of these curiosities of electric fire apparatus was the fire-fighting trolley car which served the Park Point area of Duluth, Minnesota from 1907 to 1930. Park Point, also known as Minnesota Point, is a narrow strip of land reaching out six or seven miles from the Minnesota to the Wisconsin shore of Lake Superior, toward the city of the same name, thereby forming a protected harbor for Duluth behind its 500 to 600 foot width. For many years the first few miles from the bridge connecting mainland Duluth to Park Point have been the location of numerous summer cottages as well as expensive residences. Streetcar tracks ran in the center of Minnesota Avenue, the only arterial roadway of this neighborhood. Today an airport and a seaplane anchorage are serviced by this highway.

Late in 1907 the city of Duluth purchased, for $650, an electric trolley car from the Interstate Traction Company, the streetcar company, which also donated a car barn to house the apparatus. This became Fire Station Number 5, located at 1900 Minnesota Avenue. Volunteer firemen removed the seats and installed in their place a large hose box which was open at each end to facilitate hose laying in any direction. Fully equipped the trolley carried an electric water pump, 1,500 feet of two and one-half inch diameter hose, a 35-gallon chemical tank with 200 feet of three-quarter inch chemical hose, a set of ladders, axes, and pike poles, a pair of six-gallon Babcock fire extinguishers with extra charges, and sets of turn-out gear and helmets for the firemen (FIGURE 152).

Initially this converted street car was maintained by the Traction Company, which also supplied a qualified motorman to operate it. Upon receipt of an alarm of fire in Park Point it was started along the tracks, given absolute right of way. Every few seconds an electric gong was sounded to alert the volunteer firemen, who were picked up along the route.

Finally a "mini" fire station was built in Park Point where it has been in service at 2140 Minnesota Avenue from February, 1931 until the present time. Currently a shop-built mini-pumper is housed there, along with a crew of but one firefighter.

The ultimate swan song, at least to this time, for electric fire engines can be reported with just a few additional words. In 1906 a Woods Electric Company battery-powered car was tested by the Chicago Fire Insurance Patrol. Following, shortly, the Webb Motor Fire Apparatus Company and the Couple Gear Freight Wheel Company each built battery-operated fire rigs. Then, amalgamated as Webb-Couple Gear this concern went on to construct fire apparatus in which gasoline engines powered electric generators which energized the electric motors which were built into the wheels themselves. The oldest Webb-Couple Gear electric tractor owned by the New York Fire Department was used to haul Water Tower Number 2, and it was not replaced until 1930.

FIGURE 152

CHEMICAL FIRE ENGINES AND CARBON DIOXIDE

INTRODUCTION

Affectionados of chemical fire engines all know that these were machines which used the pressure of gaseous carbon dioxide to force water from a tank and out through a small-bore hose and nozzle. This is the same system which has been used for fire extinguishers for very many years. And as a matter of fact it would not be incorrect to think of the difference between these fire extinguishers and chemical fire engines as resolving to the question of size. As a rule most fire extinguishers never exceeded about five gallons capacity, whereas chemical fire engines usually did not carry less than 50-gallon capacity cylinders or tanks.

Thus, in the classical configuration a chemical fire engine was equipped with one or more tanks of about 40 to 60 gallons volume. Each of these were filled with water in which sodium bicarbonate had been dissolved to the point of saturation. But this was not a consistent and steady state. Think about the scenario in which a seldom-used chemical engine has been sitting in its quarters, awaiting a call. With the passing of time and the changing of seasons the ambient temperature falls, as summer gives way to autumn, and autumn to winter. And as the water in the tank gradually cooled off it would lose some of its ability to hold dissolved bicarbonate in solution. As a result there would be granular deposits of sodium bicarbonate becoming caked all over the bottom of the interior of the tank. Now in order to discharge the water out of the tank sulfuric acid in its most concentrated form was added. About 80 ounces would be dumped into 40 gallons of water and bicarbonate. The ensuing chemical reaction of the sulfuric acid and the sodium bicarbonate can be represented by the formulation:

$$H_2SO_4 + 2NaHCO_3 \rightarrow Na_2SO_4 + 2H_2O + 2CO_2.$$

The carbon dioxide gas was evolved under sufficient pressure to force the water out of the container through the small-bore "chemical" hose. And generally some of the undissolved sodium bicarbonate would be expelled as well, giving the water a kind of milky appearance and messing up everything it touched.

Many fire service leaders believed that these mixtures of water and undissolved bicarbonate, together with some entrained carbon dioxide, possessed fire extinguishing properties up to 30 to 40 times that of water alone. The long persistence of this astonishing and unscientific idea does not reflect credit upon the fire officers of former generations.

The very reason for the existence of chemical fire engines was, of course, their capability to rapidly generate a stream of water. The demise of this kind of system eventually followed the creation and the introduction of the Ahrens-Fox booster method. An Ahrens-Fox brochure with the imposing title, "Nothing Compels Admiration More Than Victory and Success," probably published about 1915, included a discussion entitled, "Booster versus Chemical." Among the author's statements was this: "Affording every facility characteristic of any other known method, the Booster System has inherent advantages which are impossible of association with apparatus depending upon the result of chemical reaction. The booster idea is unique because it finds further useful work for a power plant already at hand and utilizing this at a time when it would otherwise be idle."

And having thus noted that the booster pump signaled that carbon dioxide fire engines had reached the high-water mark of their history we shift into reverse gear for a few years in order to peruse the evolution of their patenting.

A.A.C. Vignon

A retired general officer of the Corps of Engineers of the Empire of France, one Alphonse A.C. Vignon received a patent in 1862 for his "l'extincteur," a device from which water with entrained carbon dioxide could be projected onto a fire. This apparatus used a 30 to 55 pound per square inch pressure of carbon dioxide to provide the push for ejecting the water stream. Vignon's collaborator was a physicist named Philippe Francois Carlier. These inventors sold the American rights to their patent to Dawson Miles, of Cambridge, Massachusetts who, following Carlier's death, managed to have his name submitted to the Patent Office as a co-inventor. This "Miles-Carlier-Vignon" patent is illustrated in FIGURES 153 and 154 (patent number 83,844 dated April 13, 1869). In FIGURE 155 is shown an early Carlier-Vignon machine now owned by author-collector W. Fred Conway.

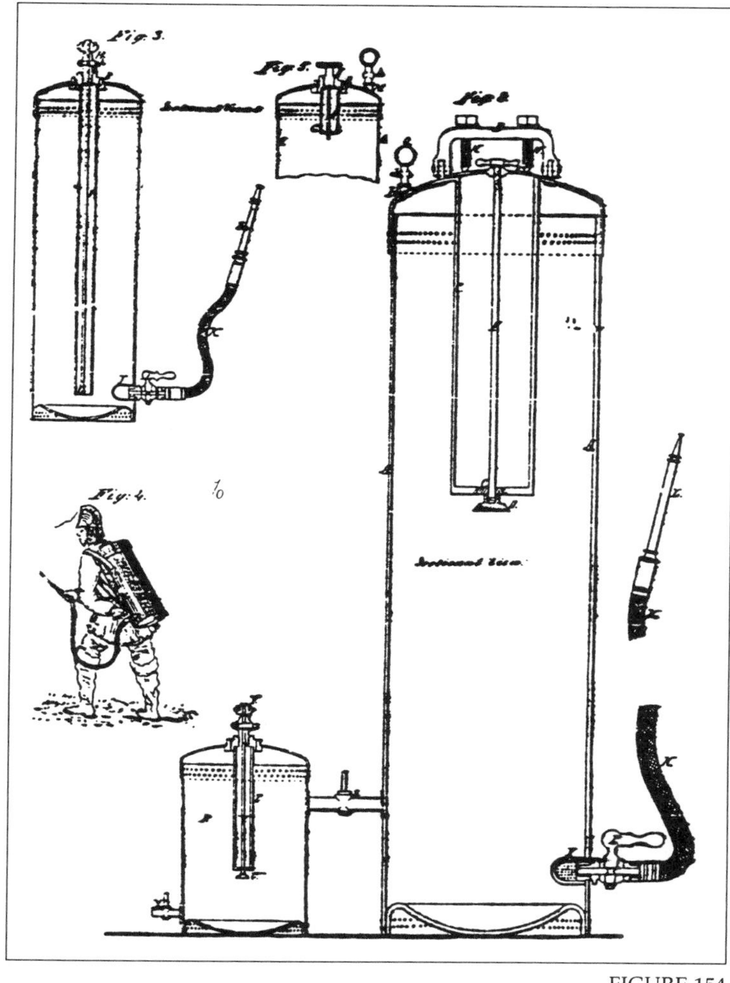

FIGURE 154

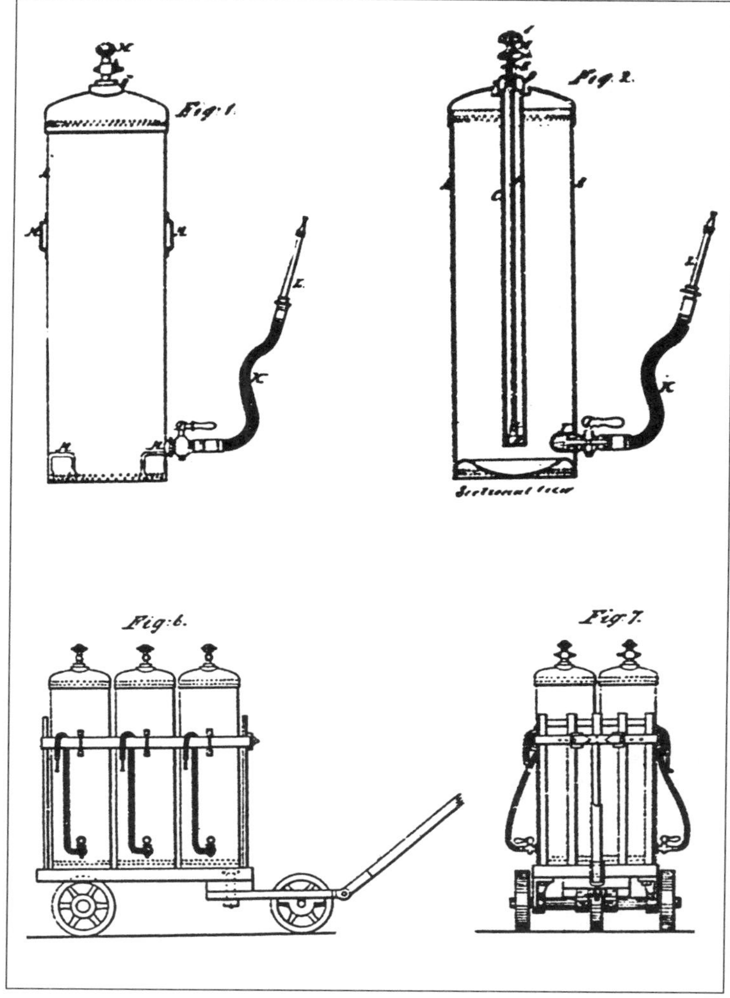

FIGURE 153

FIGURE 155

W.B. Dick & Co.

And so in a sense once one had fitted wheels and a chassis to such a device, "voila," the result was a chemical fire engine. FIGURE 156, from the October 8, 1869 issue of the British magazine "Engineering" shows another such apparatus, Messrs. W.B. Dick and Company, of Glasgow, builders. Mounted on a wheeled chassis this rig had a sheet iron tank, divided into three compartments, for its body. This hybrid rig had pumps, denoted by the numbers 5 and 6, which were fitted within the two rearmost compartments. There was a beam and rocking shaft with levers for the men working the pumps. Water and bicarbonate were used to fill one of these chambers, while a solution of tartaric acid was placed inside of the other one. Delivery pipes, designated as 16 and 17, connected the pumps to a strong vessel fitted within the frontmost compartment, and from this vessel there was a delivery pipe culminating in a hose fitting.

The builders of this apparatus, Messrs. W.B. Dick, who allegedly sold something like 9,000 of them, claimed that, "Water and carbonic acid gas combined produce a far greater effect upon a fire than an equal bulk of unmixed water–an important consideration, for it happens not infrequently that the means used for the extinction of fires are productive of as much damage as the fires themselves."

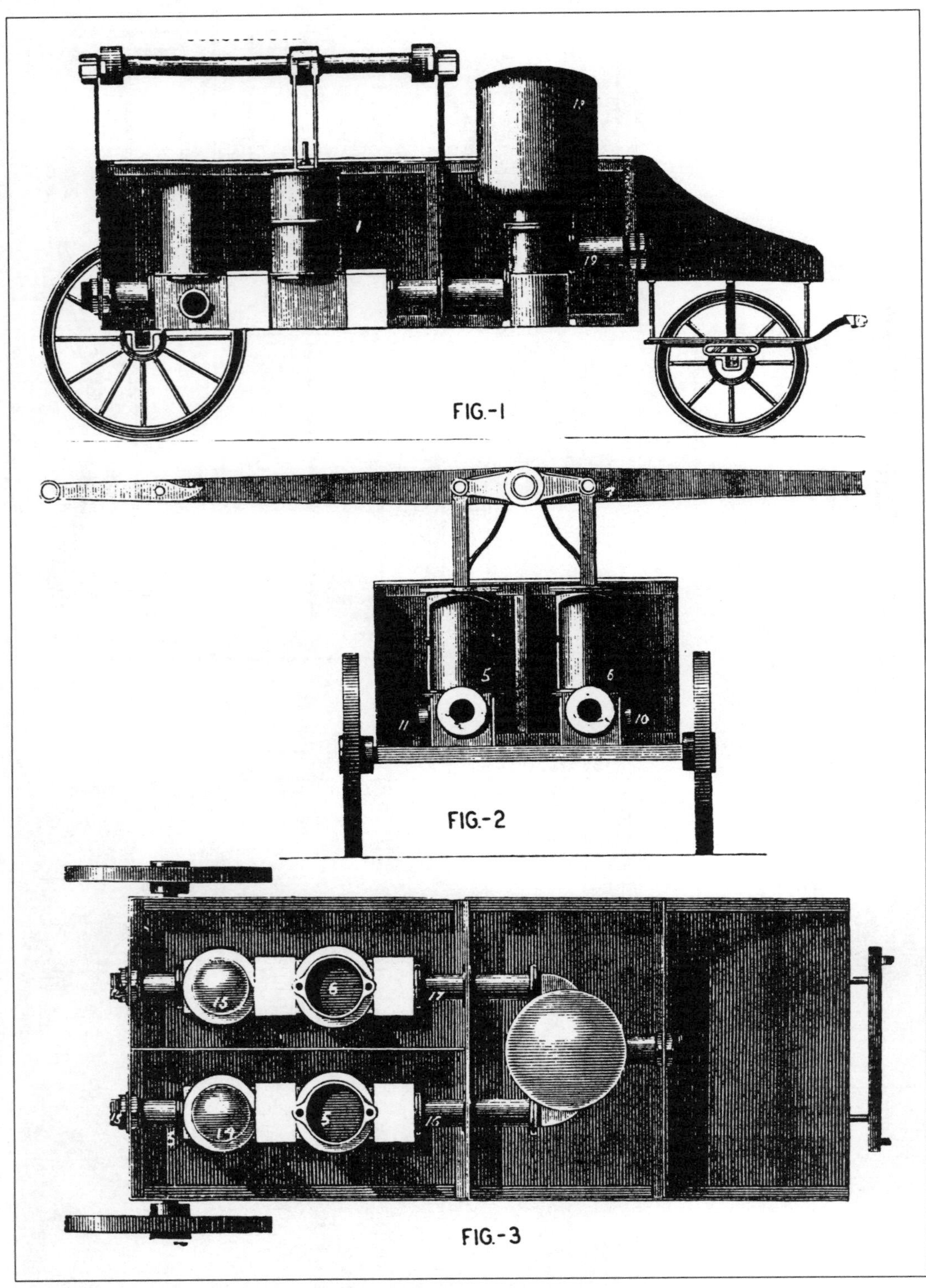

FIGURE 156

Stillson and Kley

The very first American chemical fire engine, invented and patented as such, appears to have been the one described in patent number 131,414. The date was September 17, 1872, and the inventors, Jerome B. Stillson and John A. Kley, were employees of the Chicago-based Babcock Manufacturing Company. A set of four excellent drawings were submitted to illustrate this patent document, and they are shown as FIGURES 157, 158, 159, and 160. Notice that the last three of these drawings were done to a scale, a practice which has always been decidedly the exception, rather than the rule, in patent art. In the second sketch the scale was 1.5 inches to the foot, whereas for the others it was one inch per foot. From this it can be determined that the Stillson-Kley chemical fire engine was of dimensions four feet across at the greatest width, the hubcaps, and nine feet in length. Among the essential components of this device were the upper and lower parts of the large cylinders, designed as A and B, the crank-operated stirrer for mixing the acid with the bicarbonate and water, C, the acid receptacle, D, the valve for adding bicarbonate, E, and the discharge, F. Carried on the rear step was a force pump, G, which was to have been used for filling the cylinders with water.

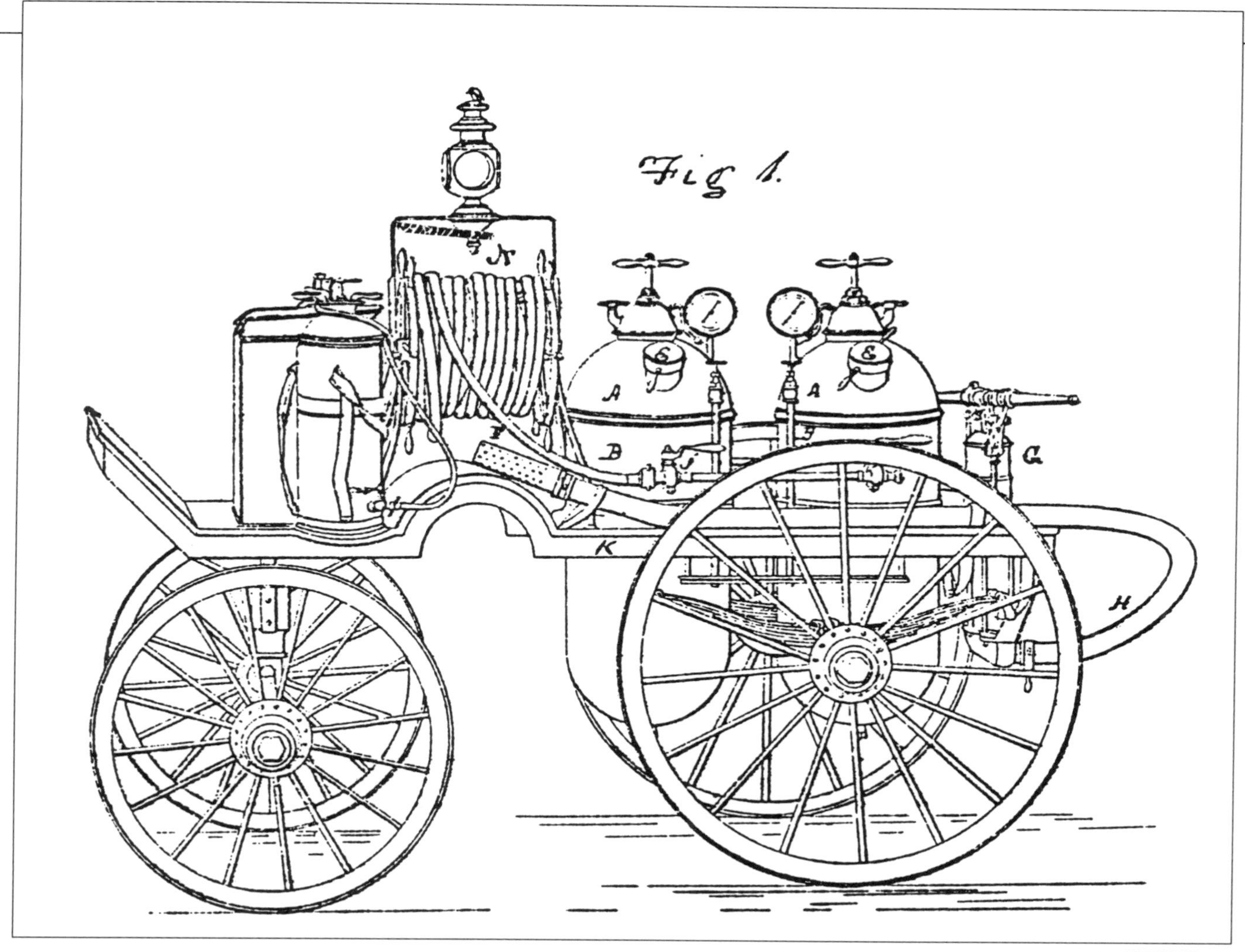

FIGURE 157

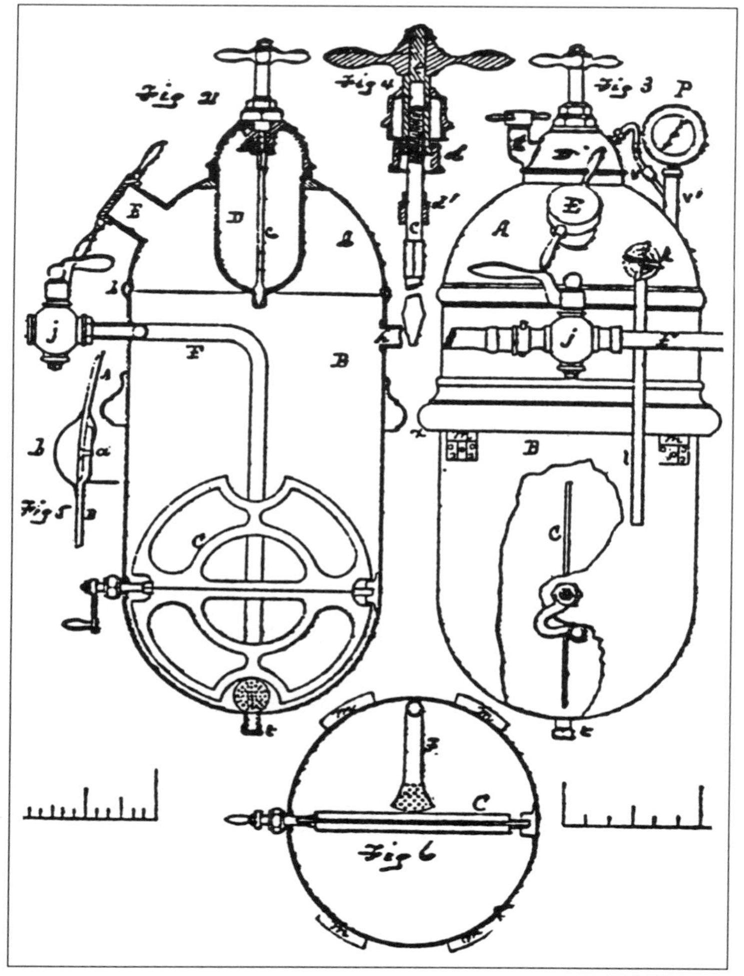

FIGURE 158

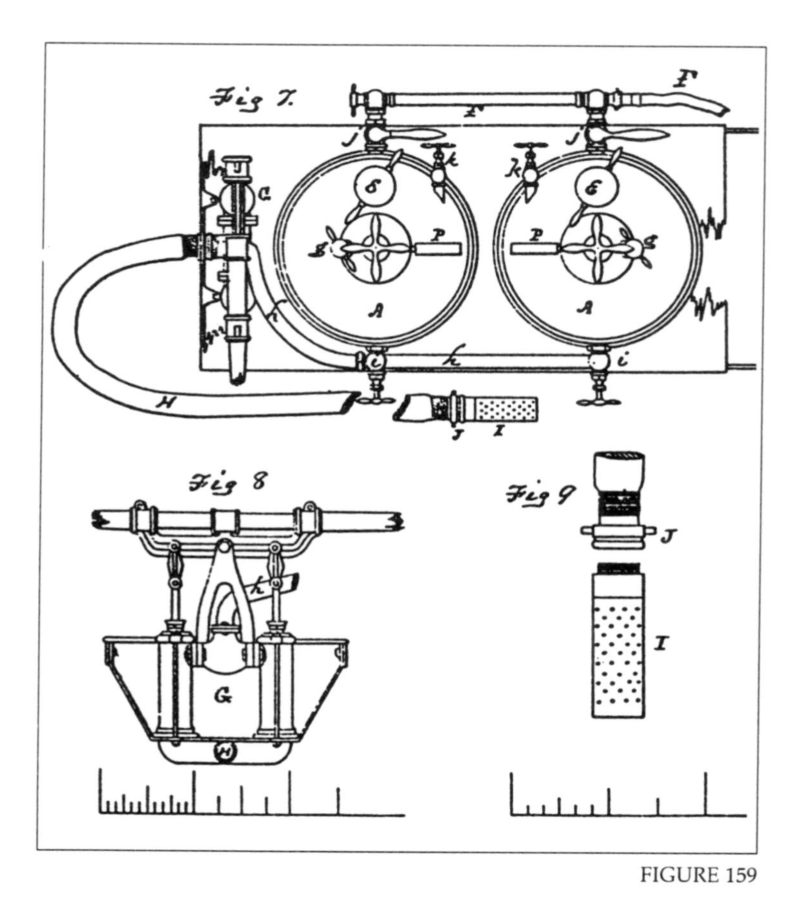

FIGURE 159

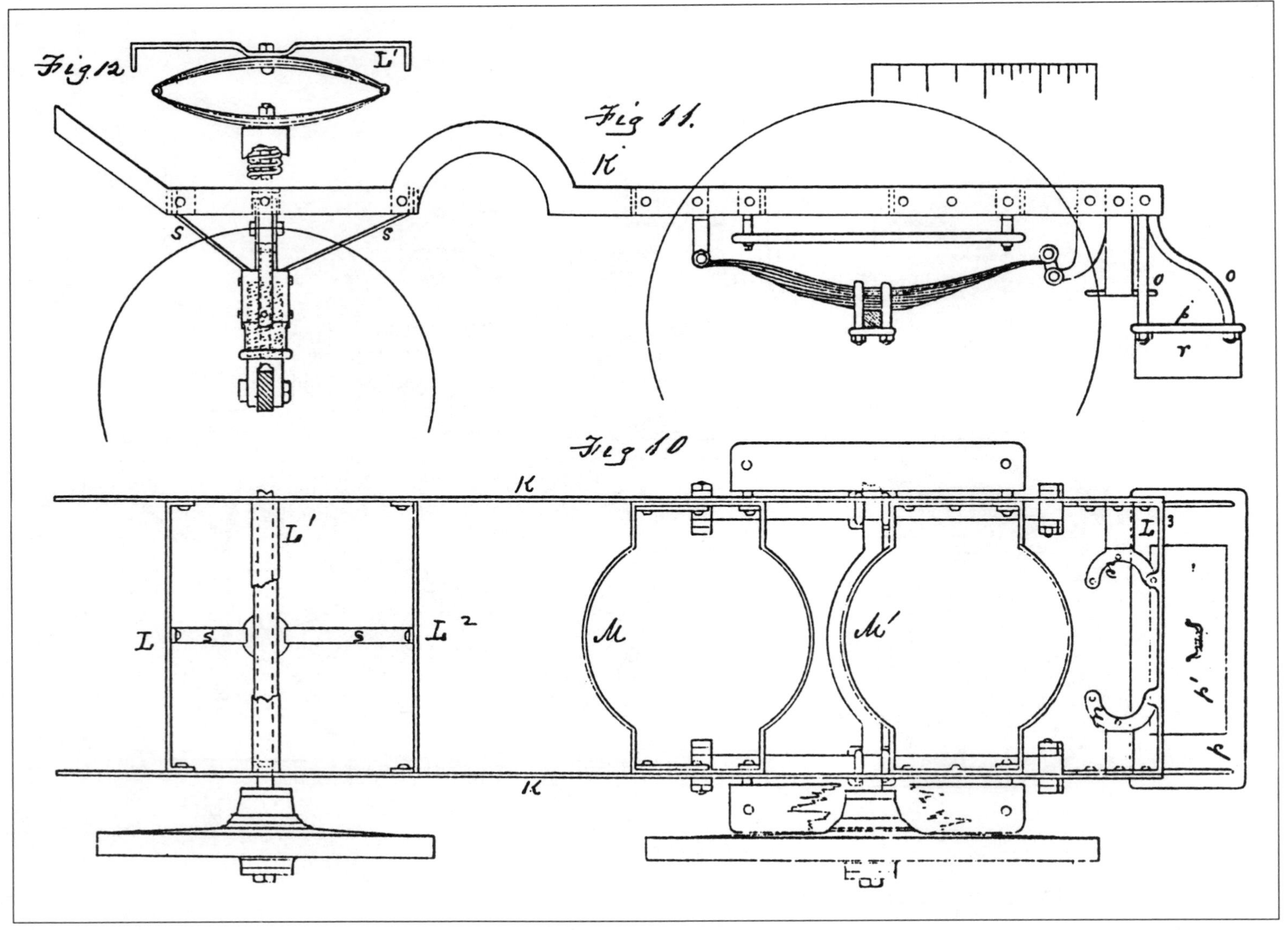

FIGURE 160

About a year later patent number 142,637 was granted to Finley Latta, one of Alexander Bonner Latta's two older brothers. This chemical engine, FIGURE 161, looked kind of like an illegitimate son of one of the popular, two-wheeled jumper hose carts of a slightly older day. This rig was operated with a length of chain, a tug upon which pulled the cover off the acid container as the water-alkali tank was rotated.

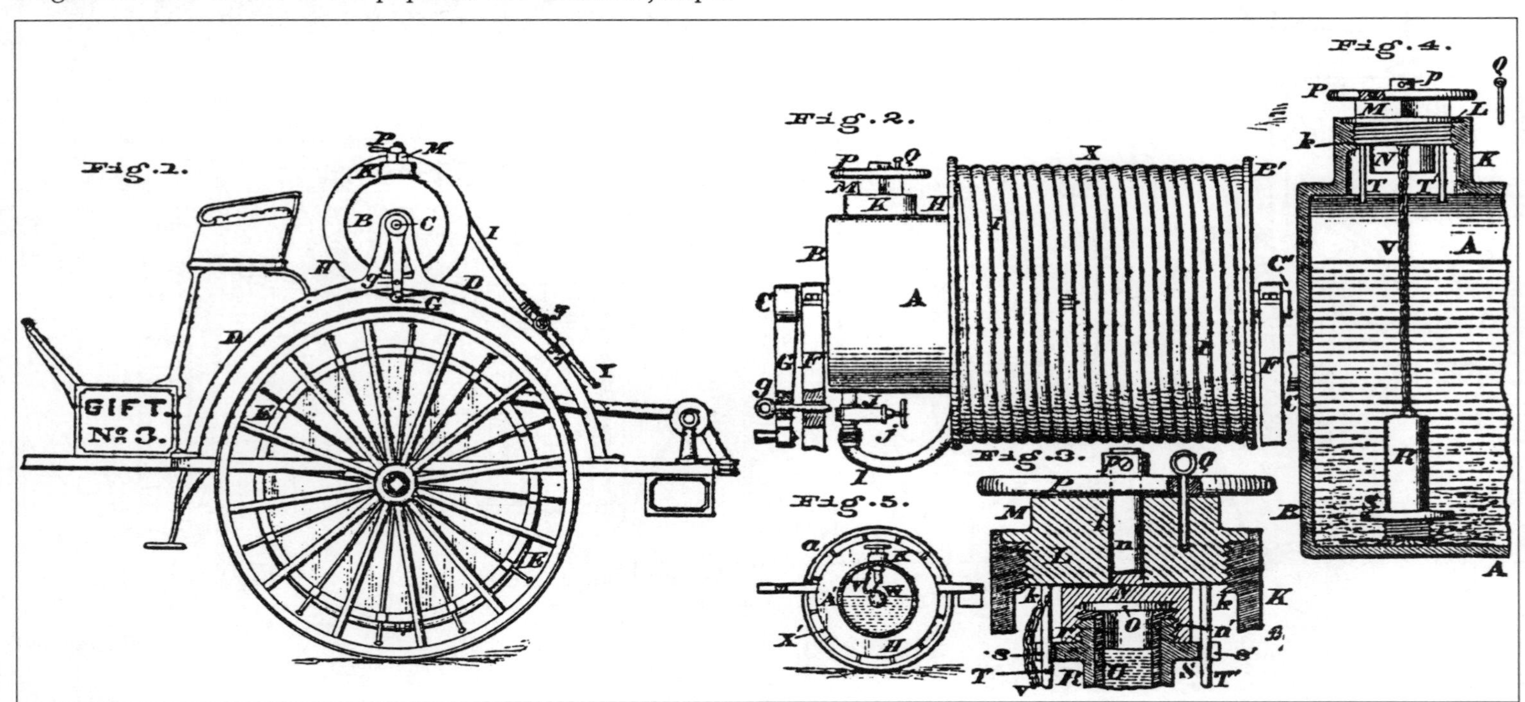

FIGURE 161

J.H. Steiner

John H. Steiner's invention (patent number 147,442 dated February 10, 1874) is shown in FIGURES 162, 163, and 164. Looking rather streamlined and somewhat semi-modern, this Albany, New York inventor had this to say about its operation: "Supposing the engine to have just arrived at a fire, with both generators filled with water having soda in solution, and with one generator in communication with the outlet-pipe, but shut off from the water reservoir. The cap, I, of that generator is removed and a bottle of acid inserted upside down, and the cap replaced. The hand screw, K, is turned so as to open the acid bottle, and then turned down, out of the way. As the acid is released gas is generated and water discharge begins. As one generator is discharged an acid bottle is inserted in the other, and upon exhausting the first, the second one is started, the cocks G and S reversed. Water is thus admitted from the reservoir into the empty cylinder." Steiner described his machine as being, "cheap, light, and simple, very readily operated, not easily deranged, quickly brought into operation, very neat and ornamental in appearance." Not at all an unwelcome set of descriptors for any fire apparatus.

FIGURE 162

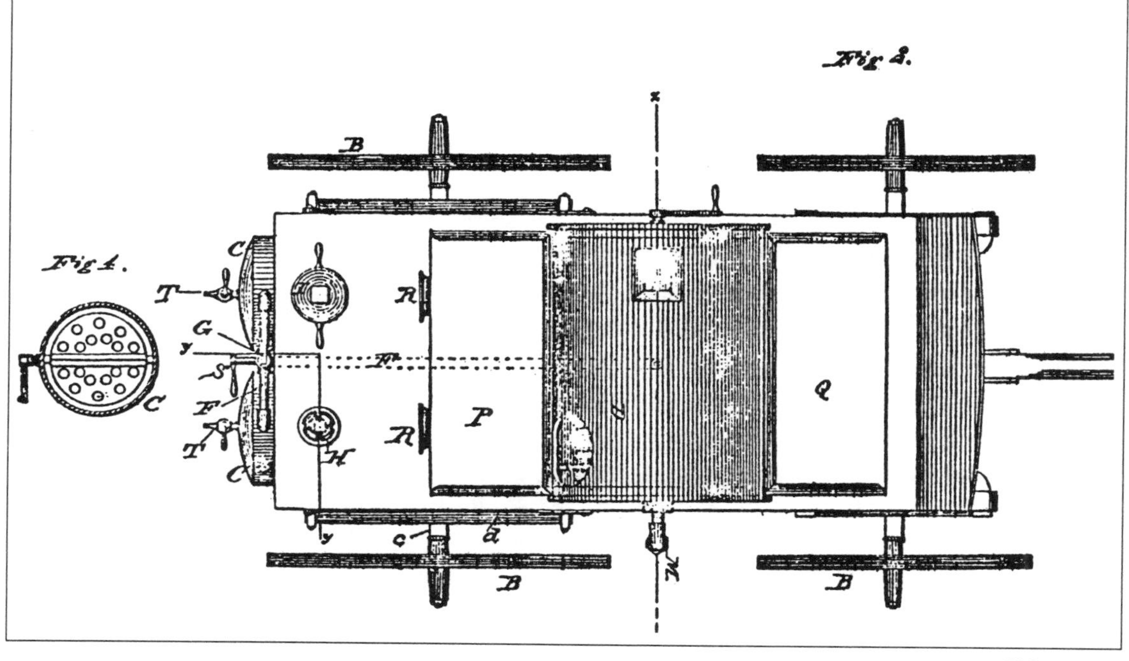

FIGURE 163

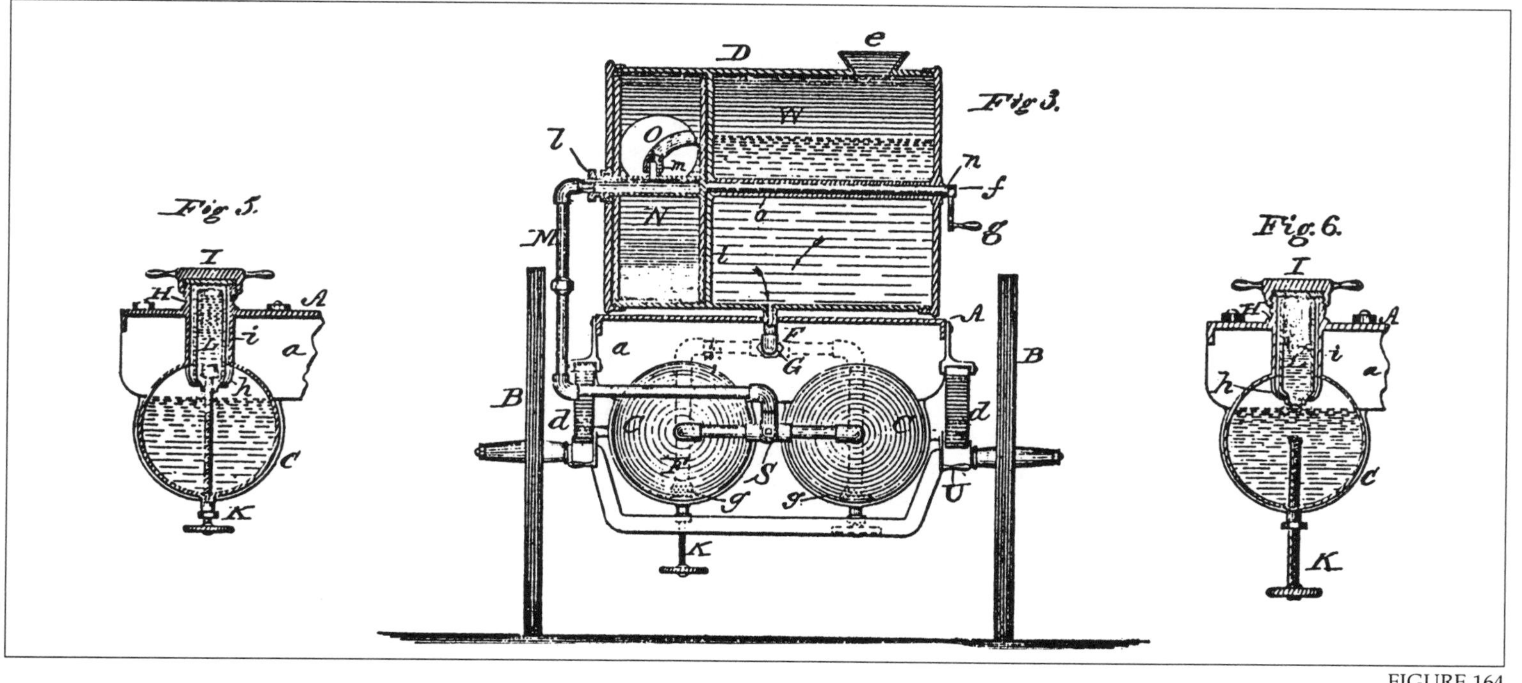

FIGURE 164

Kley and Lee

Several patents assigned to the Babcock Company had John Kley, already met, and Wellington Lee registered as co-inventors. One of their patents, number 152,850 dated July 7, 1874, was a new method for handling the acid bottles of chemical fire engines. By redesigning the mechanisms for hinging and locking the top of the acid-bottle holder they were able to replace empty bottles with freshly refilled ones faster than ever before. FIGURE 165 was the drawing which illustrated this patent. Mr. Wellington Lee, by the way, had been a partner in the firm of Lee and Larned, steam fire engine builders, established in 1856. Although the 1984 New Orleans, Louisiana, Fire Department Yearbook did not mention Lee by name, William T. King in his History of the American Steam Fire Engine placed Lee as chief officer there just before the Civil War. King described him as, "an intense Union man and an enthusiastic patriot," who left for his New York home upon the beginning of hostilities. He died in 1881 at the age of 65.

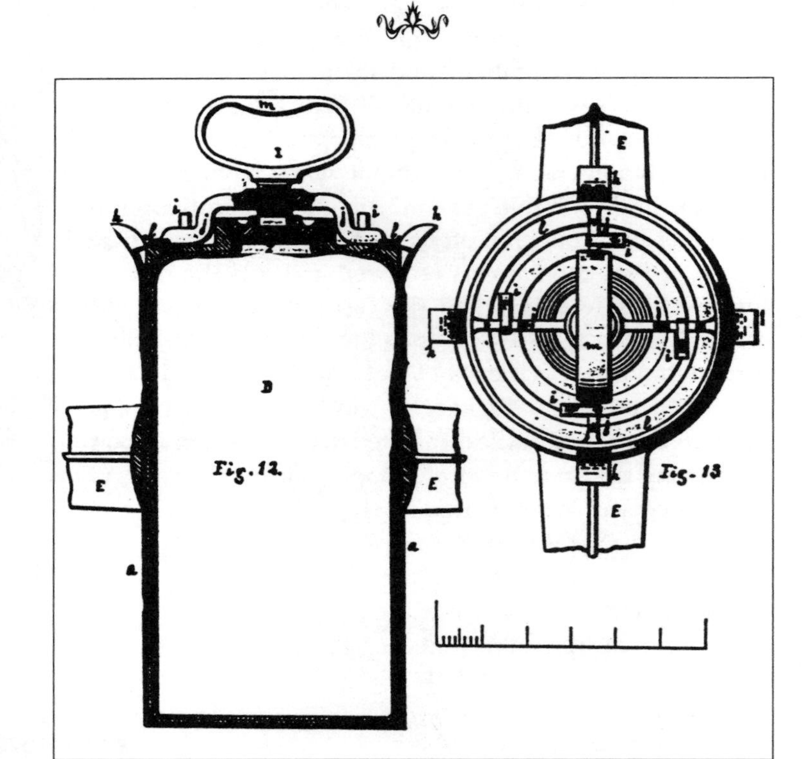

FIGURE 165

Yet another 1874 patent, number 153,355, FIGURE 166, was granted to New Yorker Samuel S. Lippincott. This invention was quite peculiar, one of those deserving its own little niche of classification. Lippincott proposed building a system which included underground pipelines for carrying carbon dioxide as well as fire engines outfitted with tanks to hold the gas. His preferred method for fire suppression was to have been by spreading the carbon dioxide through hoses to choke the fire by denying oxygen to it. There seems little doubt that such a system, at that time, would have been at least equally as effective in denying oxygen to its operators, handily asphyxiating them.

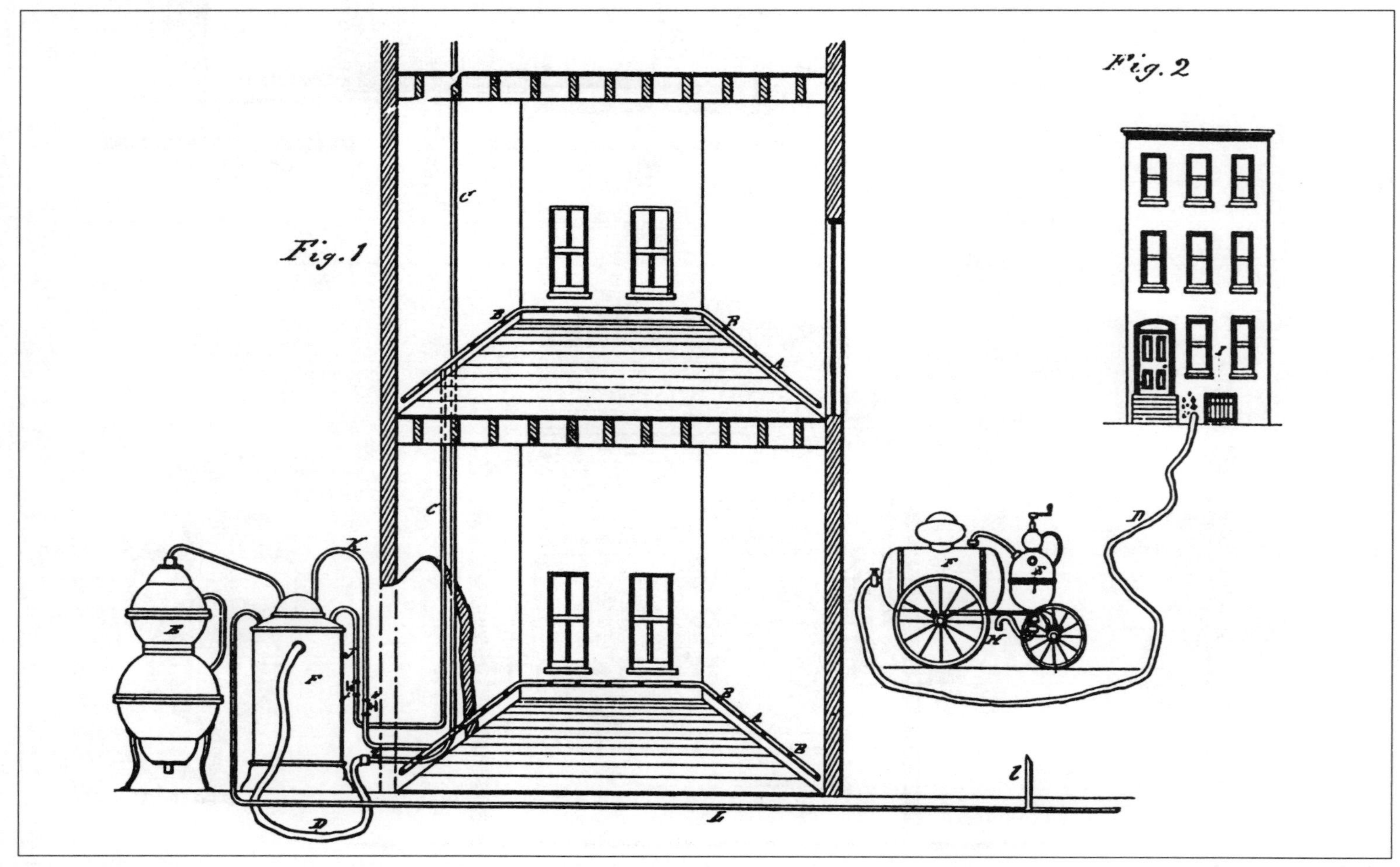

FIGURE 166

A.E. Hughes

Another inventor who assigned away his patent rights was Alexander E. Hughes. The Great American Fire Extinguisher Company, a Louisville, Kentucky concern, was cited to receive his patent number 154,395 in 1874. The patent drawing, FIGURE 167, was but sparsely labeled, with respect to parts and fittings, but as seen it was a single shot, an all or none type of apparatus. By the following year Mr. Hughes had moved to a new home just across the Ohio River from Louisville, in the area now known as New Albany, Indiana. His patent number 164,837, illustrated in FIGURE 168, showed a three-wheeled vehicle mounting a large cylinder which was divided through the center into two compartments. Each of these compartments was fitted with its own acid bottle and was fully independent of the other. June 12, 1877 was the effective date for another Hughes patent, number 191,803. By this time he had moved to Philadelphia, Pennsylvania, and he assigned these patent rights to the North American Fire Annihilator Company of that city. This waterless fire engine is shown in FIGURE 169. It used bicarbonate and sulfuric acid and generated carbon dioxide which was shunted off into storage cylinders. When needed the CO_2 was to have been directed onto the fire by a hose carried upon an on-board reel.

From Waverly, New York, a small Tioga County town just across the state line from Sayre, Pennsylvania, came a patent application submitted by George E. Barker. He received patent number 170,699 in the year 1875 (FIGURE 170). The single novel aspect of this invention, and it was an important development, was his pump which forced air through the tanks to accelerate the mixing and the agitating of their contents.

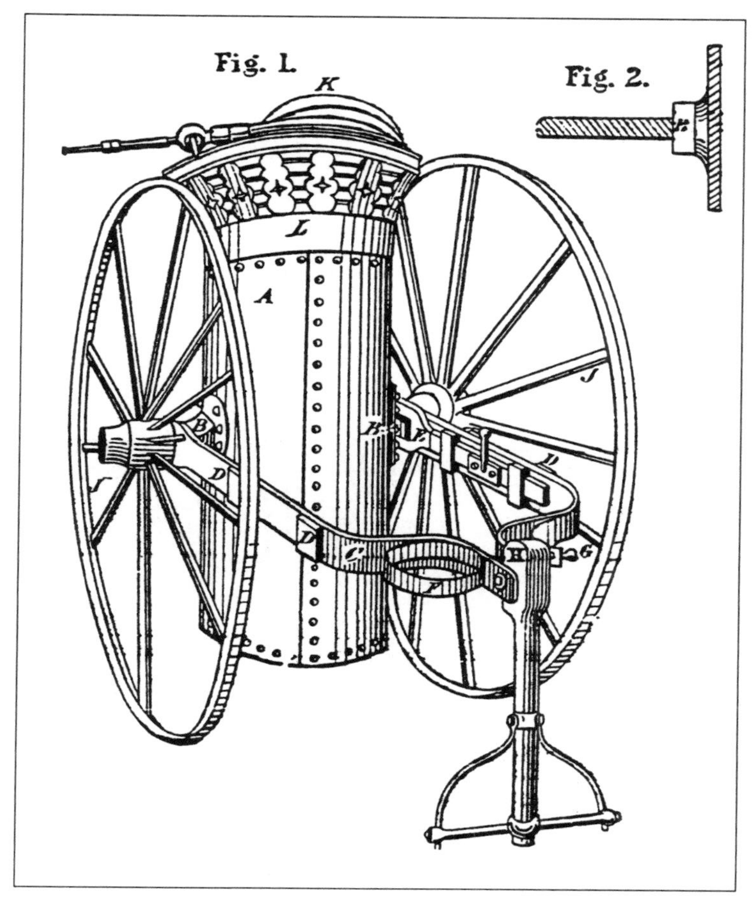

FIGURE 167

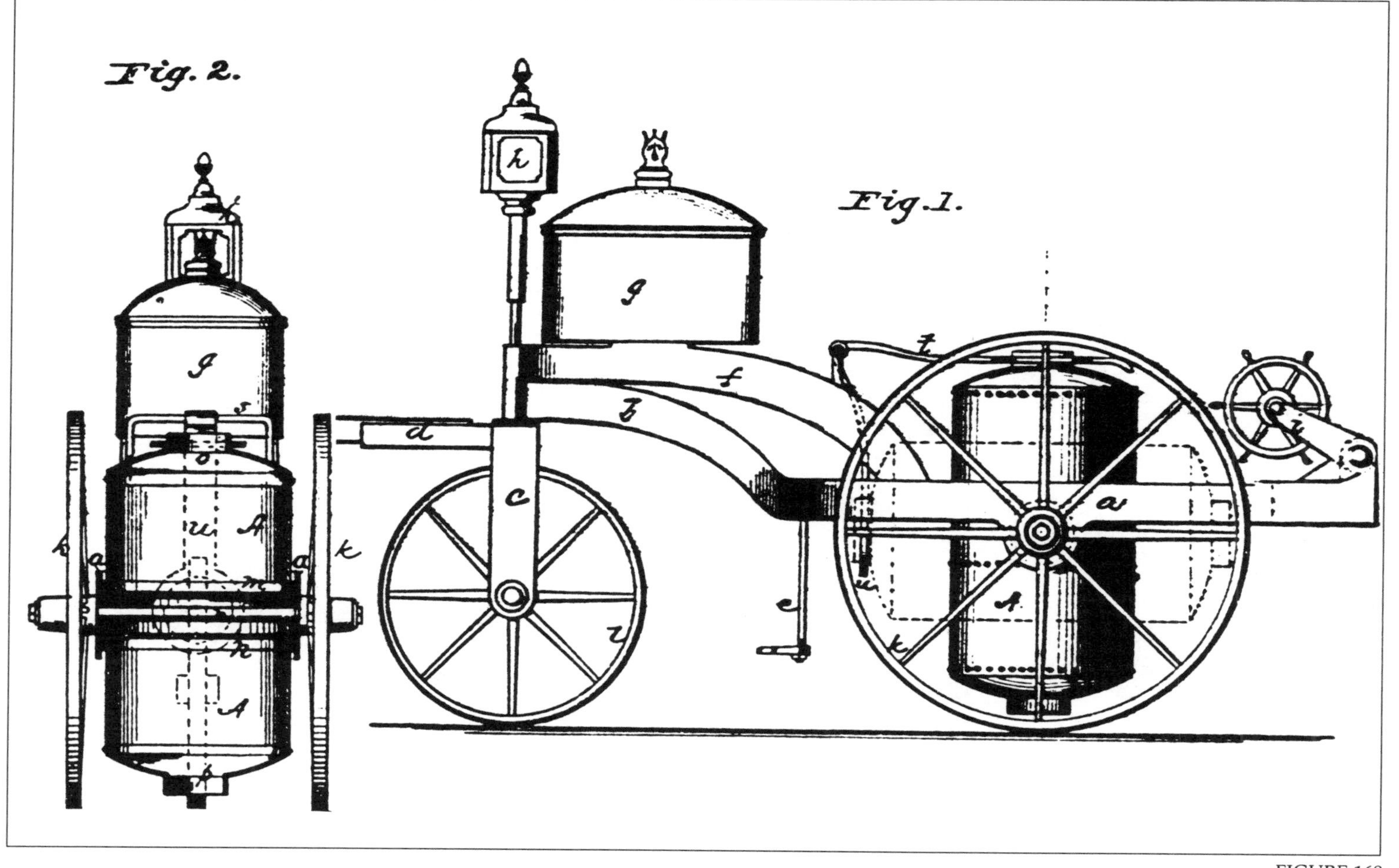

FIGURE 168

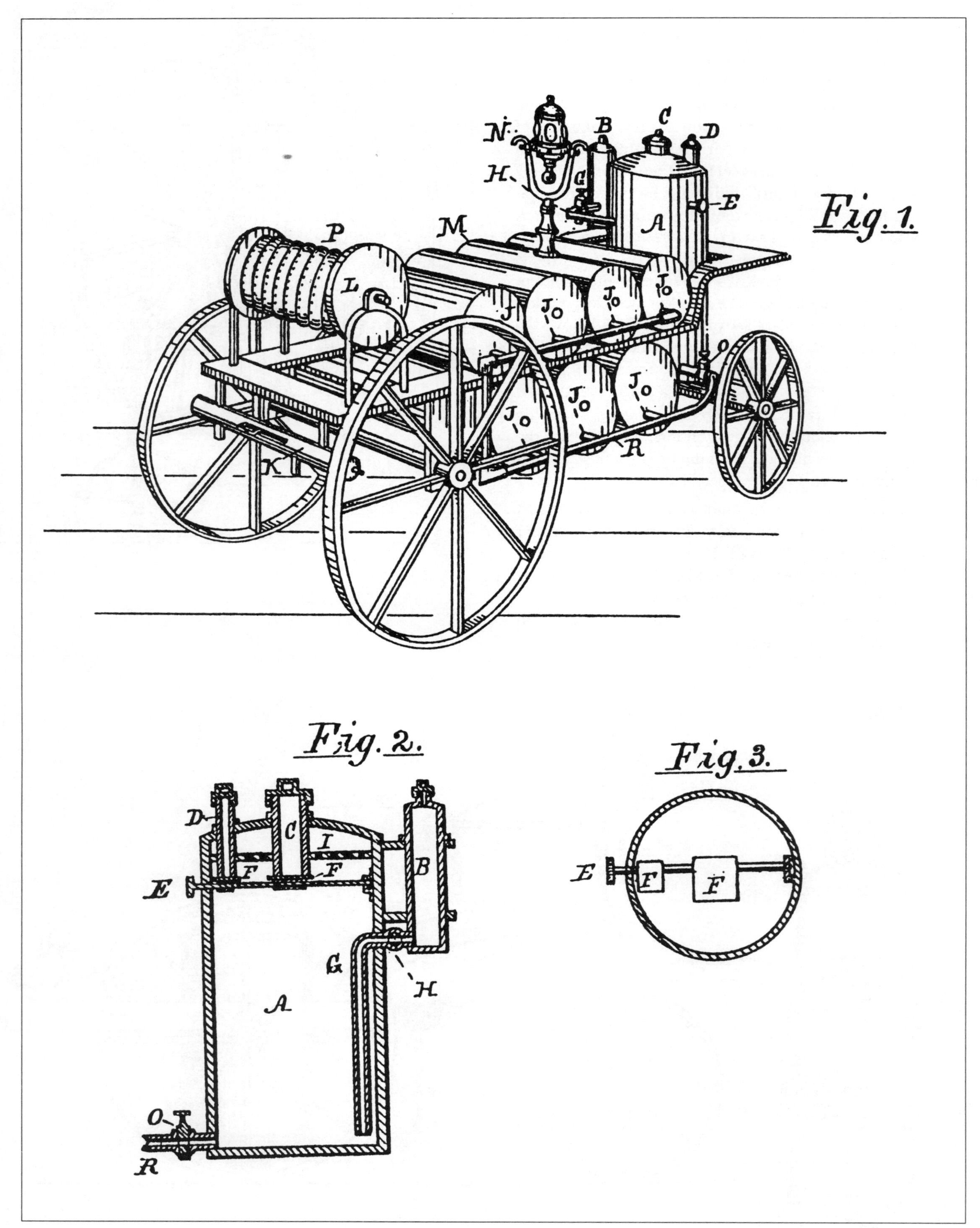

FIGURE 169

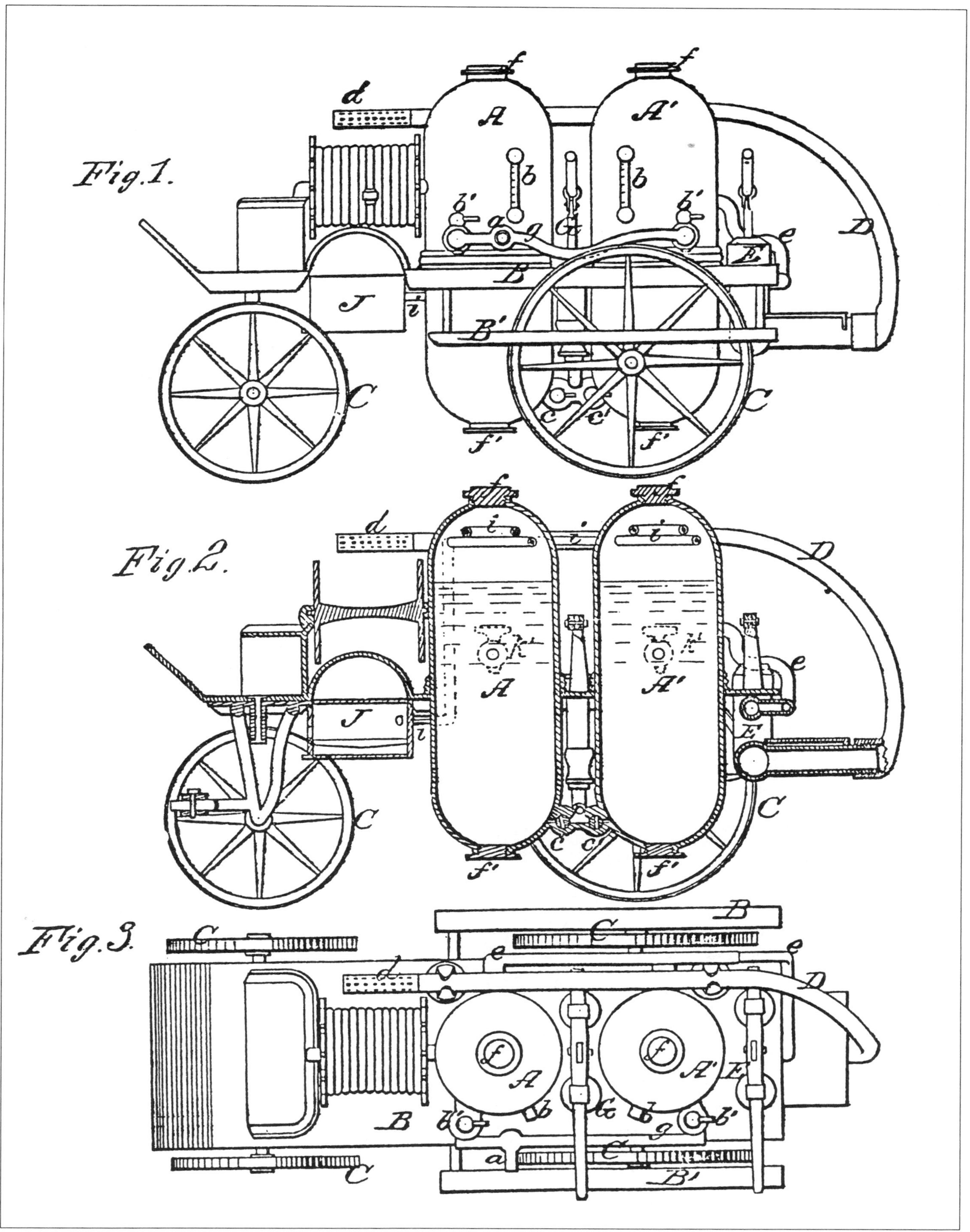

FIGURE 170

C.W. Clapp, and Others

An example of the engraver's art at its apex, at its zenith, was the beautious, slickly functional, tastefully ornamented horse-drawn chemical fire engine of FIGURE 171, from the Scientific American issue of August 25, 1877. However, compare this artwork to the patent drawing, FIGURE 172, which accompanied the extremely brief text, FIGURE 173, of patent number 174,720. This inventor was Clinton W. Clapp, from Wappinger's Falls in Dutchess County, New York. This apparatus was supposed to produce carbon dioxide and then use it to power a pump. Strictly speaking, therefore, this curious machine really was not a chemical fire engine, being equipped with a pump for throwing the water stream. One of the oddest and most uncommonly distinctive idiosyncracies of this patent document was that a mere 15 days separated its effective date, March 14, 1876 from the day the application was filed.

In FIGURE 174 is illustrated William Morrison's patent number 205,756. Notice that he suspended the large reaction cylinder using trunions so that it could be easily inverted to help ensure thorough mixing of the contents. Morrison placed a water tank, A, at the highest part of the rig so that gravity flow could refill the cylinder. The area designated with the letter D, at the rear step of the fire engine, was to have been used for carrying hose. This Canadian inventor, a resident of Toronto, Ontario, employed the seemingly strange, sort of semi-domestic term "hose hamper" to describe it.

Surely among the most eccentric contraptions receiving patents during 1878 was the "Chemical Annex" invented by Joseph H. and Thomas E. Connelly. Their patent number 208,375 is shown in FIGURE 175. Actually this was not a chemical fire engine. Rather it was a machine to be used for adding sodium bicarbonate directly into flowing water. The Connelly gentlemen obviously subscribed to the belief that this was a way to enhance the fire extinguishing properties of plain water.

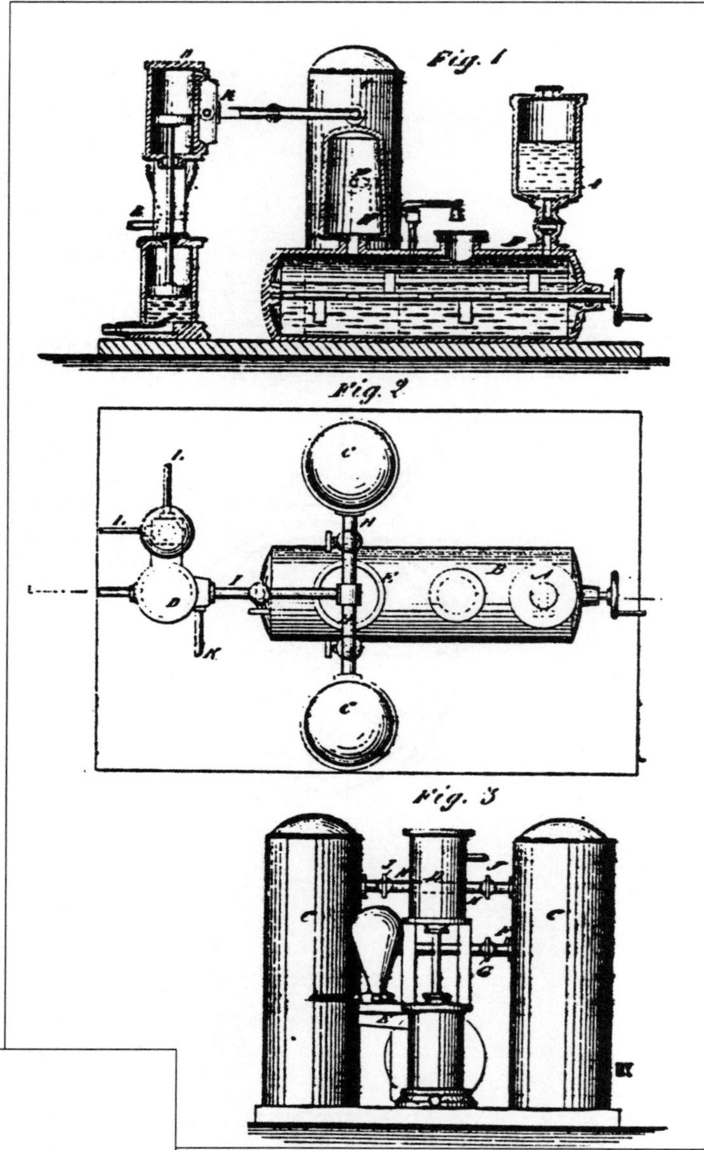

FIGURE 172

James A. Shephard's chemical fire engine is illustrated in FIGURE 176. This uncommonly attractive example of patent drawing art accompanied the papers setting out the specifications of patent number 224,109, dated February 3, 1880. The cylinder was divided into paired compartments so that one could be refilled as the other was discharging its water. Mr. Shepard of Lexington, Massachusetts provided a valving configuration such that only a single pressure gauge was required.

FIGURE 171

UNITED STATES PATENT OFFICE.

CLINTON W. CLAPP, OF WAPPINGER'S FALLS, NEW YORK.

IMPROVEMENT IN FIRE-ENGINES.

Specification forming part of Letters Patent No. 171,720, dated March 14, 1876; application filed February 28, 1876.

To all whom it may concern:

Be it known that I, CLINTON W. CLAPP, of Wappinger's Falls, in the county of Dutchess and State of New York, have invented a new and Improved Fire-Engine, of which the following is a specification:

My invention consists of a couple of receivers for carbonic-acid gas and a steam-pump, so combined and fitted with hose and nozzle for discharging the water and the gas that, by alternately charging the receiver and working off the gas through the pump, the gas can be employed as the motive agent for the pump, and, at the same time, the exhausting gas can be used separately or together with the water for extinguishing the fire.

Figure 1 is a longitudinal sectional elevation of my improved fire engine, taken on line *x x*. Fig. 2 is a plan view, and Fig. 3 is a front elevation.

A represents the charger; B, the gas-generator; C, the receivers; and D, the pump. The generator is connected by gas-chamber E and pipes F, having cocks G, so that one can be shut off while the other is filling, and the receivers are connected with the pump by pipes H and I, having cocks J to shut off one while the other is communicative with the pump. K is the exhaust from the steam or gas cylinder, to which a hose and nozzle can be attached for applying the gas to the fire; and L represents connections for the water-hose.

The gas is to be generated in the usual manner from carbonate of lime (or any of the carbonates) and sulphuric acid, by the use of strong cylinders as receivers, to be filled with this elastic gas to the pressure of two hundred to four hundred pounds per square inch, more or less. This gas is to be used to operate the engine and pump, which are of no prescribed form or kind, but are like any efficient steam-pump. The water from the pump is to be conducted and applied in the usual way for the extinguishing of flames.

The exhaust of the carbonic-acid gas from the engine is to be conducted in suitable hose, to be used in a judicious and efficient manner for the extinguishing of flames, as it can be applied in buildings, partitions, under the floors, in cellars, in attics, or in any difficult place of access, and thus confining the fires to the limits in which they originate.

The great advantage of this fire-engine over all others is its efficiency. It will throw as much, and probably more, water than any other, because of the great pressure of gas to operate the pump. But the carbonic-acid gas exhausted from the engine, properly conducted and applied, would put out more fire than the water can. It will not freeze at 100° below zero, will be instantaneous in its work, as the cylinders are always to be charged with the gas up to their proper capacity at all times; therefore the engine will always be ready to work at its greatest capacity. As one cylinder is exhausted the generator can be set in operation to replenish it, and thus a steady supply is attained. These cylinders can be made of any number or capacity, as the demand may require.

These engines are safer than steam, as they are not subject to the dangers that boilers are exposed to from low water and other causes. A well-constructed safety-valve will be a sure guard against explosion.

These engines will be lighter, consequently more portable, for manufactories, warehouses, public buildings, and ships. They may be made either stationary or portable in every and all cases, and will be efficacious and instant in their work and operation.

The carbonic gas can be conducted into the holds of ships, and, if entirely filled, would do no damage other than extinguishing any active or smoldering fires.

The carbonic-acid gas is easily generated, in retorts or generators lined with lead, in vast quantities, at a trifling cost, from marble-dust and sulphuric acid. Though it costs more than steam, yet its efficiency and promptness more than compensate for the extra cost.

Having thus described my invention, what I claim as new, and desire to secure by Letters Patent, is—

An improved fire engine, consisting of gas-generators A B, two or more receivers C, and steam pump, combined and arranged to afford a continuous supply of gas for working the pump by charging and exhausting the receivers alternately, substantially as set forth.

CLINTON W. CLAPP.

Witnesses:

KAY BROWNSON,
W. HENRY R[illegible].

FIGURE 173

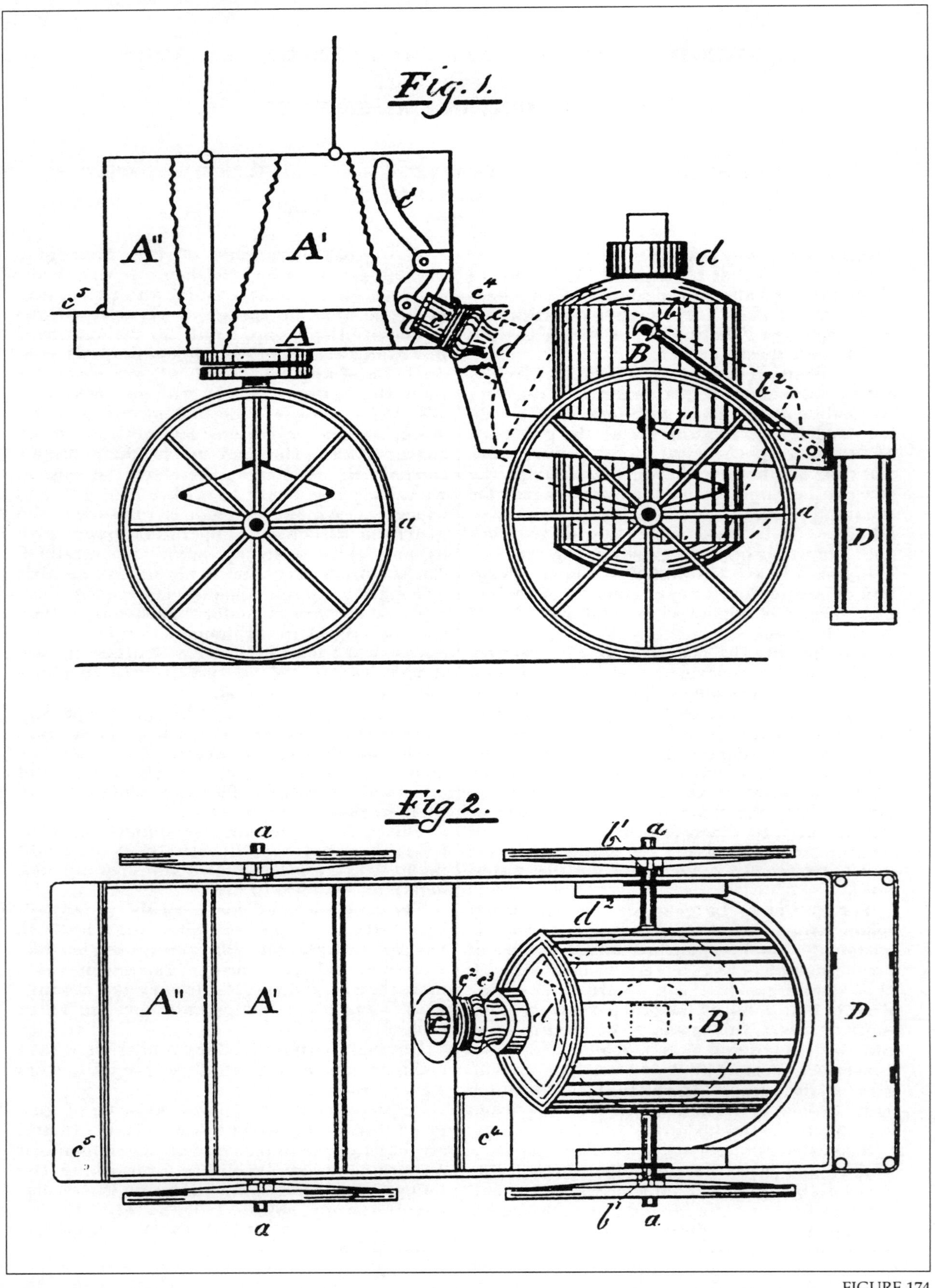

FIGURE 174

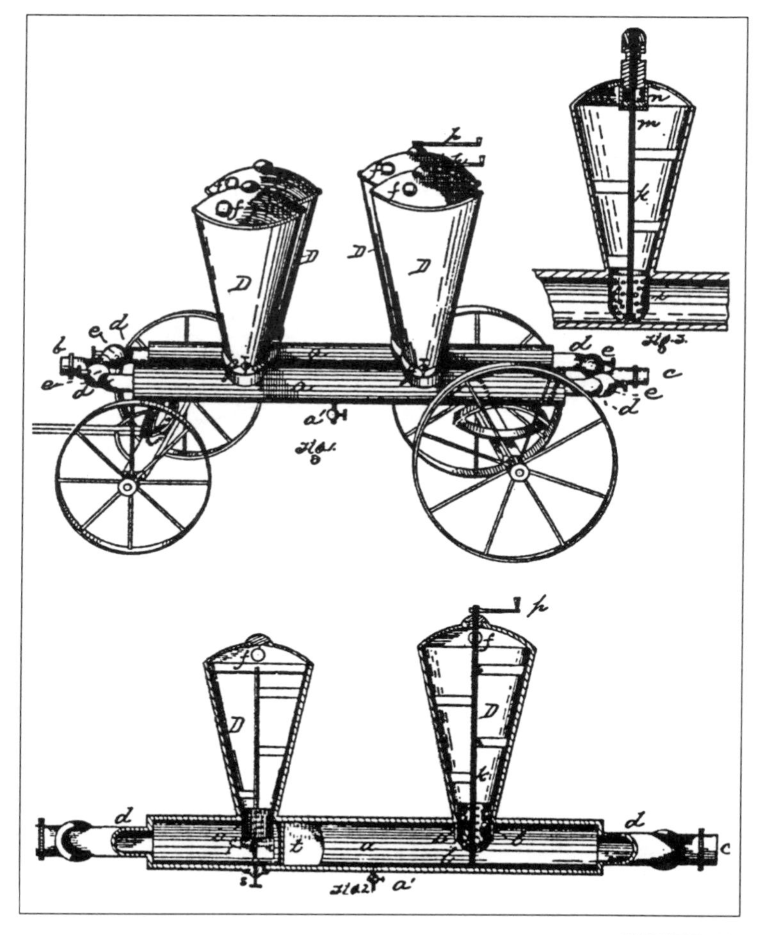

FIGURE 175

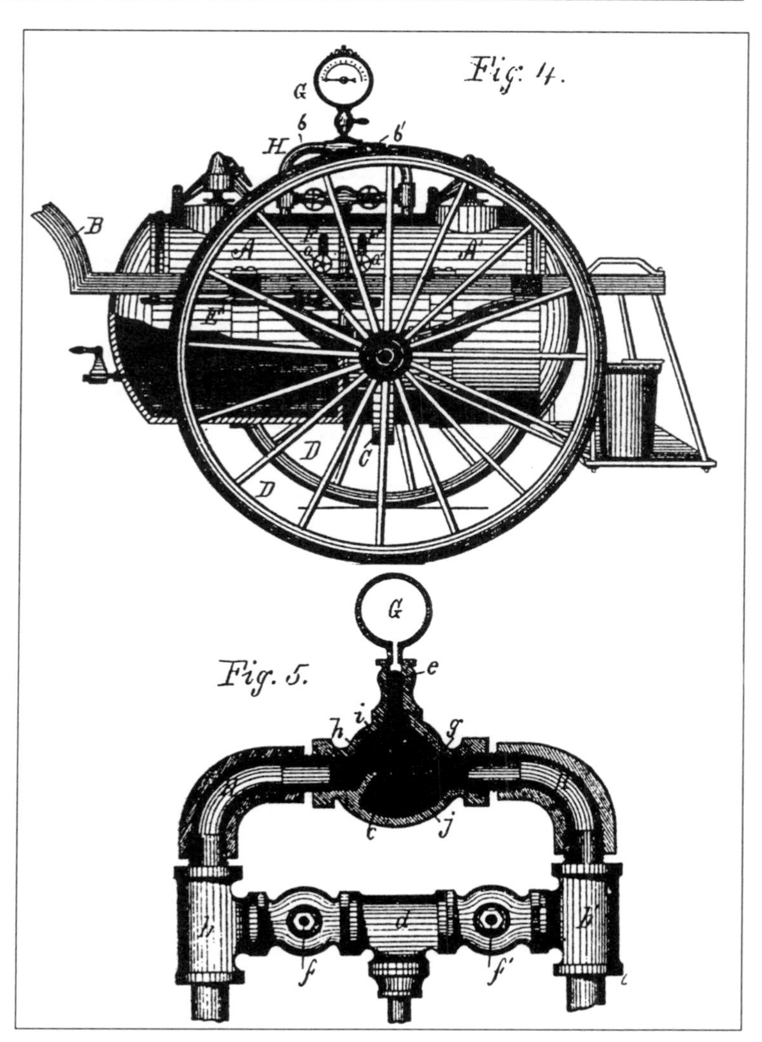

FIGURE 176

Charles T. Holloway

An inactive fire station, now a museum, surmounted by a disproportionally large clock tower is located at the intersection of Gay and Ensor Streets in old downtown Baltimore, Maryland. FIGURE 177 is an old postal card photograph of this building which was the home of the Independent Fire Company prior to becoming Engine House Number 6 of the paid fire department. The clock-maker who fabricated the tower timepiece was Robert Holloway, father of Charles T. Holloway (1827-1898), one of the really notable inventors of chemical fire engines. Robert Holloway was a lifelong volunteer fireman and Charles' firefighting interests and aptitude seem to have been almost congenital. At age 15 Charles was the president of the Hope Junior Fire Company. In 1850 he organized the Pioneer Hook and Ladder Company, the first H. & L. in Baltimore. He served this fire company as its president until 1858, when he was selected as the first Chief Engineer of the paid fire department, his commission being dated on December 21 of that year. Holloway held this position for five troublesome years, overseeing, equipping, and perfecting the new professional fire department and combating the residual evil aspects of the former volunteer system. In 1868 he was appointed as the first fire inspector, a job he held until February 6, 1894. The Veteran Volunteer Firemen's Association, nevertheless, was one of his primary interests and he was president of this group for many years. Holloway was the organizer of the Baltimore Salvage Corps; he was one of the founders of the Bureau of Fire of the city of Pittsburgh, Pennsylvania; and he was the first Fire Commissioner of the Baltimore County Fire Department.

FIGURE 177

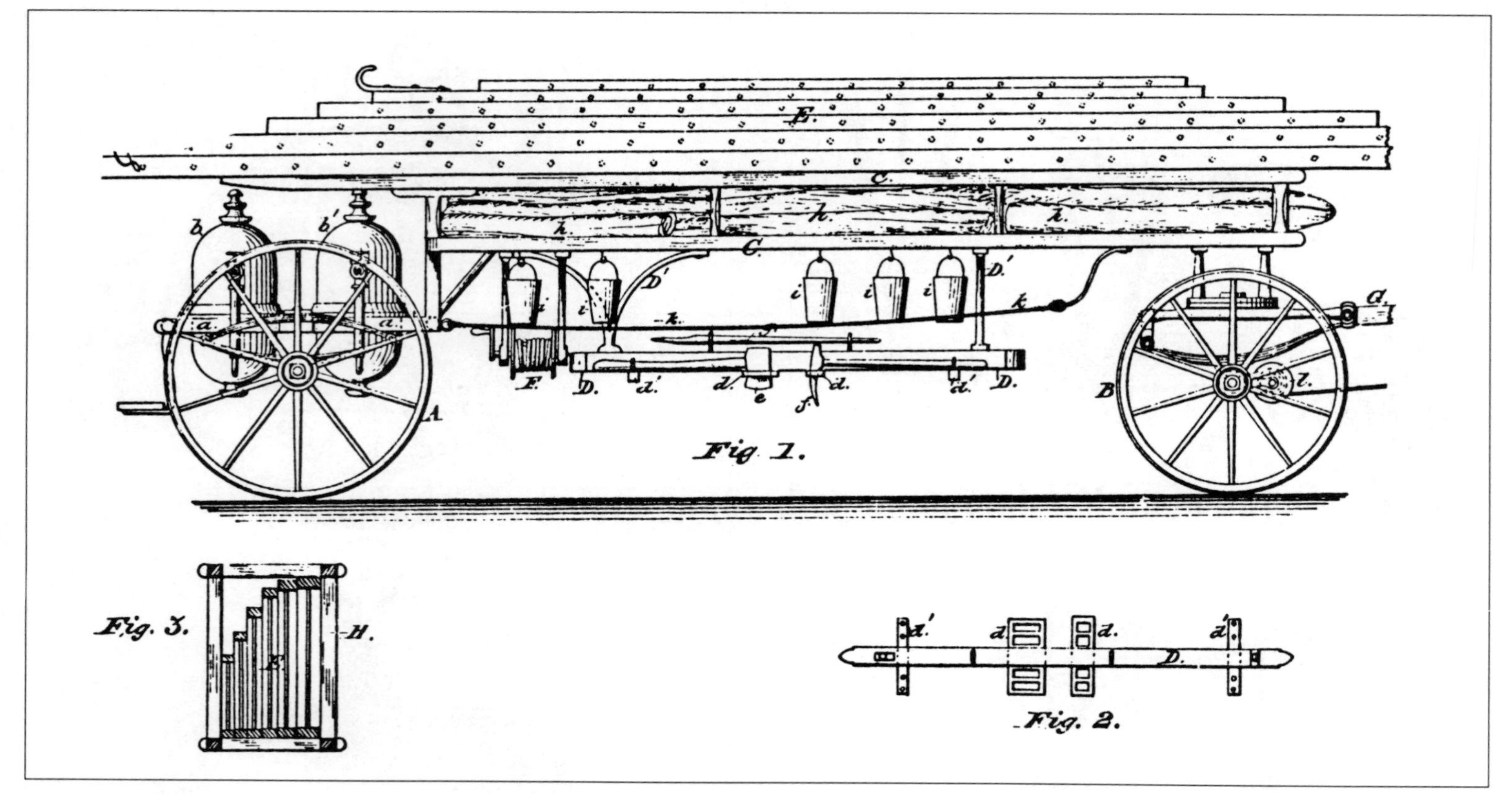

FIGURE 178

Charles T. Holloway was an altogether unusual personality, a charismatic and knowledgeable leader, an uncommonly accomplished firefighter and fire administrator, and one heck of a good businessman. He was an early advocate of steam fire engines and was responsible for their introduction to the Baltimore Fire Department. But after his 1881 appointment to head the new Baltimore County Fire Department he equipped this organization almost exclusively with chemical engines. And these were Holloway chemical fire engines, manufactured by the company he founded in 1870!

For many years, at least through 1881, Mr. Holloway continued to publicize a claim that he, and only he, held a legally-valid license derived from old patent assignments to produce carbon dioxide gas fire extinguishing machines.

By virtue of his career production of eight (or more) patents Holloway was, in a sense, a professional inventor. One of his very first patents described an integrated chemical fire engine and ladder wagon, a kind of primitive combination city service hook and ladder truck, complete with alternate fittings and hitches so that it could be pulled by men or by horses. This was patent number 224,911 dated February 24, 1880, and it is illustrated in FIGURE 178. Notice the following features in this drawing: front and rear wheels were designated, respectively, A and B; between the rear wheels platform a supported the chemical tanks, b and b. Also fastened to platform a was frame C from which projected brackets, D; which held rack D, upon which the axes and other tools were hung. The ladder E, and the salvage covers, h, were also carried on frame C, while buckets, i, hung below it. Located just behind the rack, D, was a reel of chemical hose. Ropes, k, were to be used if men hauled the vehicle, and a tongue, G, was provided if horses were to be harnessed. Holloway was prepared to outfit this multipurpose truck according to the needs of each specific purchaser, but as a rule he offered them with the following equipment: 25 or 50 gallon chemical tanks; 200 feet of rubber chemical hose; a 12 or 14-foot ladder with a scaling hook; other ladders of size 16, 18, 20, 25, 30, or 35 feet; 10 or 12 rubber buckets; 12 rubber salvage covers; 6 pike poles; paired axes, picks, and forks; a crowbar; a chain with hooks; assorted wrenches; lanterns and a bell; and side ropes and drag ropes.

Back in those good old days just as in our own time it was exaggeration and hype which characterized most advertising. Consider this description of the apparatus, according to a Holloway Company blurb: "This truck is new in design and the first of its kind ever offered to fire departments. In large cities Salvage Corps organizations, underwritten by the fire insurance companies, have been very active and very successful. On all of their wagons rubber covers or oiled cloths are carried. These are spread over furniture or merchandise for protection from fire and from the effects of water. The idea occurred to me that by combining this equipment with the gear on a hook and ladder truck, or a combined ladder truck and chemical fire engine, it could become possible for smaller communities to have these benefits. This apparatus is well equipped, well appointed, and well finished. It is built substantially, with the best materials available in the United States. I would especially call it to the attention of corporations and fire company organizations as being eminently appropriate for all of their needs." Combination ladder and chemical trucks as offered in old Holloway Company catalogues are illustrated in FIGURES 179 and 180.

An odd creation which earned patent number 259,857 (June 20, 1882, FIGURE 181) for Holloway involved an air chamber located between chemical tanks and outflow hose fittings of a chemical fire engine. According to the inventor the air in the chamber would be compressed by carbon dioxide, thus exert-

FIGURE 179

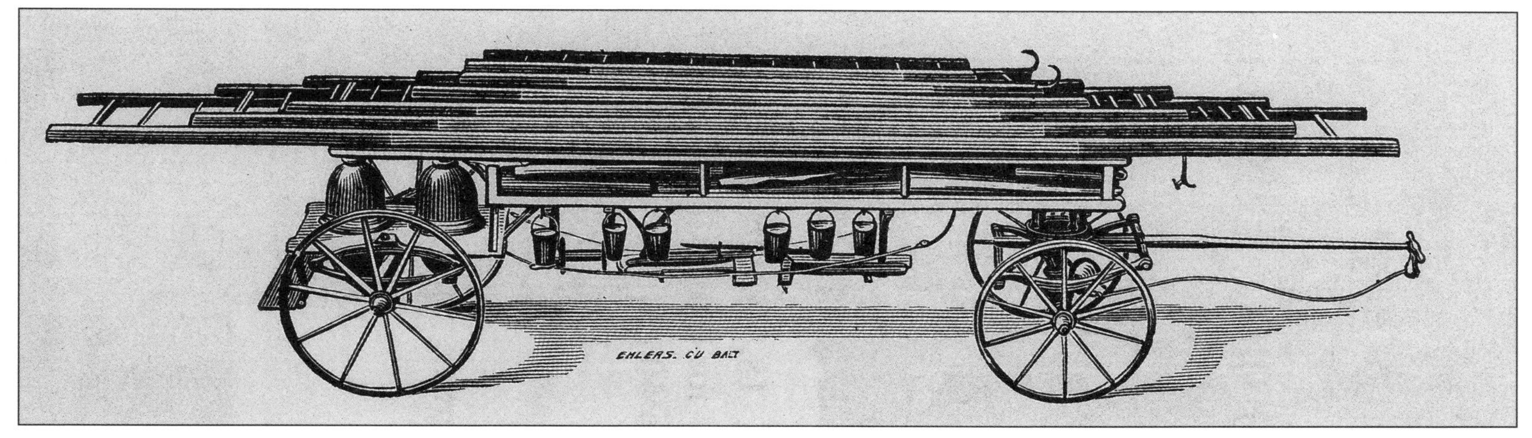

FIGURE 180

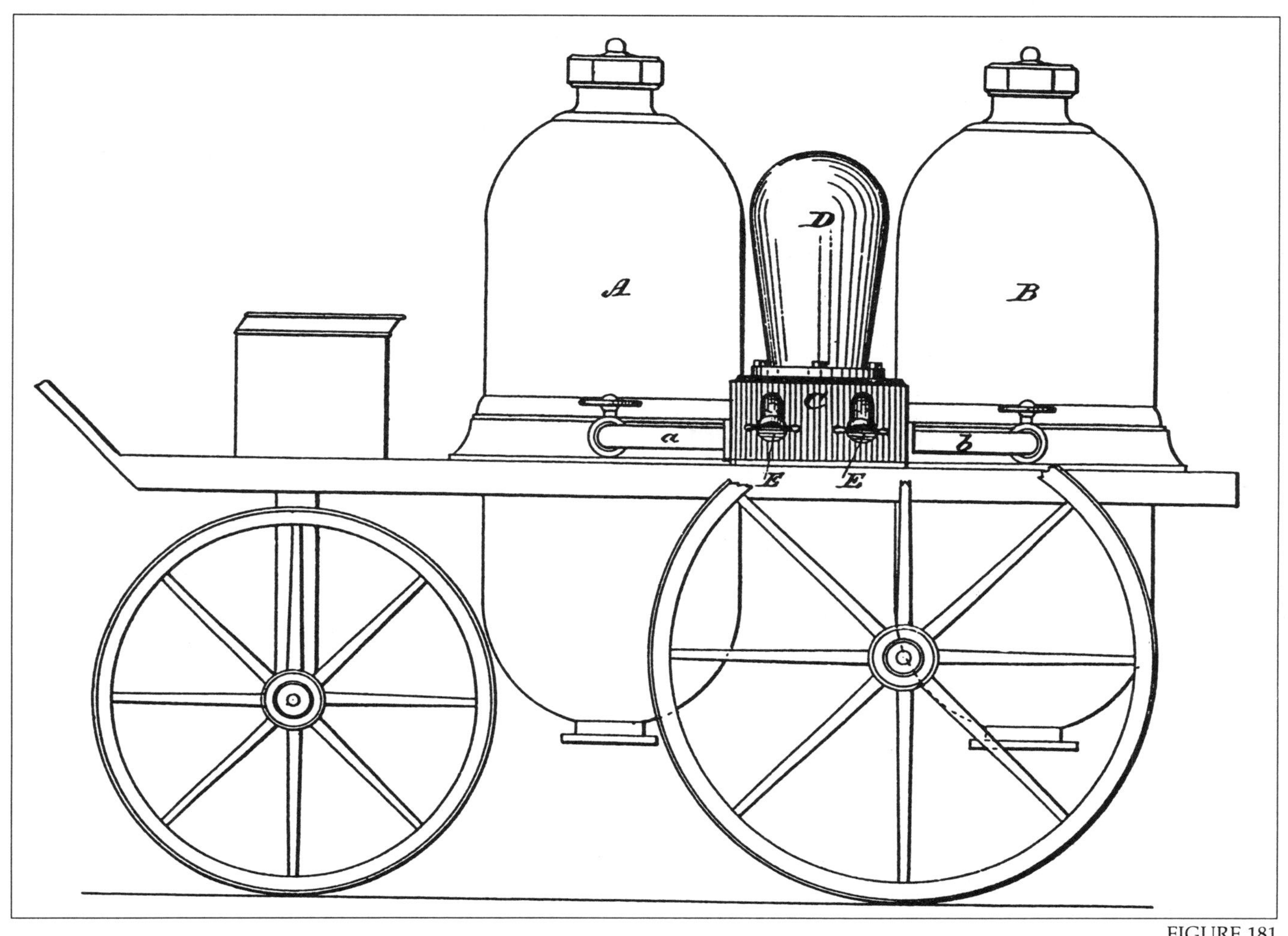

FIGURE 181

FIGURE 182

ing a force upon the gas and in turn upon the water stream, making it steadier and stronger. Such an apparatus, obviously, was never built as such.

The remaining years of the 1880s saw Holloway getting married in 1884 and receiving a couple of patents for his fire extinguisher inventions. His combination fire engine and hose wagon achieved patent recognition in December, 1891. FIGURES 182 and 183 are the drawings for this document, patent number 464,281. The elements of novelty claimed for this design related to Holloway's method of attaching the chemical tanks and to their being placed on the exterior of the truck in a horizontal alignment. Mr. Holloway's next invention was patented on March 1, 1892 and was number 469,998. This time the chemical tanks were placed under the body of the vehicle. Notice in the drawing, FIGURE 184, how the rear end of the truck was configured to avoid any obstructions to the filling and charging fittings. There was a tail gate which could be flexed to a horizontal position so as to cover and shield the tank valves whenever hose was being withdrawn from the hosebed.

FIGURES 185 and 186 served to illustrate Holloway patent number 493,028 dated March 7, 1893. This chemical fire engine had both fore and aft discharge lines, and it was equipped with a four-channel rotary plug valve of advanced design. By manipulating this valve the engineer could select any of the six flow pathways, as illustrated. U.S. Patent 596,776 dated January 4, 1898 may have been Holloway's last invention, 1898 being the year of his death. This patent described a hose wagon with latticed side walls and flooring of spaced board strips (FIGURE 187), so specified that wet hose could be exposed to currents of fresh air while being drip-dried.

One final note about Charles T. Holloway: While on duty at a fire on South Charles Street, November 20, 1870, he was severely injured when struck by a falling wall. He was trapped beneath the ruins and the hot debris for nearly four hours, with smoke and steam nearly asphyxiating him. Finally he was rescued and resuscitated. In token and expression of his gratitude to Providence for the restoration of his health and vitality he presented a marble alter to St. Andrew's Episcopal Church, which he then served as a vestryman.

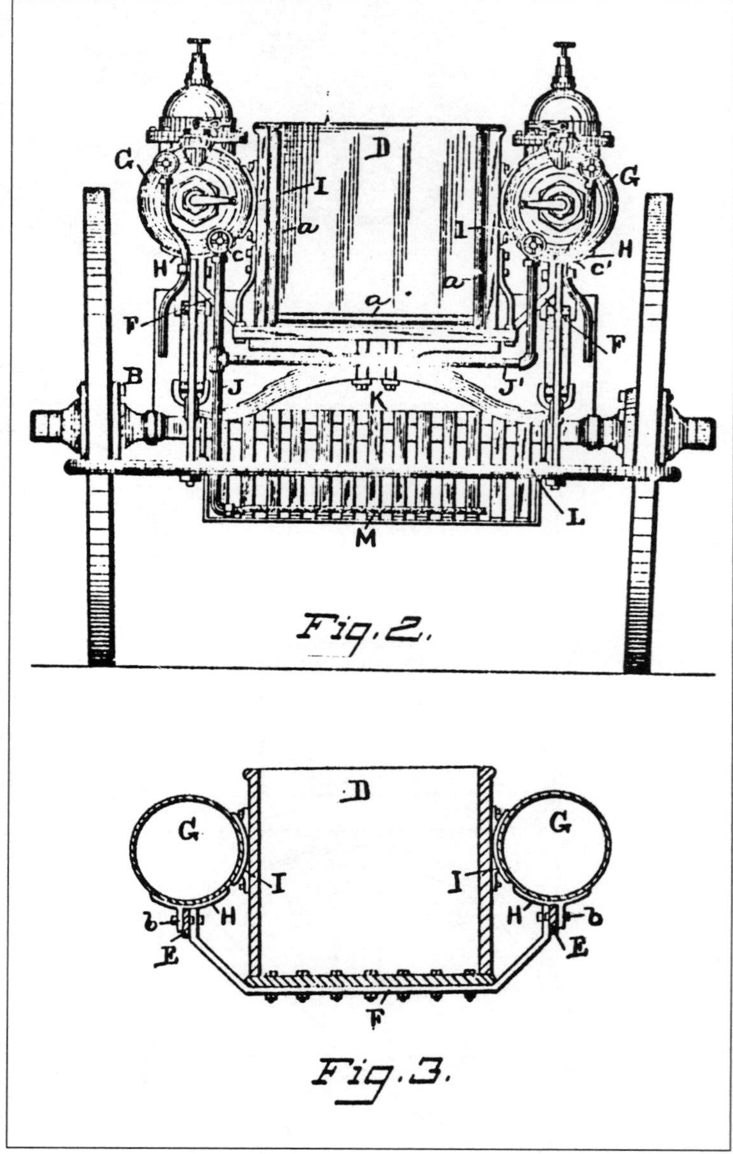

FIGURE 183

FIGURES 188, 189, and 190 are additional Holloway Company catalogue illustrations.

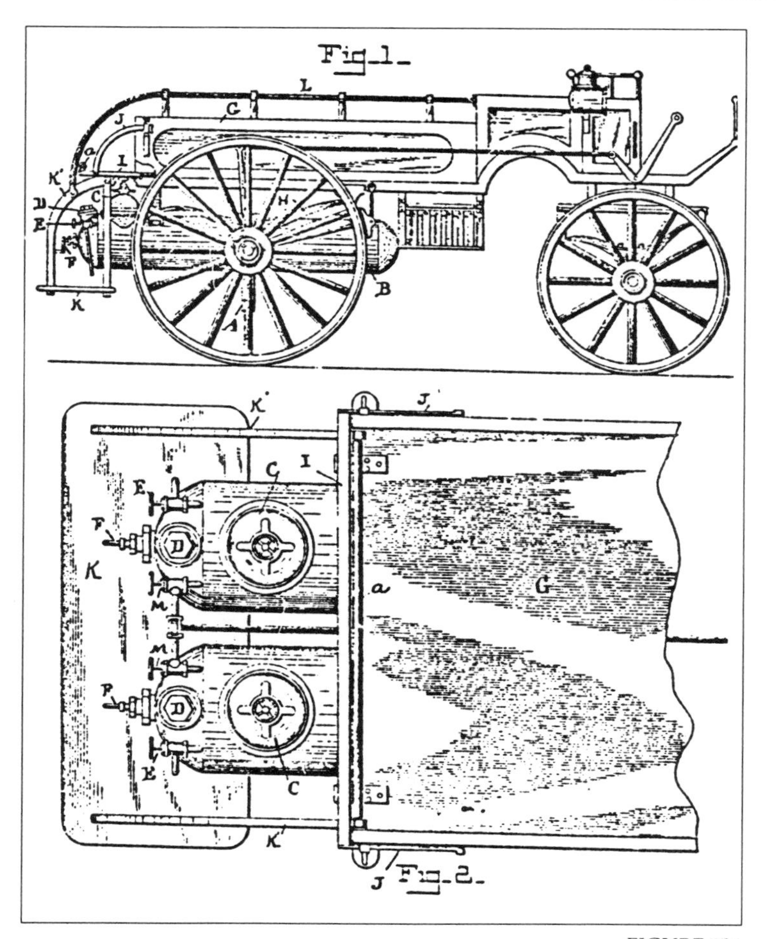

FIGURE 184

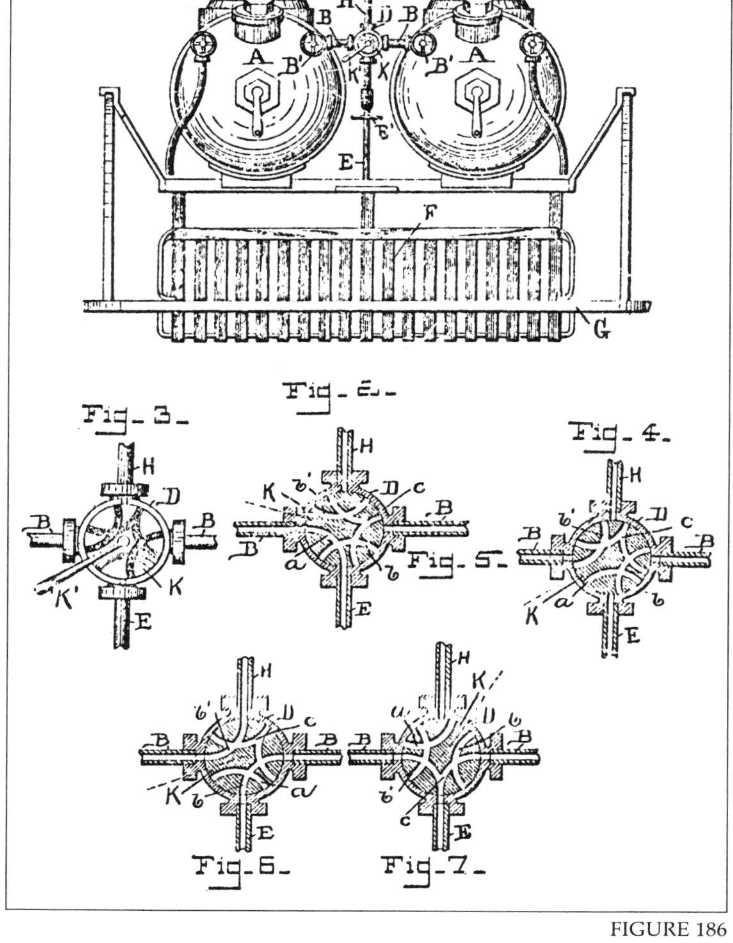

FIGURE 186

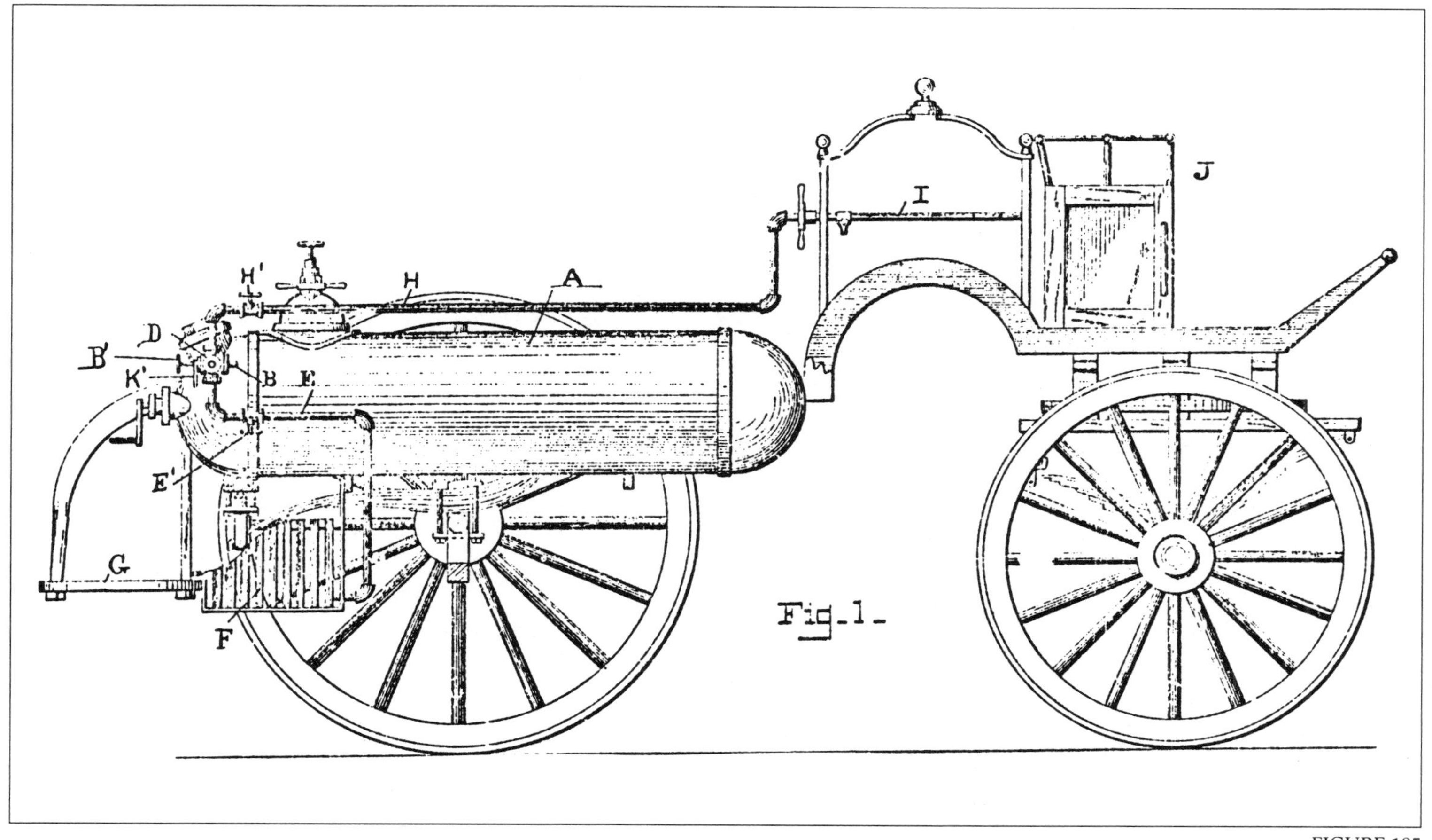

FIGURE 185

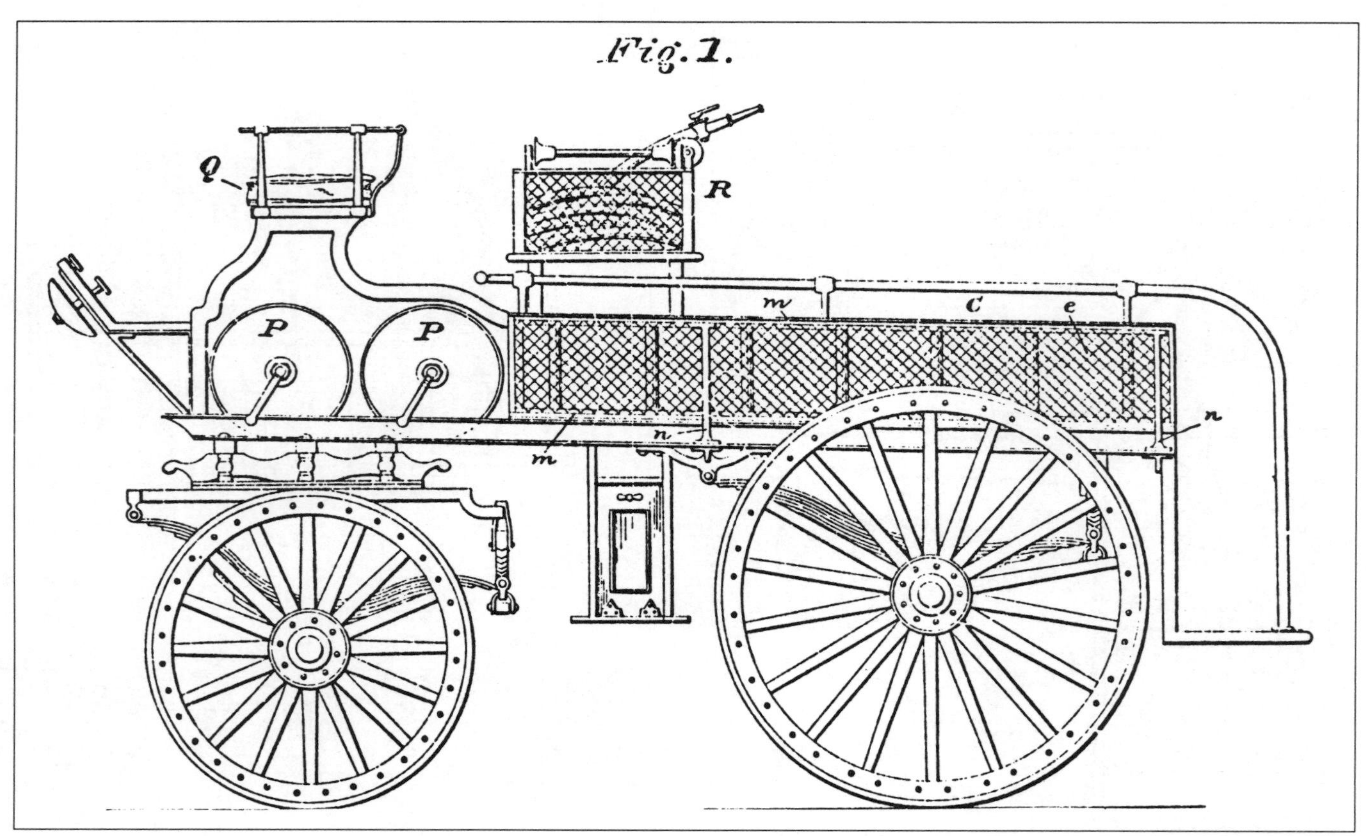

FIGURE 187

FIGURE 188

CHARLES T. HOLLOWAY,

MANUFACTURER OF

CHEMICAL FIRE APPARATUS

COMPRISING

4 Wheel Double Tank Chemical Engines, 50 to 200 gal, capacity,
2 and 3 Wheel Single Tank Chemical Engines,
25 to 100 gallon capacity.

Hook and Ladder Trucks and Chemical Engines Combined.

PORTABLE FIRE EXTINGUISHERS.

Sole Licensee under the Graham Patent,
COVERING ALL APPARATUS USING CARBONIC ACID GAS
for extinguishing fire, etc. (See Claims.)

READ IMPORTANT NOTICE ON INSIDE.

HOOK AND LADDER TRUCKS.

ESTABLISHED 1870.

OFFICE: 17 SOUTH STREET,

FACTORY: 71 N. CALVERT ST.

BALTIMORE, MD., U. S. A.

FIGURE 189

FIGURE 190

M. Cronin

A few miles down the road from Baltimore is Washington, D.C. Serving the citizenry of the District of Columbia from September 23, 1871 until his retirement by reason of disability in May, 1886 was Martin Cronin, the first Chief Engineer of the D.C. Fire Department. Chief Cronin was a holder of several patents for his inventions of firefighting gear. One of these was number 269,777 dated December 26, 1882 (FIGURE 191). This drawing shows a conventional steam fire engine strangely altered to serve as well as a chemical fire engine. It was well that this hybrid was never actually built nor used. In the illustration letter B designated a large storage tank, letter C was an auxiliary pump, and letter D was an air chamber. According to the inventor the smaller fires were to have been blackened down by wetting them with a solution called "Zaffle Fire Extinguishing Compound" which was to be sprayed through the hose carried on the reel marked by letter E. Anyhow, as we know, this rig was never built, and therefore Cronin's reputation has survived untarnished. In passing, and parenthetically: one of the major fires in Washington during Chief Cronin's tenure was the great United States Patent Office blaze of September 24, 1877.

Ernst F. Steck

Another one of the important inventors of chemical fire engines, just as he was likewise an accomplished designer of practically anything firematic, was Ernst F. Steck. In all he had received at least 27 patents. One of the hallmarks of a Steck patent was the quality artwork of the accompanying drawings, such as are seen in FIGURES 192, 193, and 194. These illustrations from Steck's patent number 327,342 dated September 29, 1885 show, respectively, a three-dimensional view of the apparatus, a top view and a side view. Inventor Steck intended to design a fire vehicle which carried the large and heavy chemical cylinders at a low point in order to avoid top-heavy instabilities when rounding corners. As the drawings show this objective was accomplished. He also wanted to position the hose reel in a more advantageous location than the conventional spot, high up and just behind the driver's seat. Again, as shown by the drawings this objective was accomplished. And from experience he had determined that the acid bottle entry valves should be placed more conveniently and directly at the hand of the responsible fireman. Again, as seen in the drawings, this objective was accomplished.

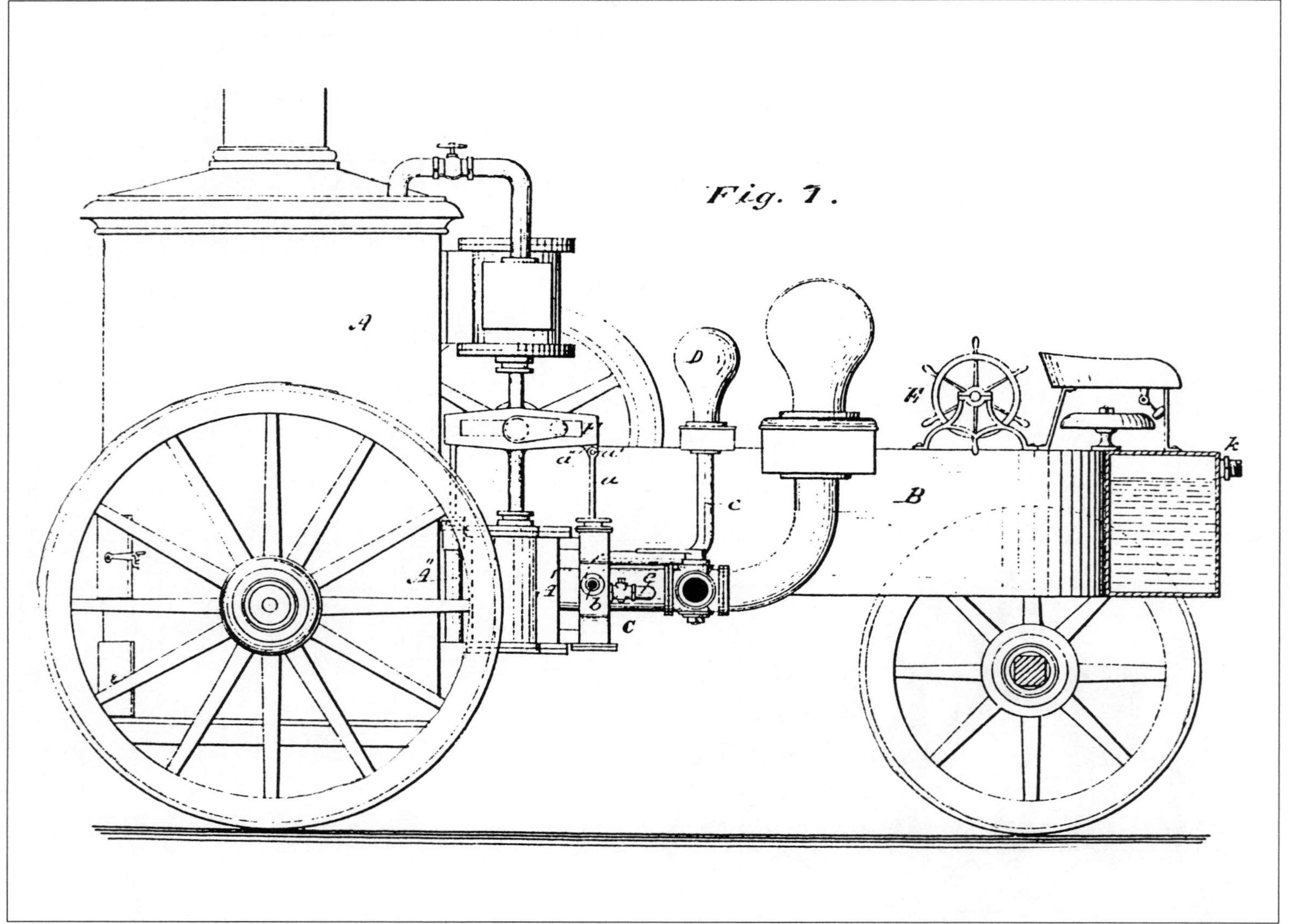

FIGURE 191

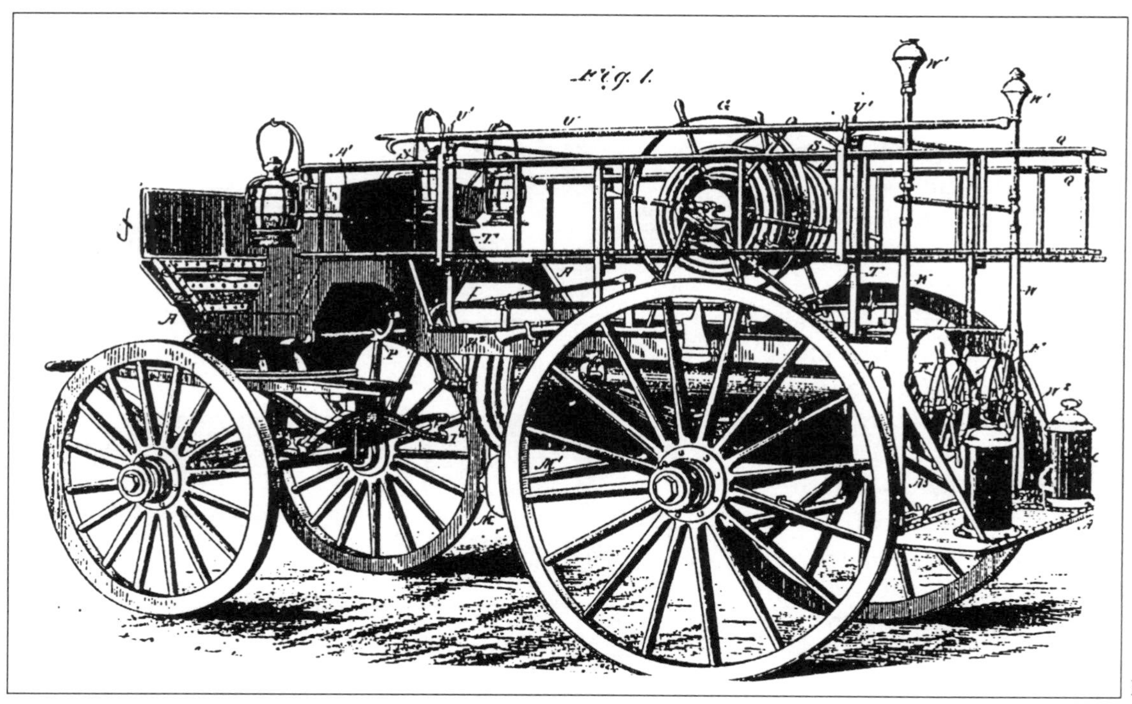

FIGURE 192

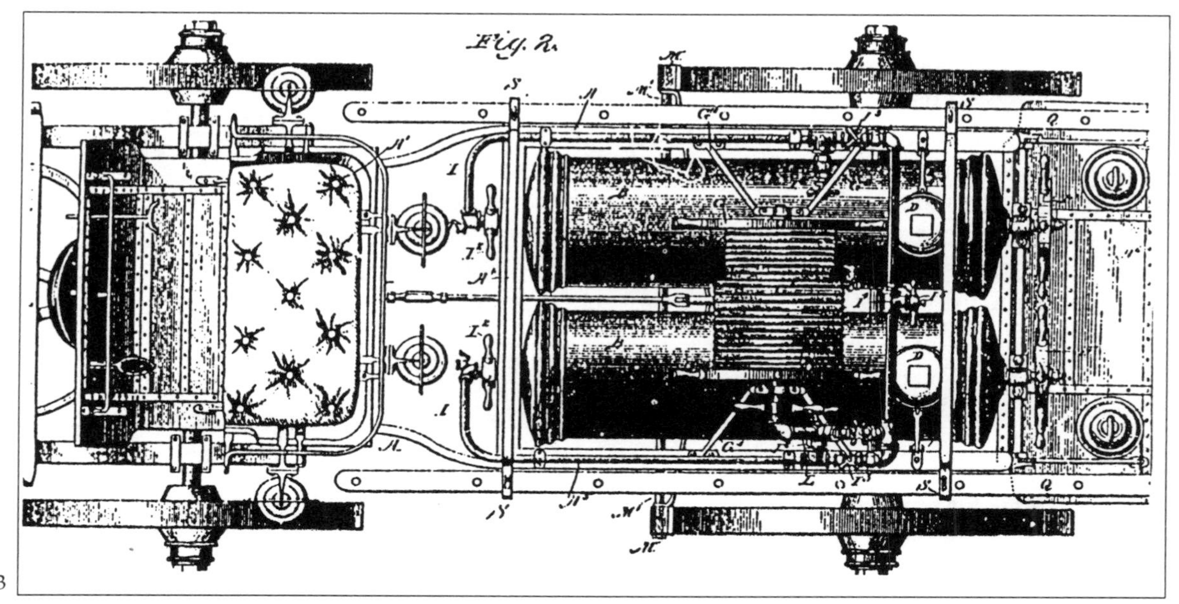

FIGURE 193

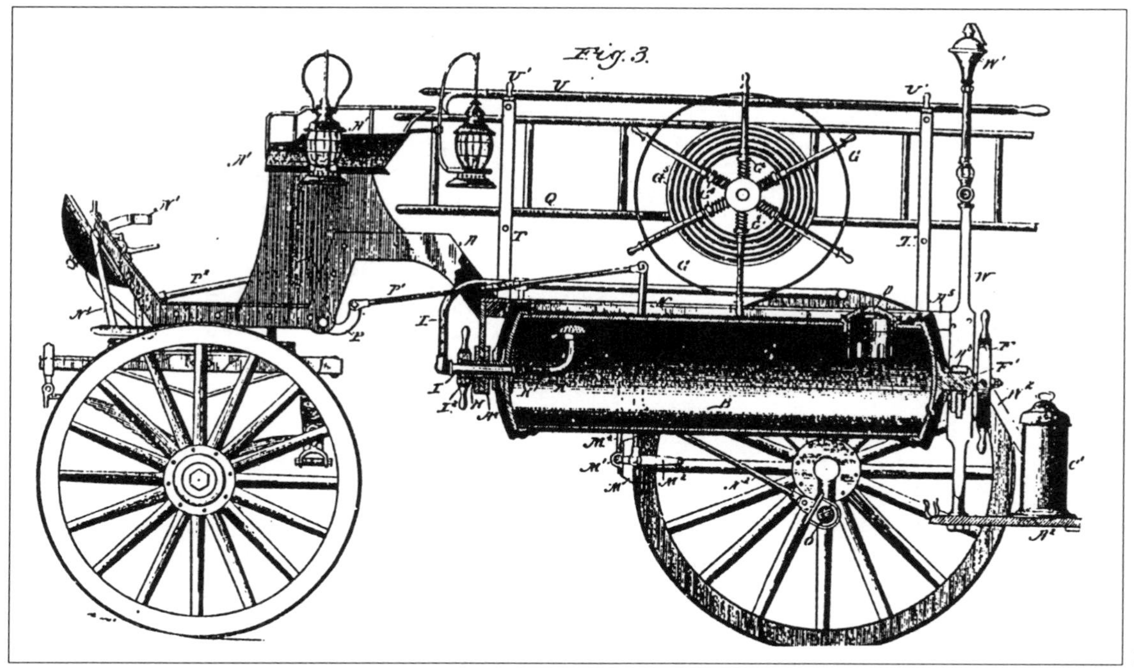

FIGURE 194

Abraham Bruegger

Assigned to the Muskegon Chemical Fire Engine Company were a number of patents received by inventor Abraham Bruegger who was, not surprisingly, a resident of the Michigan city of Muskegon, located along the Lake Michigan shore and across it from a point between Milwaukee and Sheboygan, Wisconsin. Mr. Bruegger's patents numbered 500,189 and 500,190 were both dated on June 27, 1893. The first of these set out his method of mounting the sulfuric acid container on a shaft which ran centrally through the large water and bicarbonate cylinder. By a turn of this shaft the acid could be dumped directly into the bicarbonate so that the reaction could proceed. For this to occur, however, a dome designated by the letter A in the patent drawing, FIGURE 195, had to be first opened so that the lead-lined cover plate, A', could be removed. The second of these Bruegger patents was for a complete, vehicle-mounted chemical fire engine (FIGURES 196, 197, and 198). In the descriptive text of this patent inventor Bruegger took note that, "Previously acid vessels were carried within cans upon the rear step of the fire vehicle. To charge the cylinders the covers had to be unscrewed and removed, the acid vessels taken from their cans and placed in supports in the cylinders, the covers replaced, and the entire cylinder rotated so that the acid-vessel stopper would fall out so the chemicals could mix. This was an unsatisfactory system as it required considerable adjustments and time to bring the apparatus into action. And to provide for rotation of the cylinder it was necessary to mount it in bearings, thus complicating construction and adding to the cost. In yet another type of chemical engine the acid vessel was pivoted within a dome located on top of the main cylinder. However, this configuration separated the acid vessel and the lever used to churn the mixing paddles, and therefore really required the services of two operators."

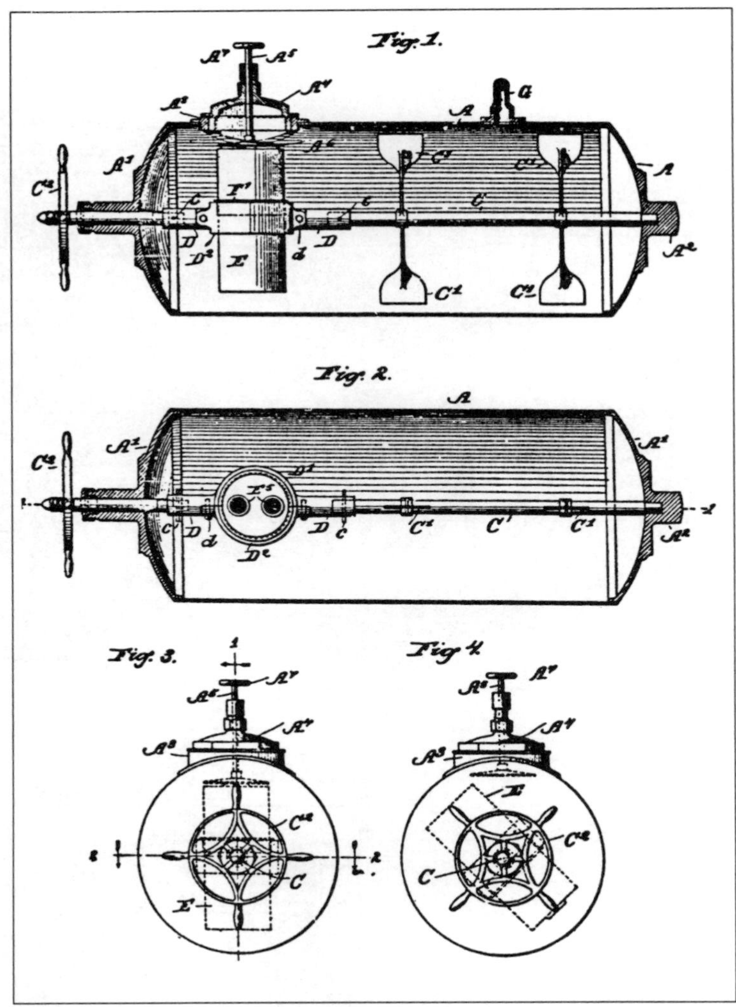

FIGURE 195

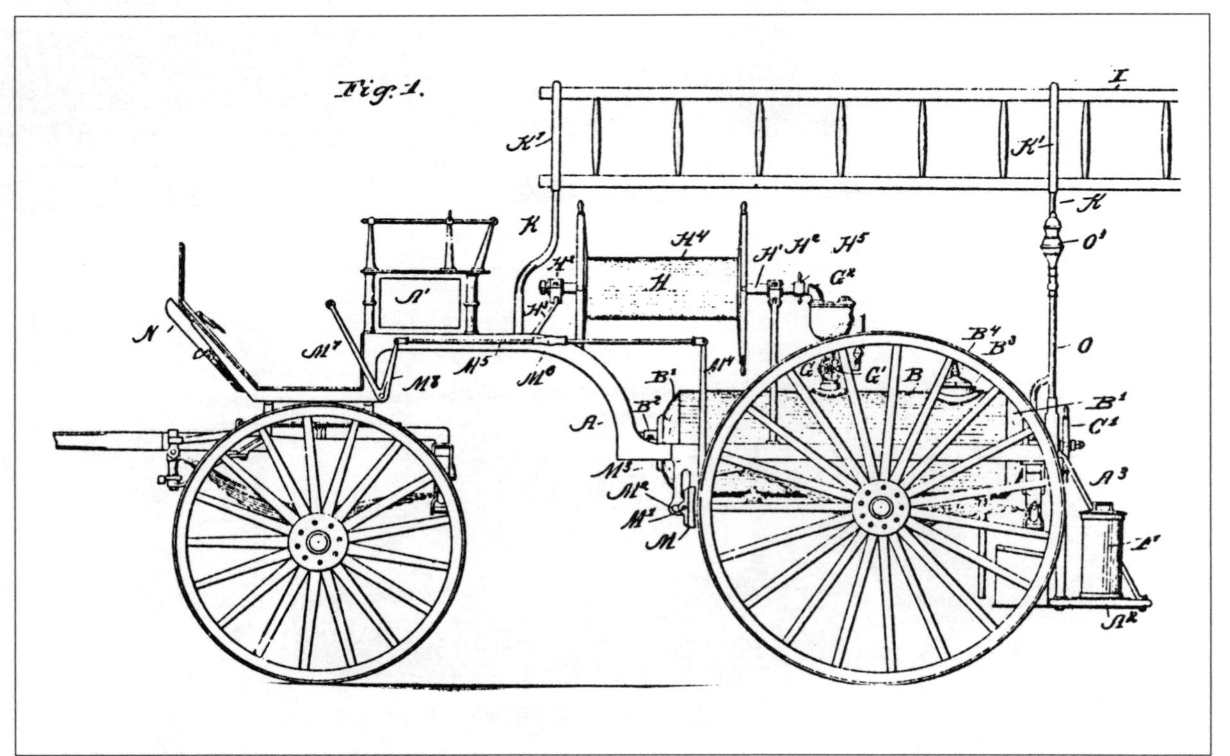

FIGURE 196

In this new fire engine Bruegger corrected all of these features of which he was so critical. In the drawings notice that, again, the acid vessel had been placed on a sleeve upon a shaft through the main cylinder, but that it was now possible to remotely remove the lead-lined cap. Bruegger had now engineered the shaft so that it, together with the mixing paddles, could be rotated without disturbing the acid bottle, which was to have been dumped by turning a shaft-mounted lever independently. The details of these maneuvers were quite simply as follows (from the fourth patent drawing): the top handle was turned to lift the cap from the acid container; lever D, was turned to dump the acid from the bottle into the cylinder; lever C, was turned to mix the acid with the bicarbonate and the water. Bruegger also designed a somewhat unusual system for operating the brakes of his chemical fire engine. By following the designators of the "M" series this mechanism can be traced out.

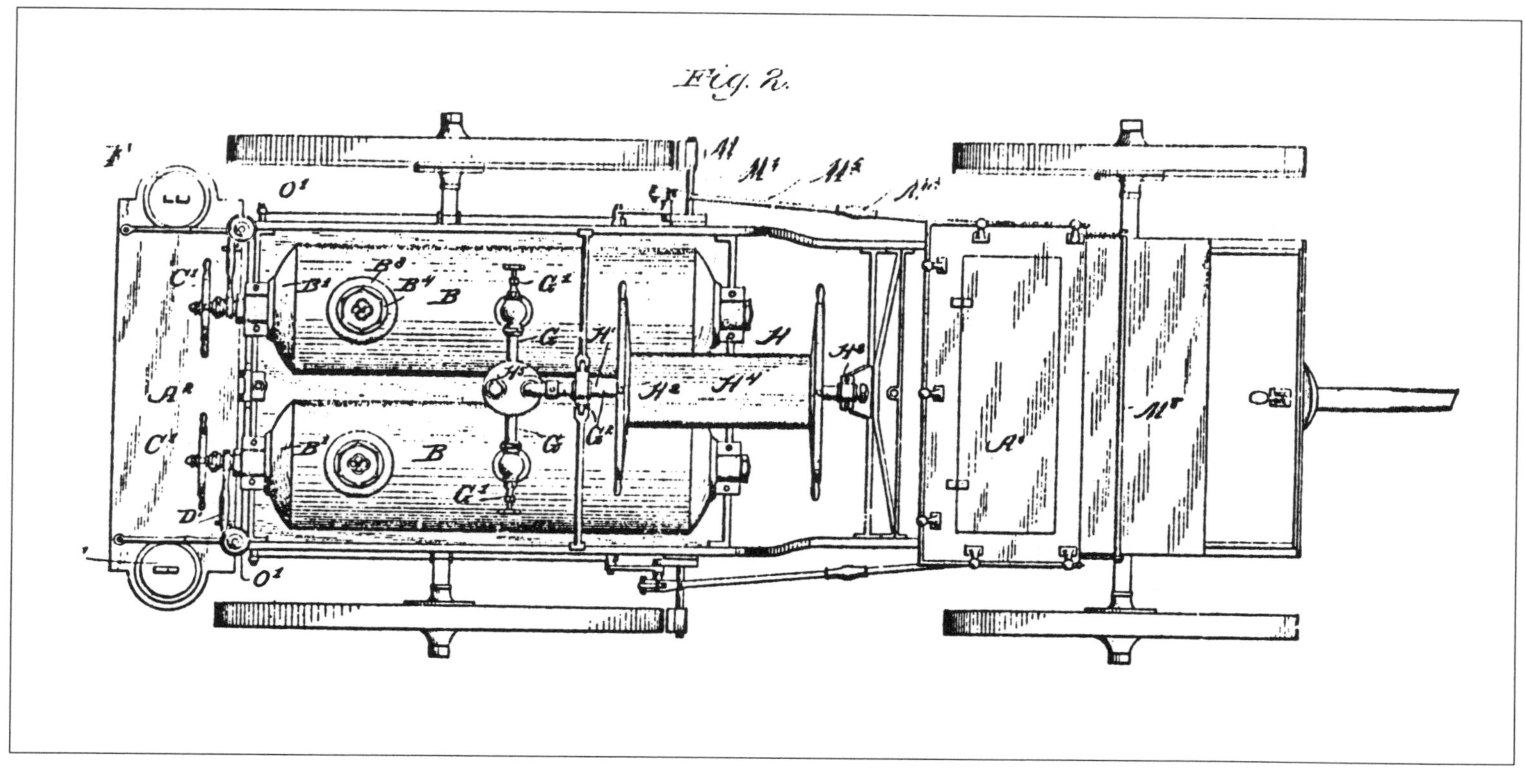

FIGURE 197

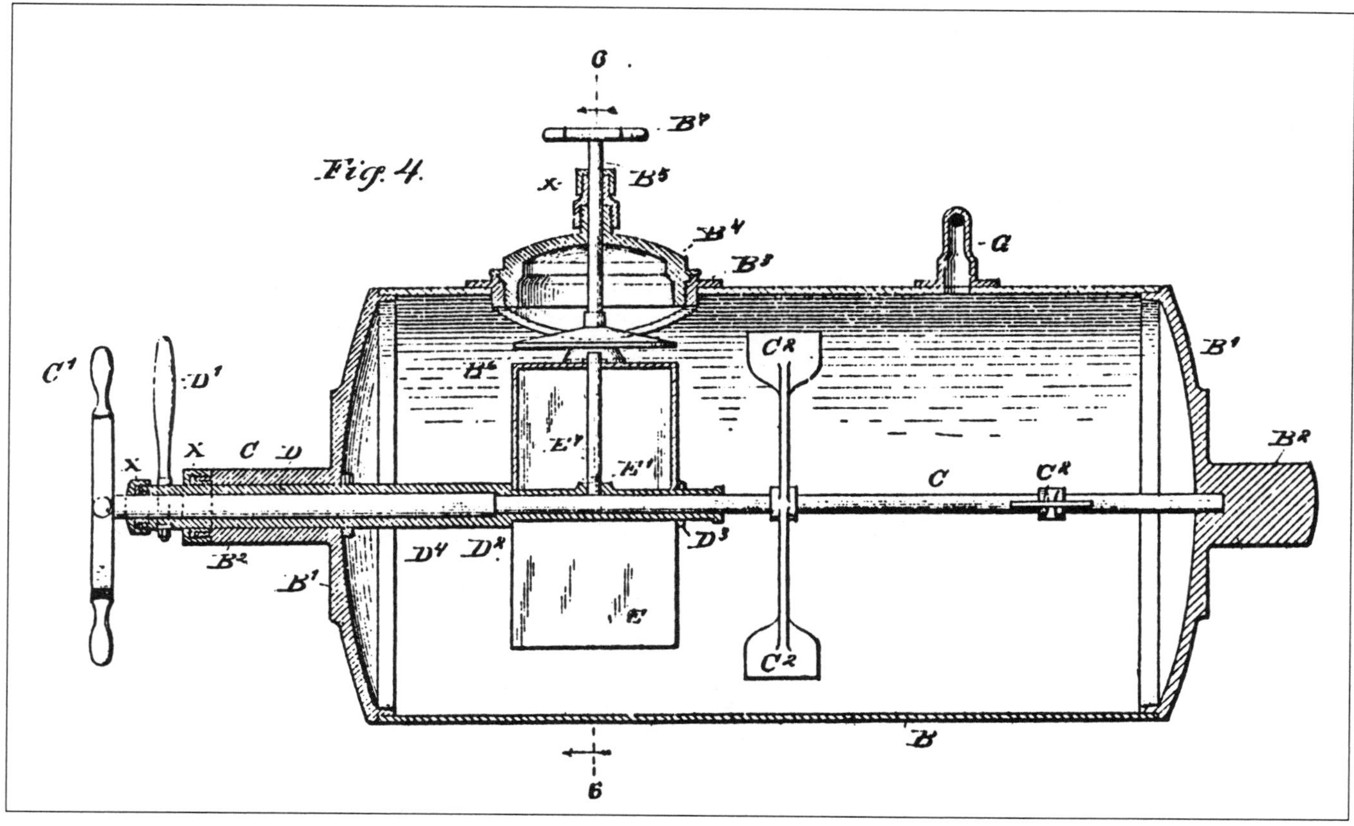

FIGURE 198

J.R. Hopkins

Combining the equipment and fittings of a ladder truck and a hose wagon with a chemical fire engine won patent number 790,839 for James R. Hopkins of Somerville, Massachusetts. Since Somerville is a part of the metropolitan Boston area one can wonder if it was more than coincidental that the Boston Fire Department was running "combination ladder" companies around that time. FIGURES 199 and 200 were the drawings for this 1905 patent. Observe the following components of this invention: the wheels were designated by the number 11, the driver's seat was 12, and the running boards were 13; the chemical tanks were 14; wire baskets for carrying tools, somewhat like the "Baltimore baskets" of a slightly later time, were 15. The ladders were designated with the number 16, and they were carried in racks which had center supports, 17, and horizontal supports, 18 and 19. Drop bars, 20, hinged at points, 21, held the ladders in place. Thus it was possible to remove the ladders equally easily from the sides as well as from the rear of the vehicle. The portable hose compartments, 25, were mounted on rollers, 26. These could be withdrawn from the back of the apparatus.

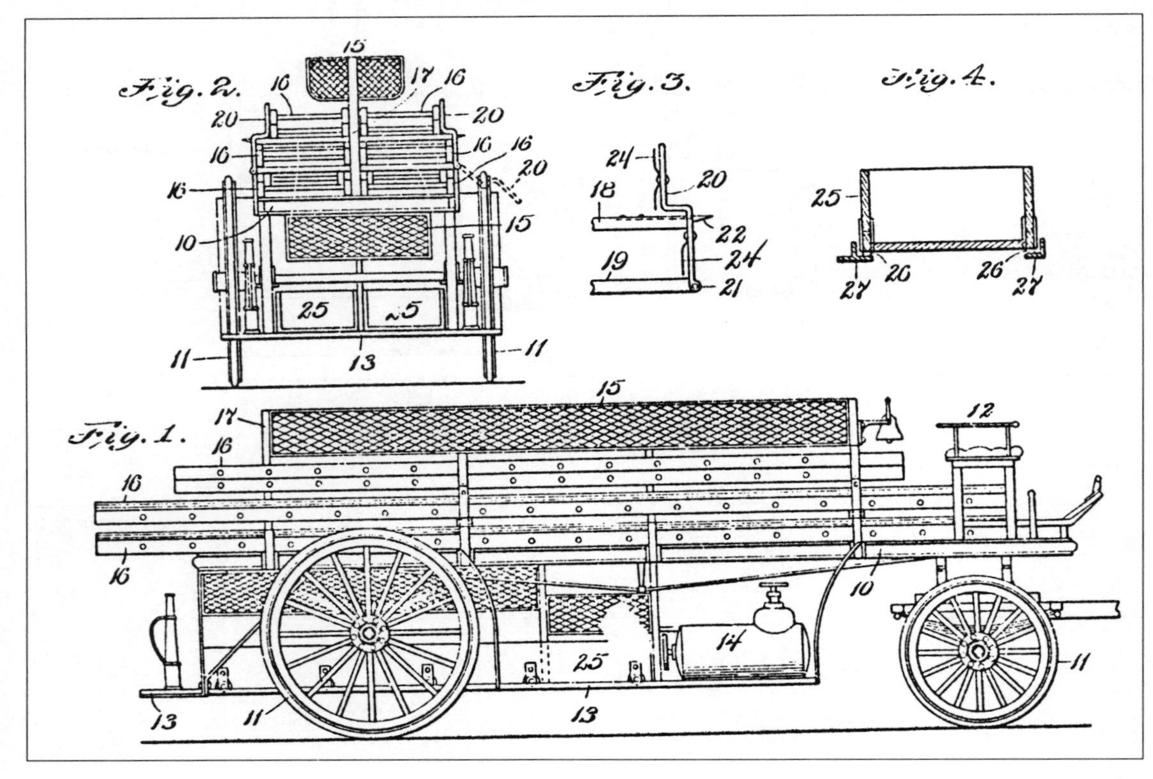

FIGURE 199

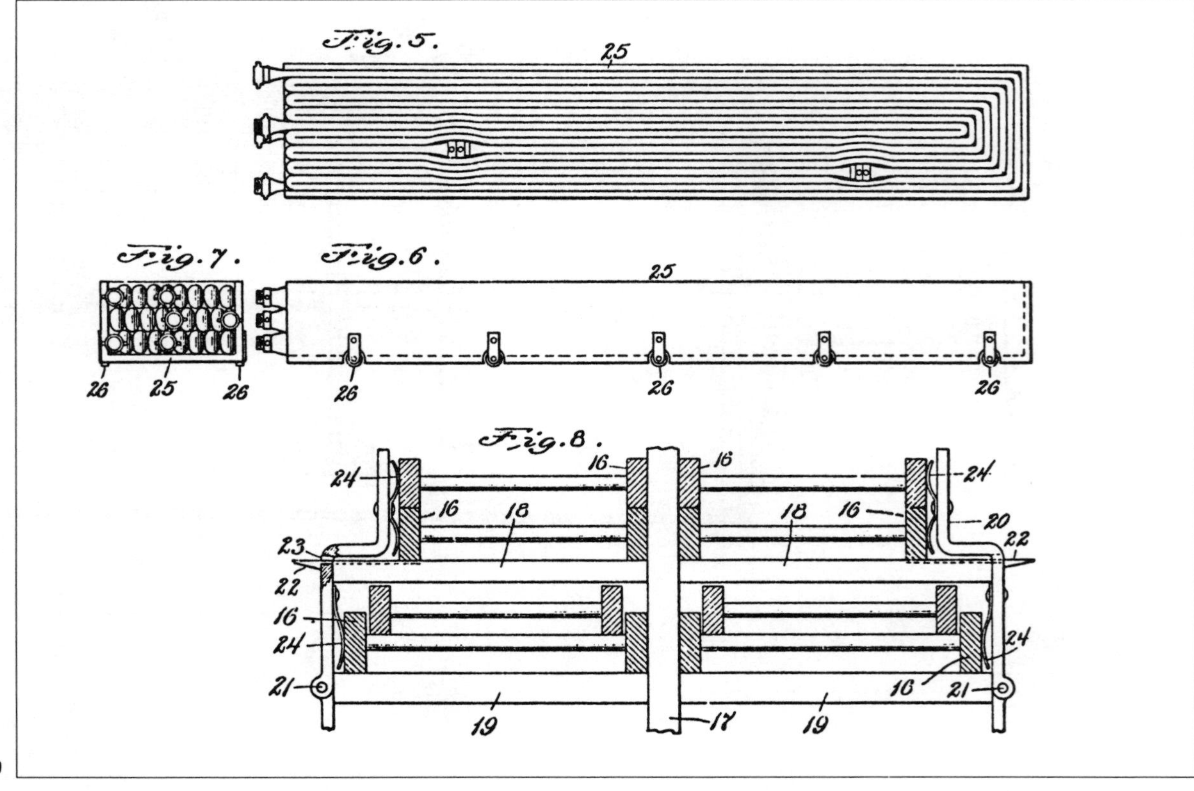

FIGURE 200

A 1905 Motor Vehicle

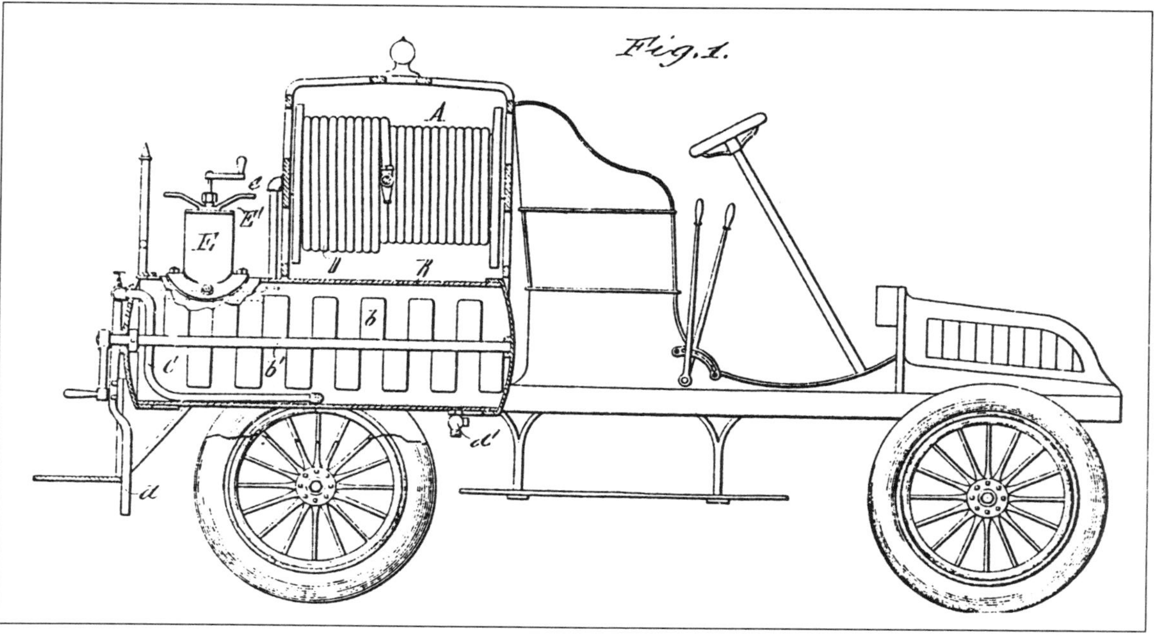

FIGURE 201

Far more than just another variation upon the basic chemical fire engine theme was Buffalo, New York resident Thomas Cochrane's invention of August 28, 1906, the application for which he had filed on November 6, 1905. This was patent number 829,629, and it is illustrated in FIGURE 201. Cochrane was quite specific to the effect that his chemical apparatus was to have been mounted on a gasoline-powered motorized vehicle. And in 1905 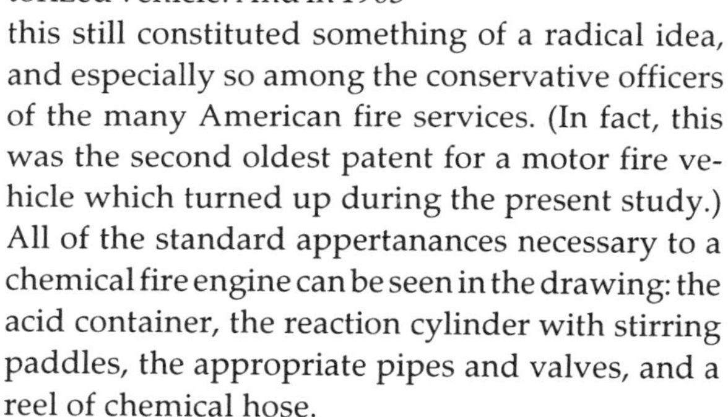this still constituted something of a radical idea, and especially so among the conservative officers of the many American fire services. (In fact, this was the second oldest patent for a motor fire vehicle which turned up during the present study.) All of the standard appertanances necessary to a chemical fire engine can be seen in the drawing: the acid container, the reaction cylinder with stirring paddles, the appropriate pipes and valves, and a reel of chemical hose.

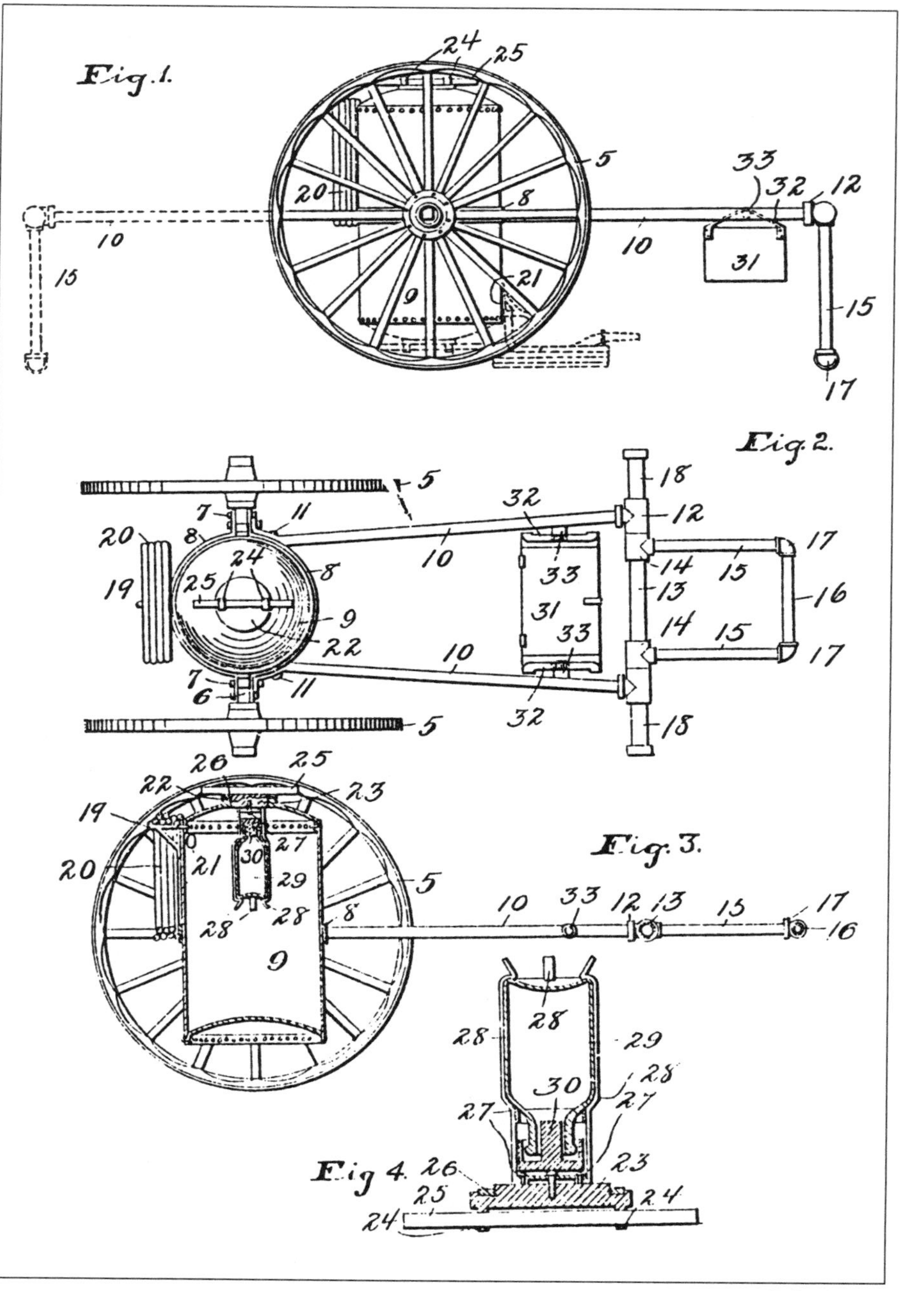

FIGURE 202

C.H. Sutphen

What surely must have been among the last ever patents issued for the invention of a purely man-hauled fire vehicle was number 852,815, with May 7, 1907 as the effective date. By 1907 even the horses, not to mention the oxen, could read the writing on the barnyard fence! The inventors of this device were Ohioans Sylvanus L. Wottring, of Prospect, and Clarence H. Sutphen, the founding sachem of the company which still bears his name, of Columbus. Actually this invention was kind of a neat design (FIGURE 202), so arranged that after pulling it to a fire by its handles the now-exhausted firemen had but to flip the handle and arm mechanism over, through 180 degrees. This maneuver inverted the acid bottle, emptying it into the large cylinder, thereby starting the reaction which generated carbon dioxide.

1908-1918 Inventions

The next ten-year period, up to about 1918, was a time of ferment and recognition that change was inevitable in the country generally, and in the fire services especially. Most of the patents issued for chemical fire engines during these years were for intriguing devices which were peculiar, hybrid machines. It remains doubtful that any of these were ever built and used. And it does not seem unfair to classify them as varying from completely useless to obviously unworkable. Just a couple of these inventions will be reported, and with brevity.

Edgar F. Sanford of Merced, California, a gentleman who was not connected with the Sanford Fire Apparatus Company, received U.S. Patents 881,872 and 881,873 on March 10, 1908. Drawings for the first only of these patents are reproduced as FIGURES 203, 204, 205, and 206. What you see is a motor vehicle with a right-sided position for the driver, and which carried ladders, which carried trays of fire hose, and which carried a drum of pre-mixed chemicals and water. The ladder bed is seen to have been located between the seats for the driver and the officer. The fire hose was carried in large trays which were caster-mounted and could thus be quickly pulled off from the back end of the truck. The chemicals, not identified but mixed with water one must assume, were within the tank, which was shown, but were without any visible means for being pumped or otherwise propelled.

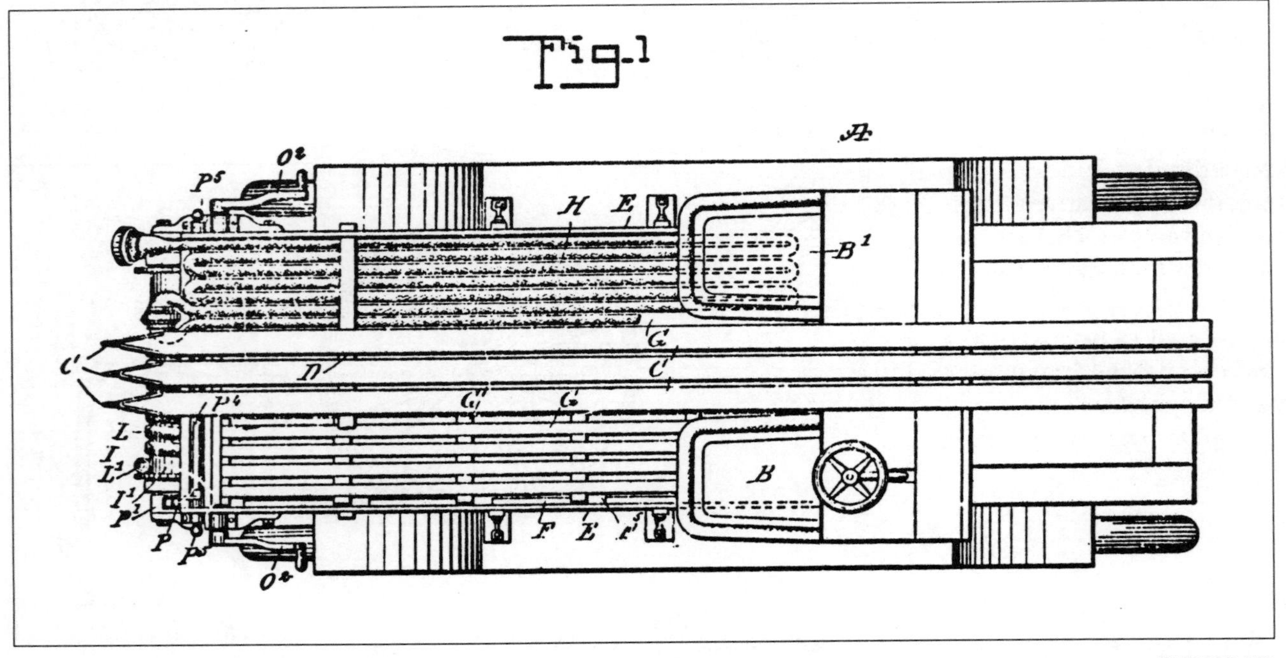

FIGURE 203

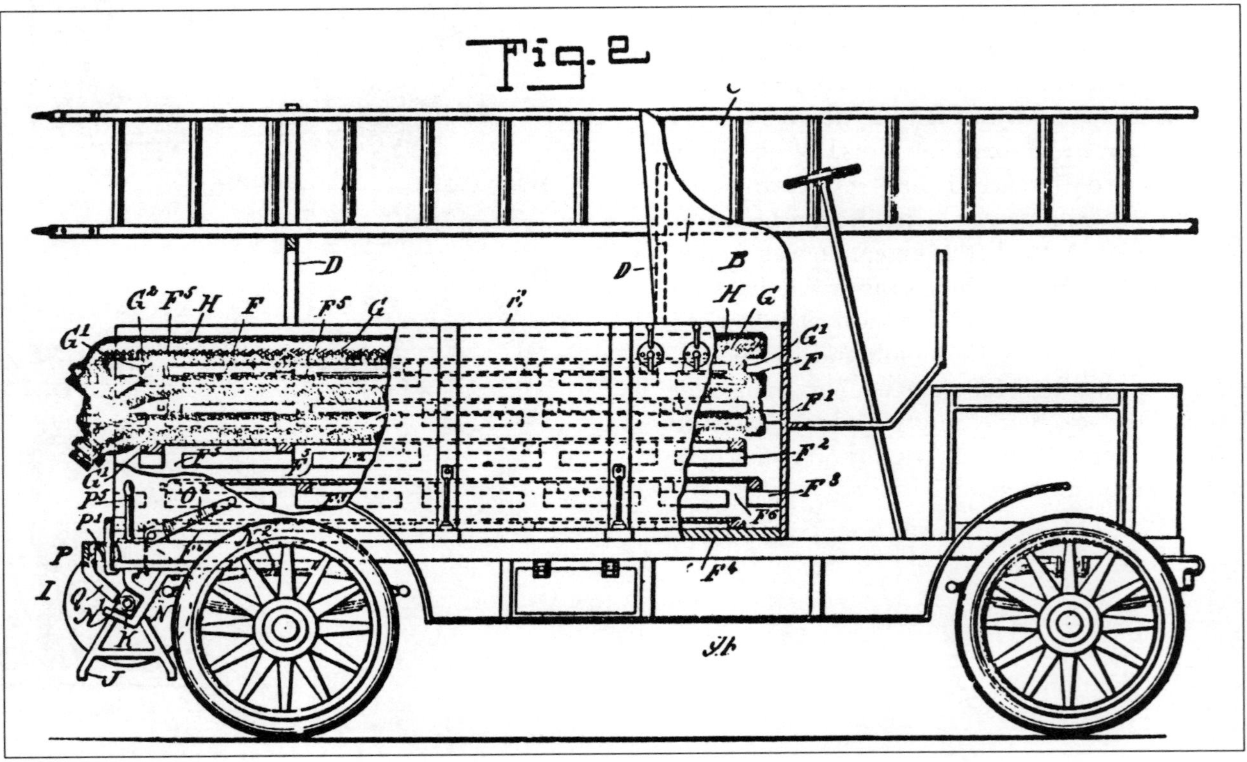

FIGURE 204

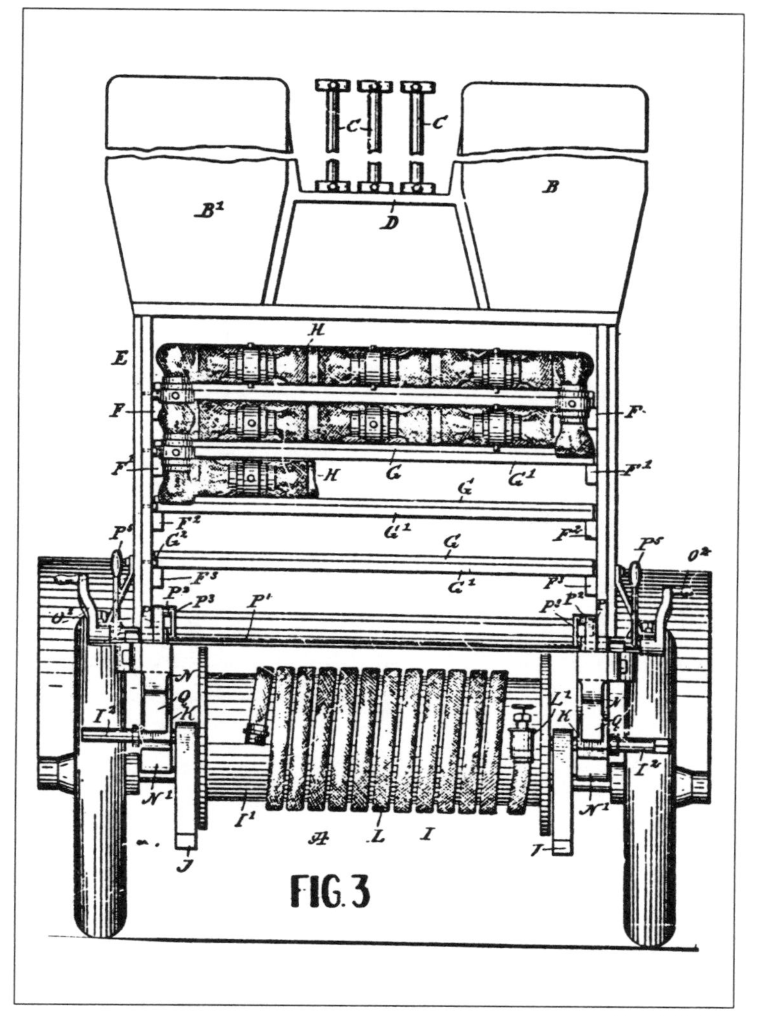

FIGURE 205

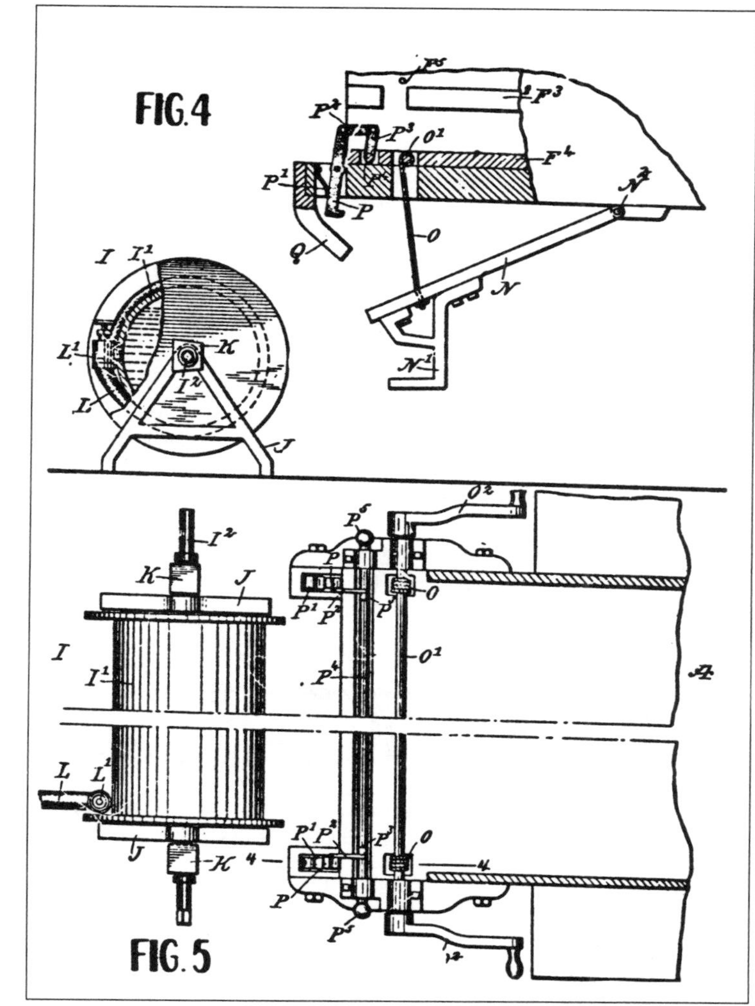

FIGURE 206

Then from the small Texas town of Wylie, near Dallas, there was Henry M. Minnis. He received patent number 976,133 in November, 1910 for his invention of a fire apparatus which carried a water tank and hand-operated, double-acting pumps geared to be workable with minimal power input. Minnis made provision for a fly wheel in case a motor was added to operate the pump, and there was a method for introducing chemicals into the water stream. This invention is illustrated in FIGURES 207 and 208. It was the inventor's intention that some "suitable chemicals," whatever that might mean, were to have been injected into the water stream just before it was discharged from the machine. The chemical was to have been stored in a cylindrical hopper, number 45 in the drawings, and a metal plate, number 46, was to have been used to push it into the water. The plate was pressed downward by the action of two rods, number 52 and 53, which were in turn operated by a cam on the shaft of the pump, number 25.

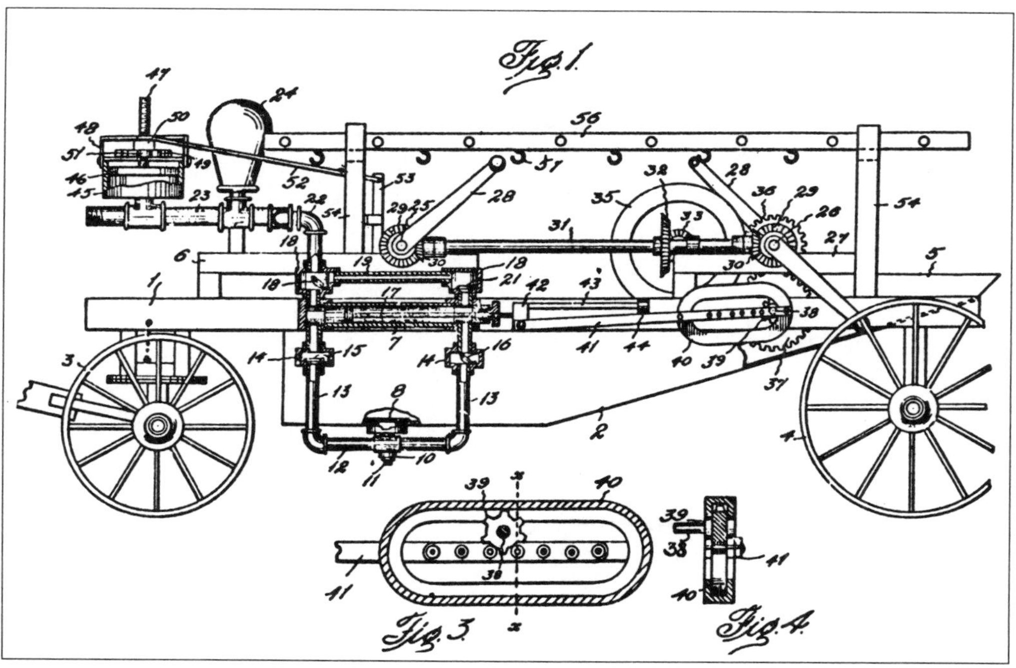

FIGURE 207

There was a second Minnis patent, number 984,231, FIGURE 209, which he received about four months later. This invention was a real chemical fire engine. It was supposed to have worked in this manner: water and bicarbonate were fed into a hopper, number 9; they were mixed and pumped into a tank, 2; acid was added from a container, 15. The evolved carbon dioxide then expelled the fluid. Inventor Minnis, strangely, ignored some very important details, such as the corrosive effects of concentrated sulfuric acid upon check valves!

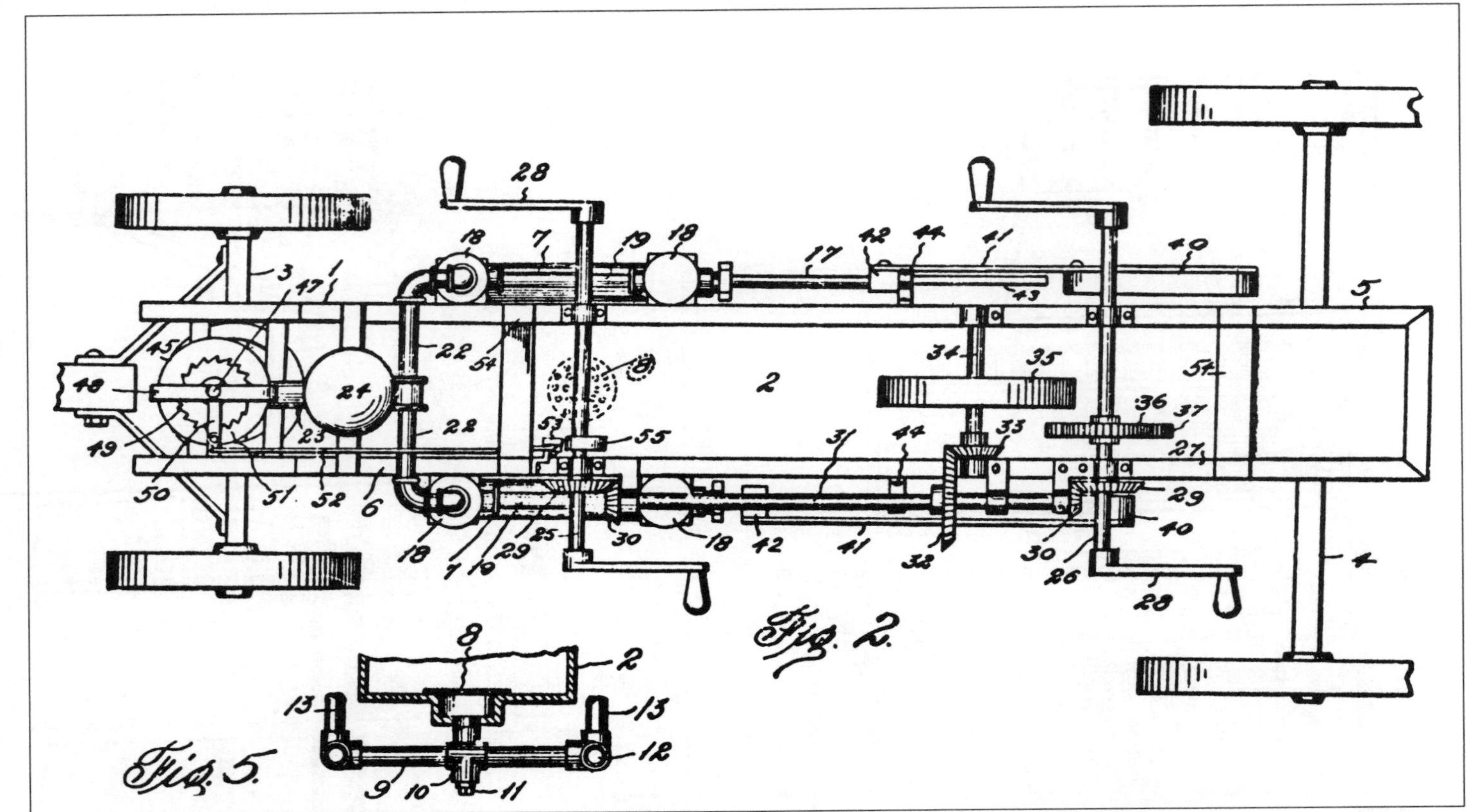

FIGURE 208

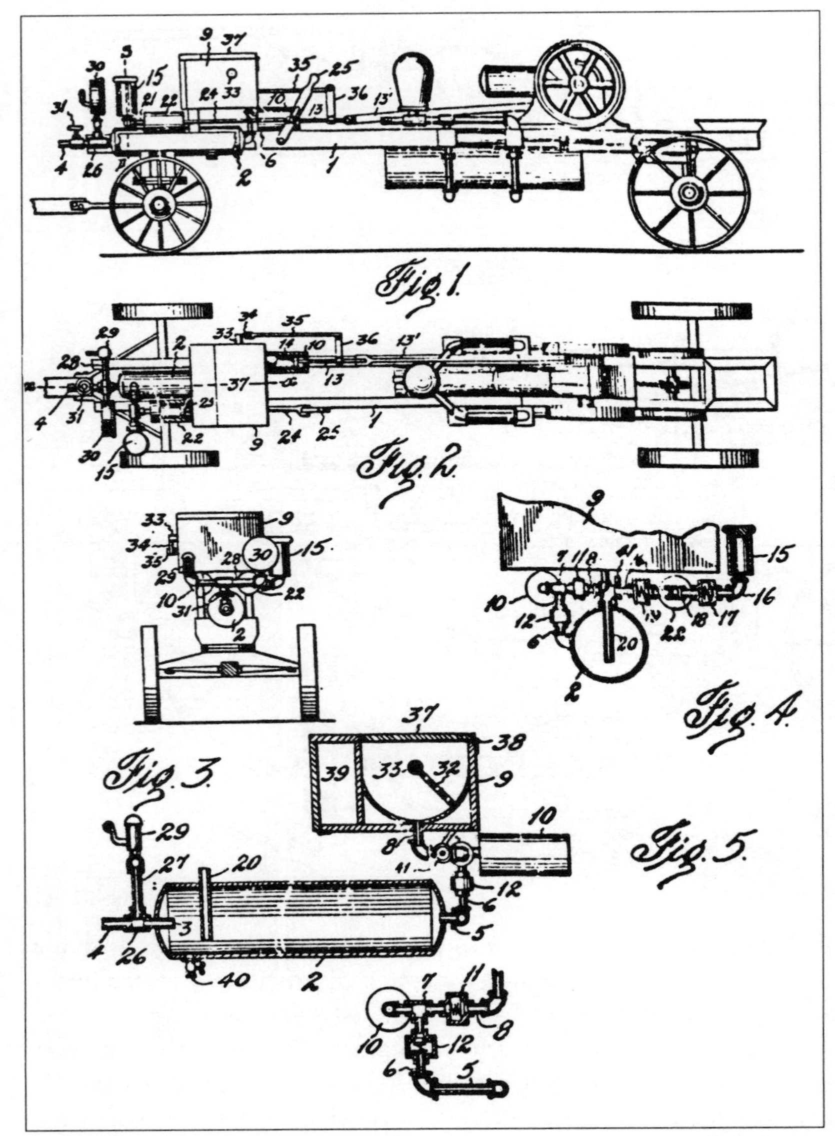

FIGURE 209

Howe and Lambert

In 1917 the Howe Fire Apparatus Company moved from Indianapolis to Anderson, Indiana, presumably to be closer to the Buckeye Manufacturing Company, builders of the Lambert automobile, the chassis of which was to be used for Howe fire vehicles. Perley G. Howe, who was one of B.J.C. Howe's sons, and Alvan Ray Lambert received patent number 1,272,956 in July, 1918, for their method of adapting a piston pump to a Ford automobile. A chemical tank was one of the components which they described. The patent drawings are shown in FIGURES 210, 211, and 212. Notice, first, these particular parts of the machine: the transmission, a; the drive shaft, b; the pump-powering shaft, g; a chain for turning the pump crank shaft, i; the pump crank shaft, j; the pump pistons, k; the pump cylinders, l; the inlet manifold, m; the outlet manifold, n; the air chamber, t; and the chemical tank, v. According to the inventors the chemical tank together with its associated lines and valves was furnished for use in case water was not immediately available at the fireground. The capacity of the small-diameter chemical hose was significantly less than that of the pump, a situation from which an advantage was derived, the surplus water being passed back into the chemical tank to agitate the mixture, keeping it well stirred.

After building this fire engine Howe and Lambert must have noticed that with the loaded water tank on its left side there was an inherent lack of balance. So, for counter-balancing the inventors added auxiliary supporting springs and these maintained an overall level posture for their car.

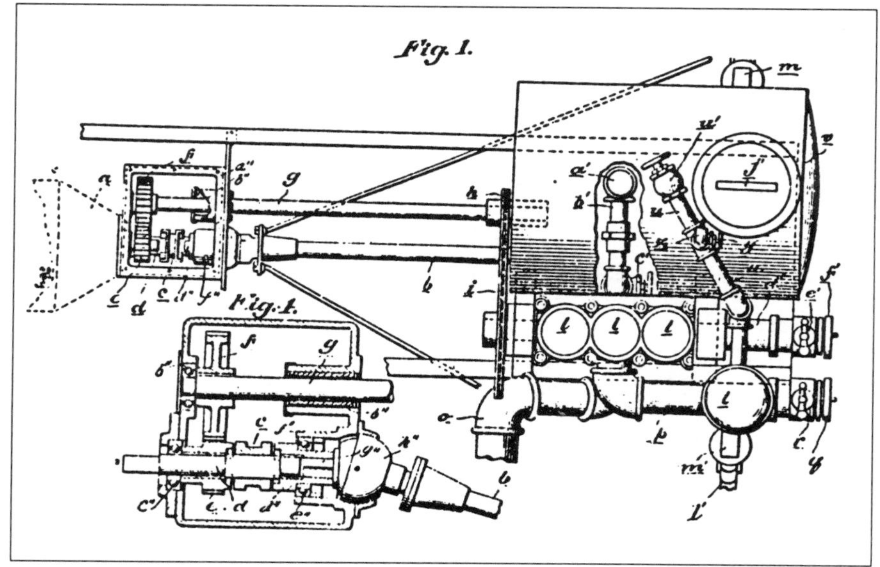

FIGURE 210

FIGURE 211

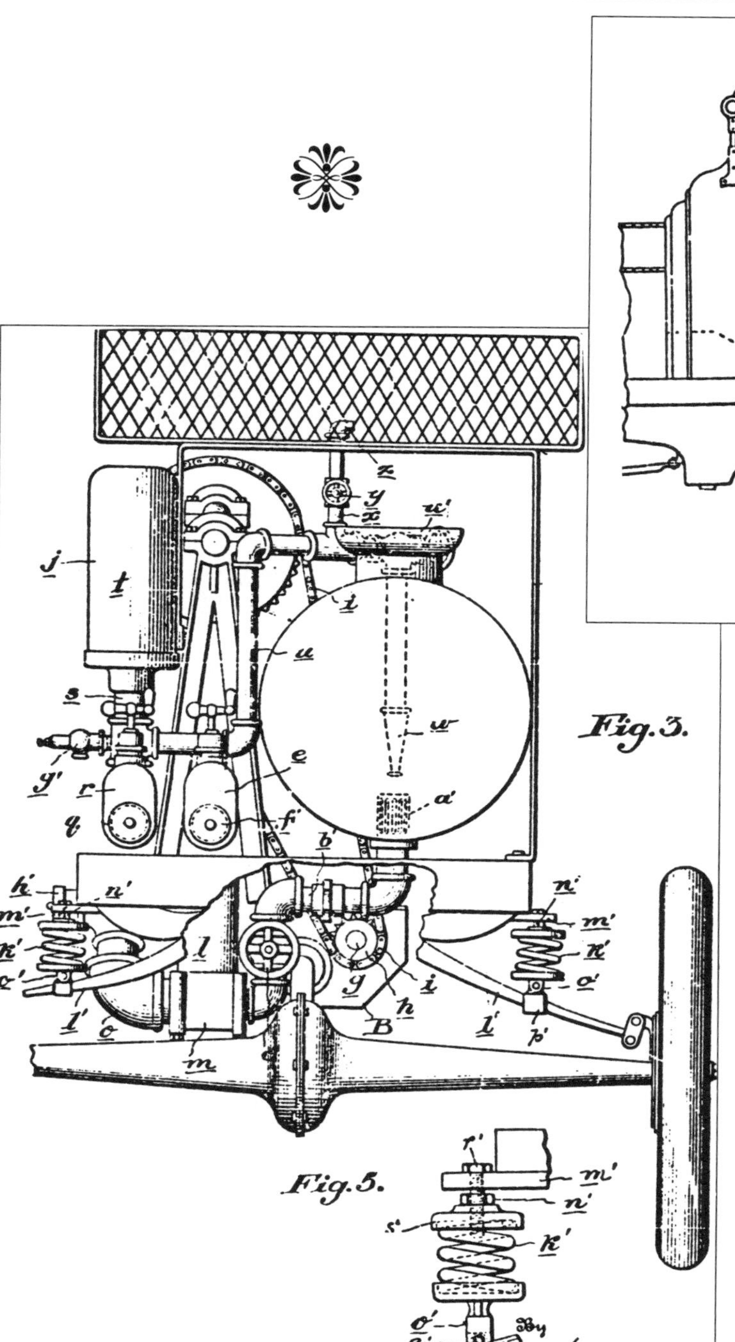

FIGURE 212

M.M. Connor

Another of the inventors who subscribed to the chemicals-are-better-than-water school of thought was Maurice M. Connor of Buffalo, New York. His 1918 patent, number 1,287,238, was actually a large, fancy, truck-mounted seltzer machine (FIGURES 213, 214 ,and 215). In these illustrations one sees a right-sided chauffeur's position, a power take-off (PTO), a pump, a water tank with mixing paddles, and flasks of carbon dioxide. Connor's procedure called for bleeding off the carbon dioxide into the water, mixing well, and then pumping this carbonated water onto the fire. Imagine the consequences if there had been a fire at an ice cream factory: production of the world's largest ice cream soda! At least this method would have been clean and neat in comparison to the operation of a real chemical fire engine spewing out gobs of unreacted sodium bicarbonate along with the water, staining and ruining whatever it failed to drown.

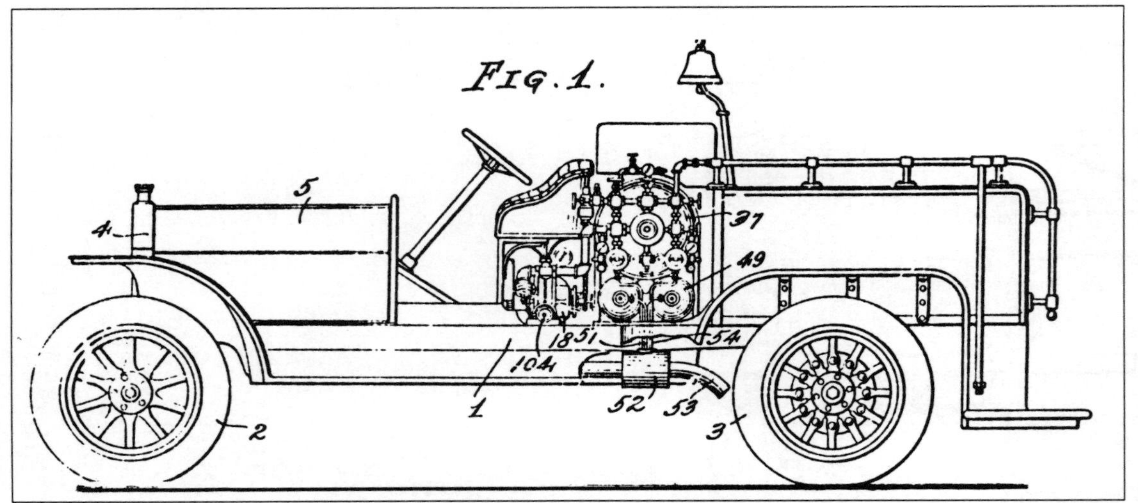

FIGURE 213

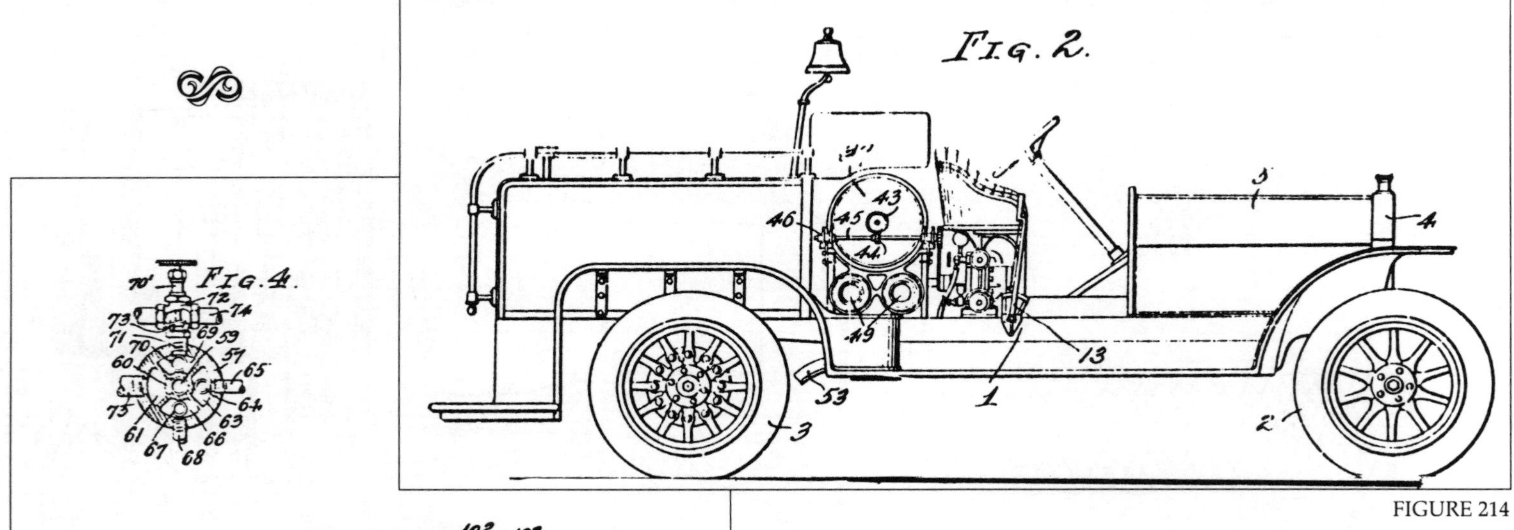

FIGURE 214

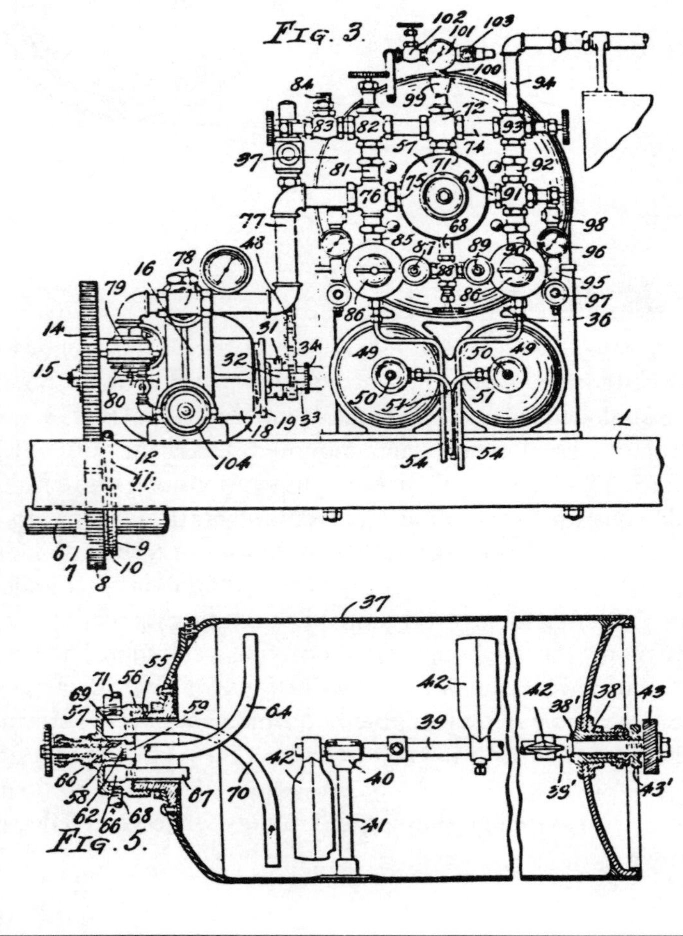

FIGURE 215

D.W. Adams

And finally, just to show that using chemicals to fight wildfires is not at all a recent idea, consider Daniel W. Adams invention, patent number 1,003,853 dated September 19, 1911. Adams, of Glendale Springs, North Carolina actually built and used this apparatus as shown (FIGURES 216, 217, 218, and 219) carrying it upon horses into the woods. The second drawing illustrates the construction details of the mixing tank, the acid supply tank, the water tanks and the air cylinder. Adams described his step-by-step method of using compressed air to move water and dissolved bicarbonate into the mixing tank, and of next adding concentrated acid. He routinely produced pressures up to 300 pounds per square inch as carbon dioxide evolved, and such a pressurized stream could throw water to a height of fifty feet. No doubt basking in that warm glow which one feels when something works well, inventor Adams proceeded to redesign the system by introducing servomechanical valves, as shown in the last two drawings. This configuration was by far the more elegant of the two types, but it was easily the most failure prone as well.

These days chemical fire engines are to be found at firemen's musters, and in the line of march at parades, and not working at fire scenes. This is good, since they look terrific with their tanks, valves and pipes all chromed and shined up, glistening in bright sunlight. It would seem, somehow, to be blasphemous, and a desecration, if dirt, cinders, or ashes were ever again to fall upon these lovely antiques.

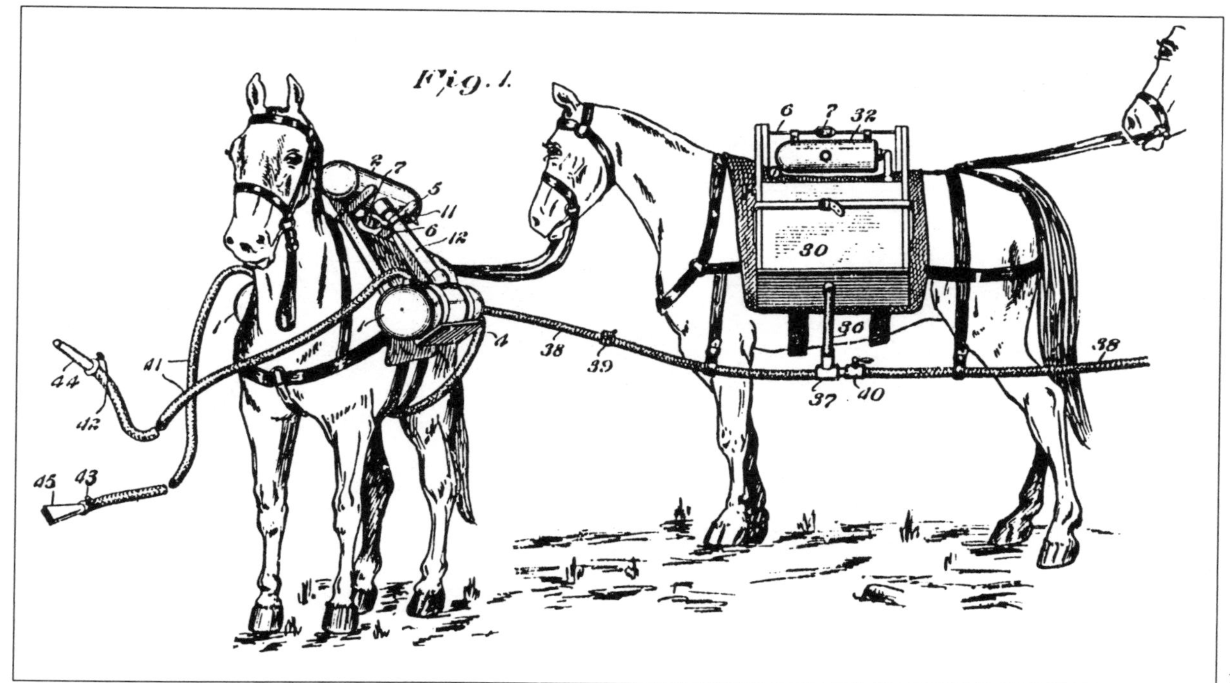

FIGURE 216

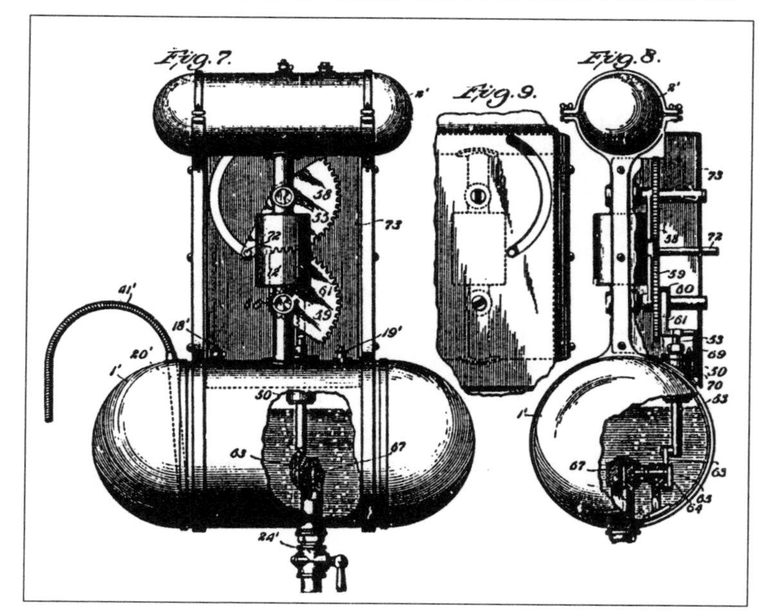

FIGURE 218

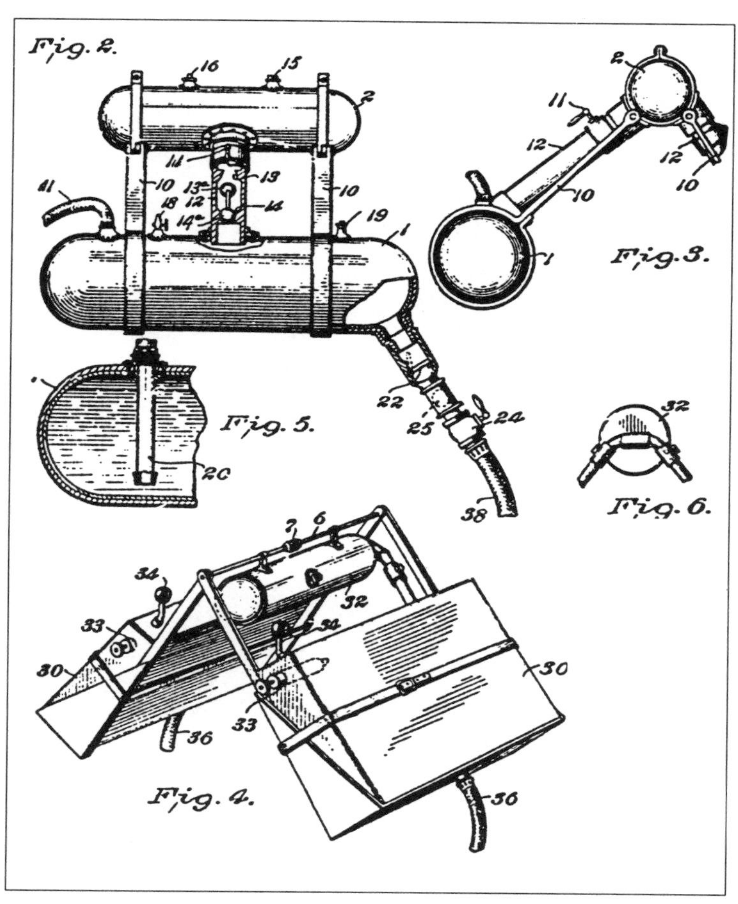

FIGURE 217

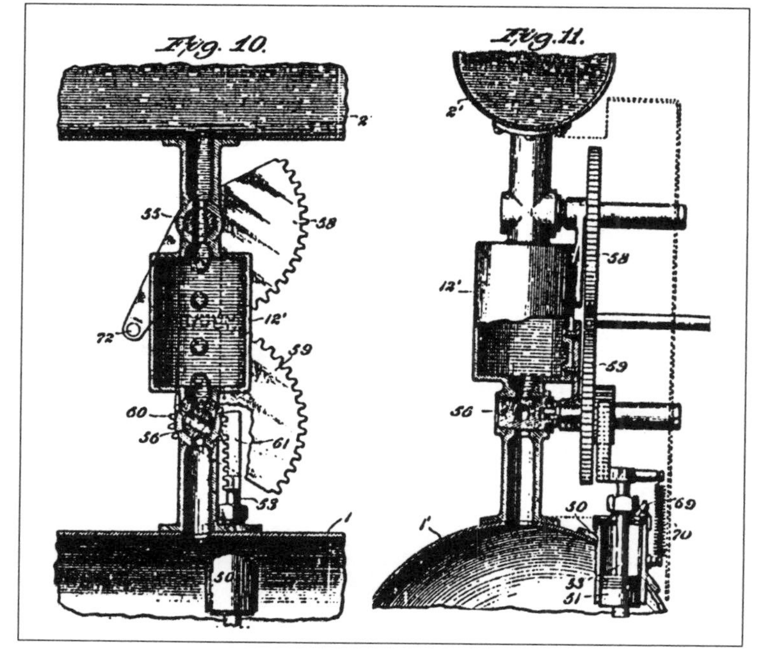

FIGURE 219

Chemical Fire Engine Photo Album

As before we close this chapter with a photograph album, this one being quite brief. The Pioneer Fire Company of Ephrata, Pennsylvania, regularly hauls its wheeled reminder that suckers are born every minute to parades, showing off their Gibbs and Gordon chemical fire engine (FIGURE 220). The year, 1867, inscribed on each side of the apparatus was the year in which the Ettla Fire Company was founded. This was a puny rig, as was soon realized by firemen following their 1874 purchase of it.

FIGURE 220

FIGURE 221 shows an 1875 model chemical cart owned by the Wilmington Manor Volunteer Fire Company of Delaware. Notice the plaque which identifies the rig as made by American LaFrance, an obvious impossibility, since 1875 was the founding year of the LaFrance Fire Engine Company. "American LaFrance" came into existence as such following the collapse of the International Fire Engine Company cartel, just after the turn of the century.

FIGURE 221

The Oceanic Fire Company, Long Branch, New Jersey, is shown, FIGURE 222, in the line of march during a parade, proudly showing off their 1876 dual-tank Babcock chemical fire engine. Inventors Stillson and Kley, patent number 131,414, designed this rig in 1872. (Their patent drawing is FIGURE 157.)

Another single-tank chemical fire apparatus is shown in FIGURE 223. This machine was built by Boyer Fire Apparatus Company of Logansport, Indiana, successor to the Obenchain-Boyer Company about 1929.

Photographed in Bethany Beach during a Delaware Volunteer Firemen's parade was the Ford vehicle shown in FIGURE 224. From Lynbrook, New York, this antique belongs to the Tally Ho Chemical Engine Company. Notice the container for an acid vessel on the rear running board, just to the front of the fire extinguisher.

Another parade piece is the three-tank chemical fire engine of FIGURE 225. The photograph of this old rig from the York New Salem Community Fire Company was taken during the York, Pennsylvania 250th anniversary parade. And the last of this group is a 1927 Foamite-Childs apparatus, FIGURE 226, equipped with two chemical tanks, a hose bed, and a rotary pump.

FIGURE 222

FIGURE 223

FIGURE 224

FIGURE 225

FIGURE 226

THE INTERNAL COMBUSTION ENGINE AND THE DEVELOPMENT OF MODERN FIRE PUMPERS

INTRODUCTION

A vertiable army of inventors, designers, and manufacturers of gasoline and diesel-powered fire engines have come and have gone during this century. Fire vehicles bearing the nameplates of Ahrens-Fox, Crown, Hahn, Howe, Knox, Maxim, Pirsch, Robinson, Waterous, and Webb are examples of all of those which have responded to their last calls. The roster of actively practicing manufacturers of fire engines seems to change from time to time, no doubt influenced more by decisions made in the financial boardrooms of the country than by the needs of the active firefighting forces. Right now, in the years 1993-1994, fire pumpers are being built by American Eagle, American Fire Apparatus and American LaFrance, by Becker and by Boardman, by Emergency One and by Emergency Vehicles, by Ferrara, Firemax, Firewolf, and Four Guys, by Mack, Marion, Mason Dixon and Murray, by Panama and by Pierce, by Ranger Fire Apparatus and by Rescue Technologies, by Saulsbury, Seagrave, Slagle, Smeal, Spencer, Summit and Sutphen, and by 3-D, HMC, and KMC. And there are other manufacturers as well.

And so the discussion within this Chapter will recognize inventions which have been clearly identified with specific builders, of course, but otherwise the presentation is intentionally general. Fire engine patents, inventions of apparatus parts and accessories, and inventions of fire pumps and their accessories will each, in turn, be addressed. But since it was the gasoline-fueled internal combustion engine and the automobile and truck industries which made possible the fire vehicles of today, it is with these–fuel, engine, and industry–that one begins.

Gasoline, Petroleum, and the First Dreamers and Builders

Gasoline comes from petroleum, and interest in petroleum was derived from the need for fuel for nocturnal illumination lamps. By the middle of the 1800s it was evident that the price of whale oil, then used in lamps at night, was escalating toward unacceptably high levels. In response to increasing demand the whaling industry of New England had become efficient, so efficient in fact that these great marine mammals were scarce, having been hunted nearly to extinction. Clearly, a substitute for whale oil was needed. There had been some efforts to derive oils from coal, and by 1860 a number of refineries dedicated to this purpose had been established. Lamp oil was also being prepared from crude petroleum which was harvested in areas of surface seepage. One such operation was located in northwestern Pennsylvania. As is well known even to schoolchildren a gentleman named Edwin L. Drake was hired to drill there for subsurface oil, and on August 27, 1859 oil began flowing from a depth of just under 70 feet. From this key event at Titusville, Pennsylvania there has arisen the entire history of petroleum-based fuels, the internal combustion engine, and all of highway transportation and American civilization as we experience it to this day.

During its years of infancy petroleum producers concentrated on lamp fuels, and they did not consider gasoline to be a desirable product. In fact gasoline was seen as a dangerous and highly flammable byproduct, to be dumped as soon as possible. Then, as now, disposal of unwanted liquids often involved flowing them into any convenient creek, a practice which sometimes provoked dramatic, fiery consequences downstream. But because gasoline was so cheap and so available it was seen as an ideal motor fuel by such visionaries as George Seldon, the Duryea brothers Charles and J. Frank, John William Lambert, Elwood Haynes, Ransom Olds, Alexander Winton and James Packard, and others in this country, and by Gottlieb Daimler, Wilhelm Maybach, and Karl Benz in Europe. All of these gentlemen were infatuated by the idea of the automobile. The Duryeas sold the first motor car in America, in 1896. But prior to Henry Ford, of whom more later, all of the builders of automobiles were, really, assemblers. They procured engines, chassis, and bodies from a variety of sources and from these they fabricated their vehicles. Ford introduced those concepts which we have come to know as vertical integration and the assembly line, and he made it possible for people of modest means, such as his own employees, to purchase his vehicles.

In addition to an engine, a chassis, and a body a successful and useful motor vehicle requires only a few additional items of equipment. Probably the most important one is the clutch mechanism, as this allows the engine to run while the vehicle is not moving. Another necessity is a gearing device to compensate for heavy loads and for hills and to make it possible to drive in reverse. A gear-box to wheels linkage is essential, and pneumatic tires and a starter device are also highly desirable. Curiously, despite the introduction and the refinements of all of these useful devices, and despite the ready availability of cheap gasoline, trucks powered by steam engines were still being built as recently as 1935. But for this entire century, even to this day, it has been the four-stroke internal combustion engine which has most profoundly affected our lives and has sculpted, forged, and fashioned our environment. For fire vehicles the compression-ignition engine originally devised and patented by Rudolf Diesel has become the prime mover. And at this time in our history there do not seem to be any alternatives likely to supplant it. The Stirling hot air engine, the Wankel trochoidal engine, the fuel-injected stratified-charge engine, and the divided-chamber engine, by way of examples, have not demonstrated any applicability as fire engine power plants. The era of electric and gasoline-electric powered drive trains for heavy vehicles has come and has gone, although this file ought not yet to be stamped as forever closed and forgotten. Practical vehicular power derived from photoelectric reaction cells, or from fuel cells, or even from batteries or by burning hydrogen or propane or alcohol still requires years of testing and development.

The birth year, it seems, for the American motor car was 1894, and the Duryea brothers were the obstetricians. In 1899 Ransom Olds sold his company to S.L. Smith, while two years later he introduced his single-cylinder, tiller-steered, "Curved Dash" Oldsmobile, destined to be the first American car built in large numbers. Thirty-five of them were produced each working day by 1905, the year which saw Olds retiring from one company and going on to found another, Reo. But these were only the years of the infancy and the early childhood of this industry. It was Henry Ford who presided during the troublesome teens and early maturation years of large-scale automobile building.

The Ford Model T was designed by Ford, together with J. Galamb and G.H. Wills. This car was equipped with a four-cylinder 2,892 cc side-valve engine which could develop 20 horsepower at 1,500 revolutions a minute. It had a detachable head, a magneto and fly wheel low tension ignition, and the epicyclic gears had pedal control. Ford produced 12,292 of these cars in 1909, 40,402 in 1911, and 182,809 in 1913. The price of a Model T fell, consequently, by reason of scale economies, from $950 to $690 to $550 during those years. Beginning in 1914 the time needed to build a Model T was reduced to just 90 minutes. And by 1915 at least 355,000 cars were produced, with the cost to buyers now $440. In 1923 when annual output was approaching 2,000,000 vehicles the price was down to $290 each.

The inventors who worked to develop fire pumpers, and whose creations constitute the substance of the following pages, fall more-or-less into three groups. Some were obviously preoccupied with motors and with functioning vehicles. These inventors were, generally, the earlier ones. Later on when power plants and vehicular chassis were off-the-self items, the patents were more directly related to innovations of structural details. Finally, there were the inventors who were the specialists in pumps and their fittings.

There are of course many components, many parts, and many sub-assemblies which are essential in a modern fire pumper, but the inventors of most of these items did not create them with only this purpose in mind. While all are obviously important none lie clearly within the current purvue: Allison transmissions, Archibald wheels, Bosch ignitions, Cummins engines, Delco batteries, Detroit diesels, Fuller transmissions,

Lance Neville alternators, Michelin tires, Rockwell axles, Ross power steering systems, Taylor-Wharton air cylinders, and Timkin roller bearings, to cite but a very few, and not to mention such contemporary firematic components as the Federal Firehawk, the Hannay reel, the Onan generator, the Roll-O-Matic door, the United Polytank II, and the Whelen strobe lamp, and more.

The Porteu Fire Engine

Now the very first ever "automobile" fire engine was one designed by Porteu, and built by the Lille, France manufacturer Cambier. This apparatus had its initial demonstration in public during the French Heavy Autocar Trials in Versailles in 1898. FIGURE 227 shows a sketch of the Porteu fire engine, taken from Scientific American Supplement Number 1231 of August 5, 1899. It had a gasoline-fueled four-cylinder engine rated at 22 horsepower, and it could propel the vehicle at a maximum speed of about nine miles an hour. This same motor also operated the pump. (In that year, 1899, there were at most fifty gasoline–powered vehicles–the term "automobile" was yet to be coined–in the entire United States. By 1902 this number had increased to about 12,000, and France was soon to lose its place at the head of the automobile line. American pre-eminence in the engineering and production of motor vehicles was not to fade away until our own time.)

The Eisenbise Fire Engine

Harvey W. Eisenbise was the first American inventor of a motor fire pumper. He was a veteran fireman and fire company officer in Reading, Pennsylvania, where he served the Friendship Fire Engine Company Number 4 as President in 1897, in 1903, and later during 1917 and 1918. At one or another time he also filled the positions of vice president, assistant secretary, and trustee of his volunteer fire company. ("Eisenbise" must have been, and perhaps still is, a common family name in Reading and Berks County. Numerous Eisenbises have been listed on the membership rosters of Reading fire companies.) FIGURES 228 and 229 are from this first Eisenbise patent, number 846,835, dated March 12, 1907, but filed in June 1904, when there were no more than 700 miles of paved roads in the entire United States. (The historian who hands out the awards and the salutes for these "first" motor fire pumpers must be careful lest he stick his thumb into his eye. By the year 1906 the Waterous Company had supplied a pumper to Wayne, Pennsylvania–more on this later–and was poised to sell to Alameda, California, a unit which used the same engine for both pumping and propulsive power.)

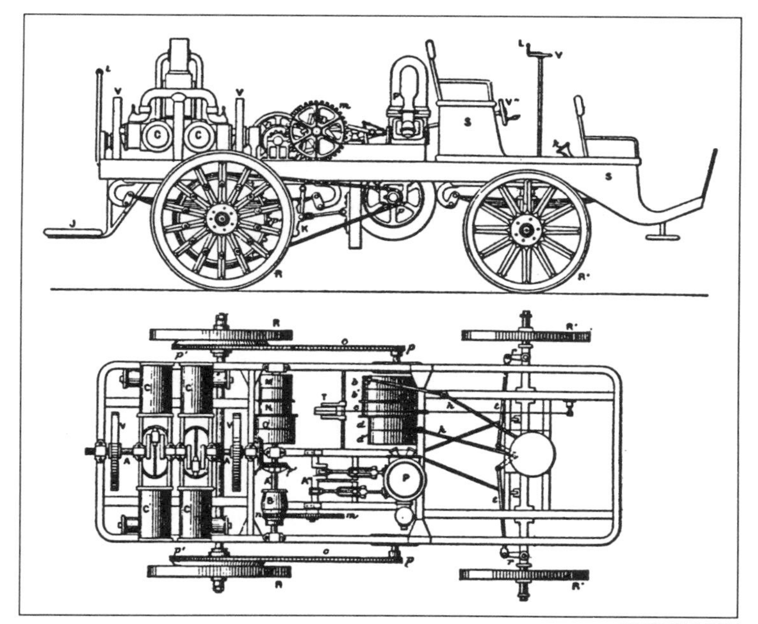

FIGURE 227

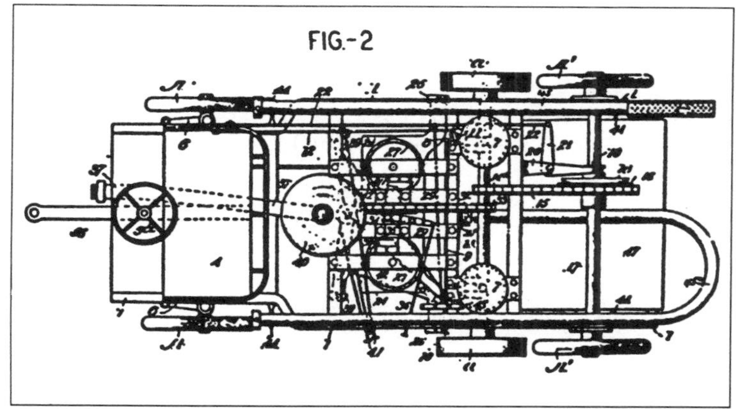

FIGURE 229

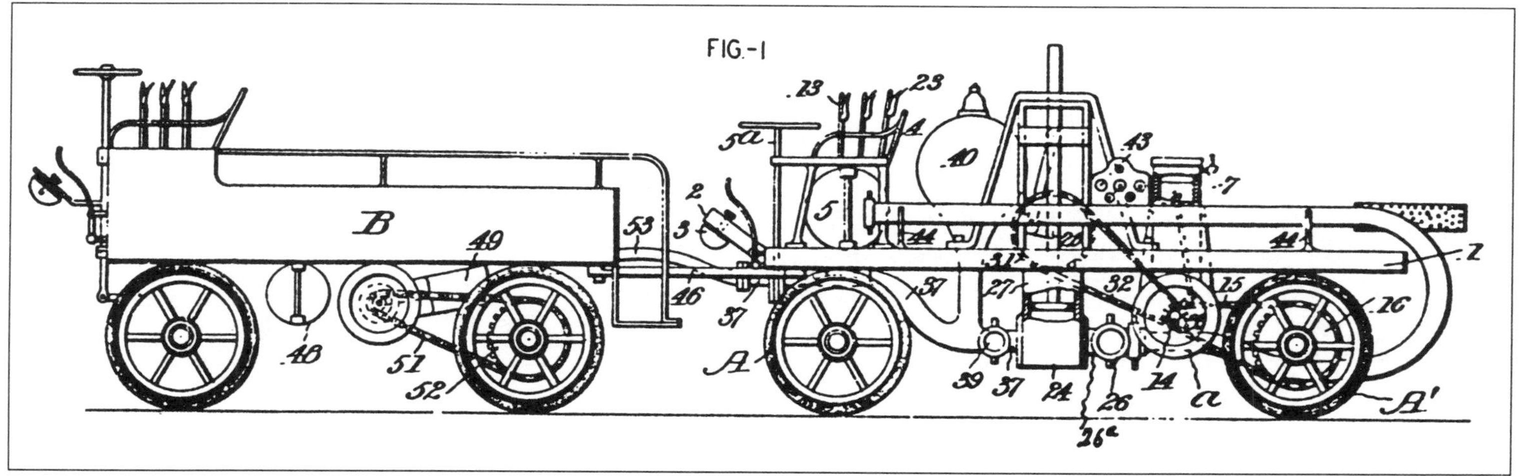

FIGURE 228

Perhaps it was his experiences with steam pumpers and hose carts that predisposed the inventor to continue to look upon the hose carrier as a distinct vehicle, but the drawings do show that he wanted it to arrive at the fire simultaneously with the engine. Notice that in his design the hose wagon was linked to the pumper by firehose, while being detachably coupled to it as well. Upon reaching the fireground the two vehicles were to have been unhooked one from the other so that the pumper could be connected to a water source while the wagon was moved to the most effective tactical position. Both the Eisenbise hose wagon and pumper were sketched in the same "buckboard" configuration as was conventional for the horse-drawn vehicles of that day. For the pumper the vehicular frame was designated by the number 1; the "motorman's" foot rest was 2, the bell was 3, and the motorman's seat was 4. The gasoline (in those days it was spelled "gasolene") tank was 5, the engine cylinder was 7, the engine shaft was 14, the chain-drive mechanism was 15 and 16, and the rear axle was 17. The pump chamber was 24, the pump cylinder was 27, the air chamber was 40, and 45 was a suction hose. A platform for the firemen was designated as number 47. On the hose wagon the gasoline tank was number 48, the engine was 49, and the drive chain was 51. Number 53 was a permanent hose coupling. Not clearly explained in the text of this early patent was just which one of these two vehicular motors was supposed to power the rig on the road to the fire.

A second and much more sophisticated Eisenbise patent was number 1,023,141 received in 1912 (figures 230, 231, 232, and 233). This fire engine, which looked like a fire engine usually looks, was really a very simple apparatus, and one which could squirt a chemical stream as well as water. The chemical supply flow was pump-powered and did not employ the usual gas-generated pressure expulsion. Eisenbise, as will be seen, came close to the concept known as the "booster system," but not quite close enough to qualify as its inventor.

Since this Eisenbise patent, just like his earlier one, has an uncommon historical significance attached to it, the drawings again deserve close examination. The inventor designated the vehicular chassis as number 4, and it was carried by "steering" wheels, 5, and "driving" wheels, 6. Power was from a "hydro-carbon" motor enclosed within a casing, 7, from which was protruded a shaft, 8, which carried a bevel gear, 9, which was to mesh with a second bevel gear, 10. This mechanism was to power a countershaft, 11, upon which were to be mounted sprockets, 12, which were to connect with driving wheel sprockets, 13, by chains, 14. The front wheels were controlled by a "pilot" wheel, 15, and there was a "danger" bell, 16, which was to be sounded by a foot lever, 17. Water was to enter the pump, 20, through a suction line, 35, which was branched, 36 and 37. A length of hose, 40, with a perforated cap, 41, was attached to the suction inlet and was to have been used for obtaining water from cisterns or other non-hydrant sources. Located below the floorboards and terminating adjacent to the front bumper was the pump outflow line, 45, which gave off a valved branch to each side of the fire engine. Paired reels of one and one-half inch diameter hose also were reached by valved branches from the main outlet. There was a basket, 56, for stowing the three-quarter inch chemical hose, and a roller, 57, was provided to ease its playout pathway. A partition, designated as number 86, divided the hose bed so that hose loads could be varied, and the hose bed was open at the rear to facilitate playout while the vehicle was moving. Among the remaining structural details were the hand rails, 88, a ladder, 89, and side and rear steps, 90 and 91, for carrying firemen.

Keyed to the main drive shaft of the vehicle was a piston pump, 68, which was to have been used for ejecting a stream of water and dissolved chemicals from the "chemical tank," 64. Had Eisenbise not been preoccupied with this belief in the usefulness of chemical solutions this auxiliary system could well have been given some other name, such as, "booster." The inventor did remark that since carbon dioxide was not being generated, "the chemical system was not pressurized, and the danger of explosion was eliminated." An Eisenbise "chemical booster system" was fitted to the hose wagon illustrated in the third patent drawing, even though it was not equipped with a standard water pump. And finally, note that these vehicles were complete with right-sided driving positions.

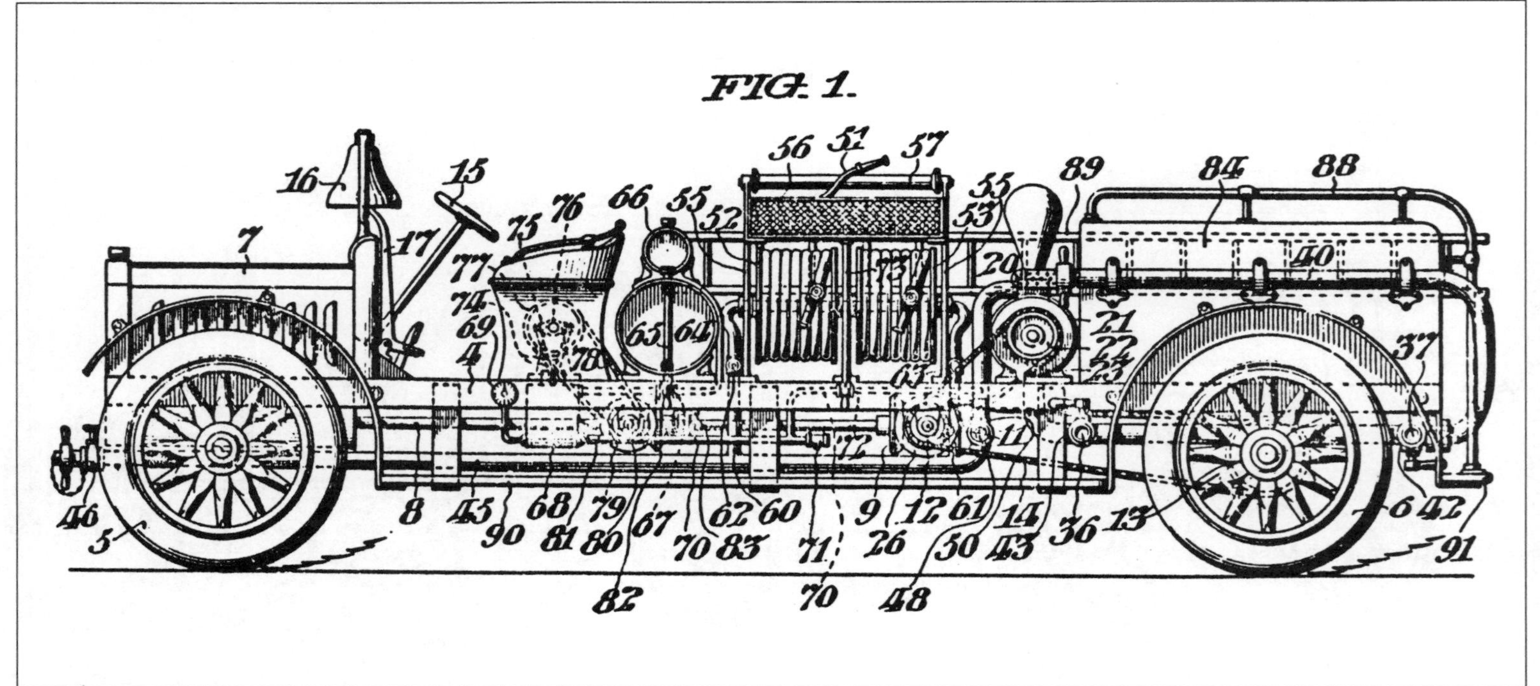

FIGURE 230

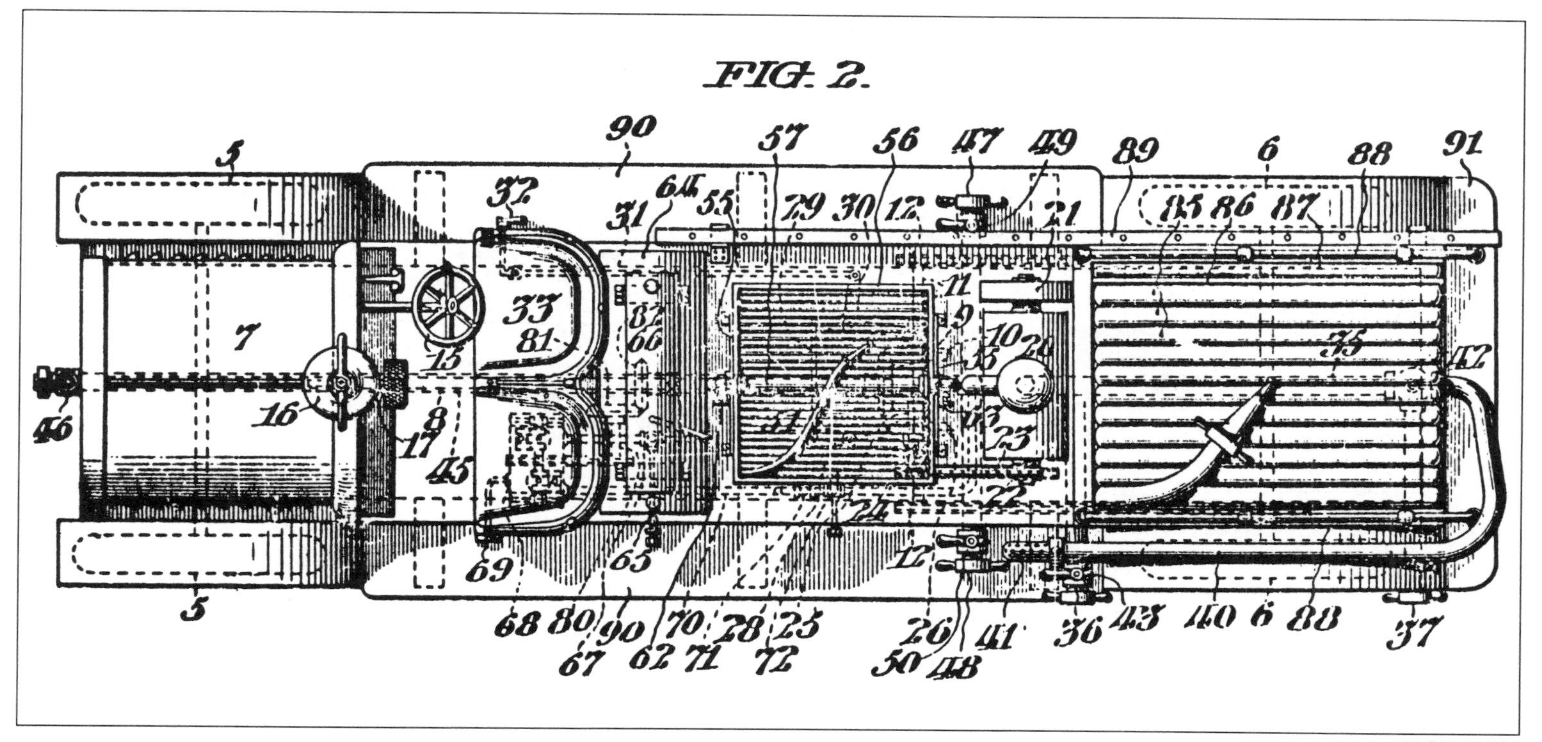

FIGURE 231

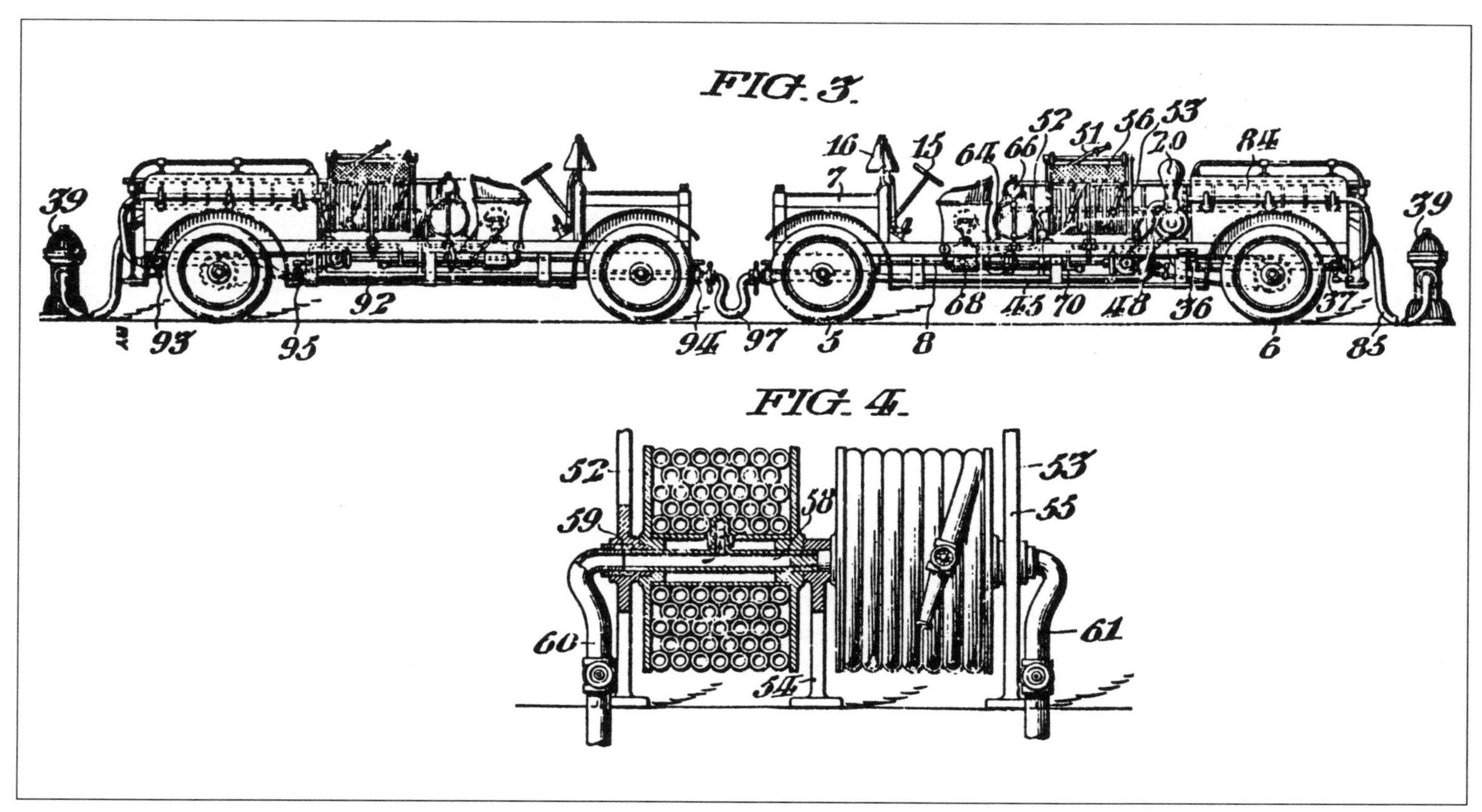

FIGURE 232

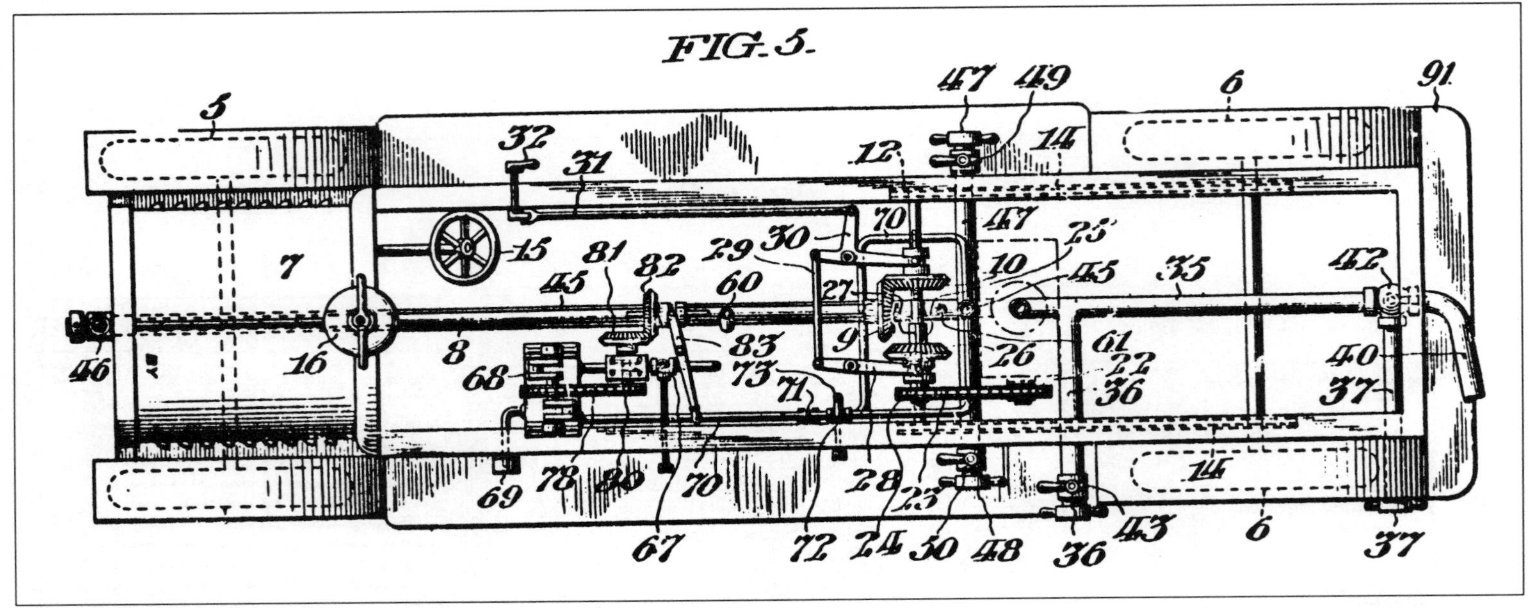

FIGURE 233

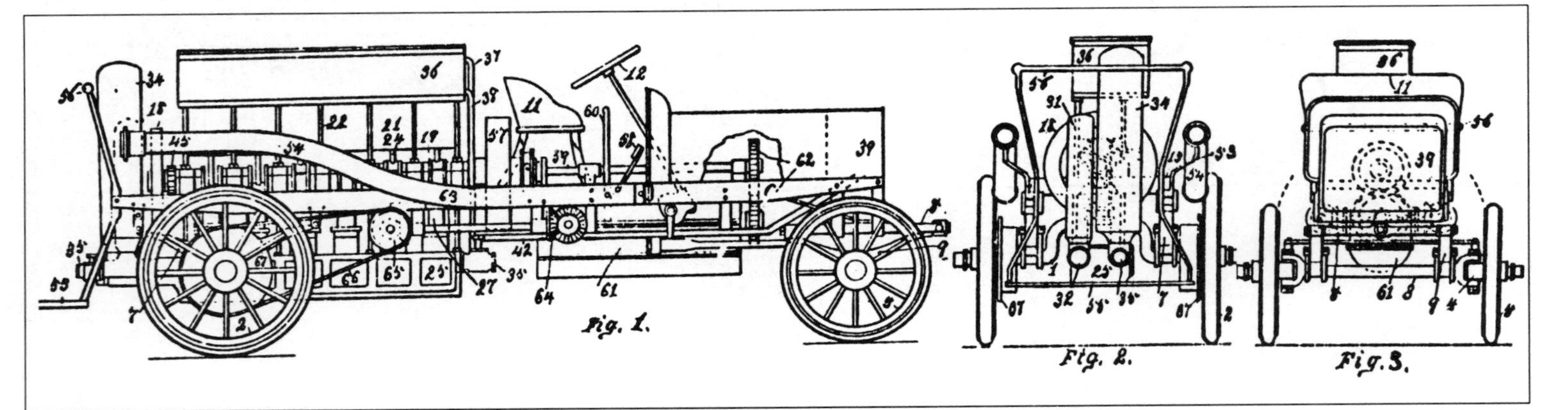

FIGURE 234

The Shafer Patent

An inventor with a total of eleven patents bearing his name was William H. Shafer (October 7, 1864-August 5, 1933) who served the Ahrens-Fox Fire Engine Company in a number of positions of responsibility for many years. During his most inventively productive period, the first decade of this century, he patented a uniquely peculiar, primitive, gasoline-powered fire pumper (patent number 876,404 dated January 14, 1908). A couple of Shafer's features designed into this never-built vehicle are worth mention, and the drawings for this patent are shown in FIGURES 234 and 235. The inventor used the words "explosive motor," a well-worked contemporary phrase, to indicate a cylinder mechanism, and he created a truck in which a series of explosive motors were each directly connected to a corresponding series of force pumps. Specifically, he wanted each of six in-line, four-cycle cylinders to be placed directly above a similar number of single-acting force pumps, with the plunger of each one aligned with and attached to its respective piston. In such a configuration the pump plungers would have been constantly in motion with the engine, whether or not they were actually at work pumping water. This was indeed a quaint idea, and a most eccentric one, since the Shafer record was otherwise marked by rather more important patent creations.

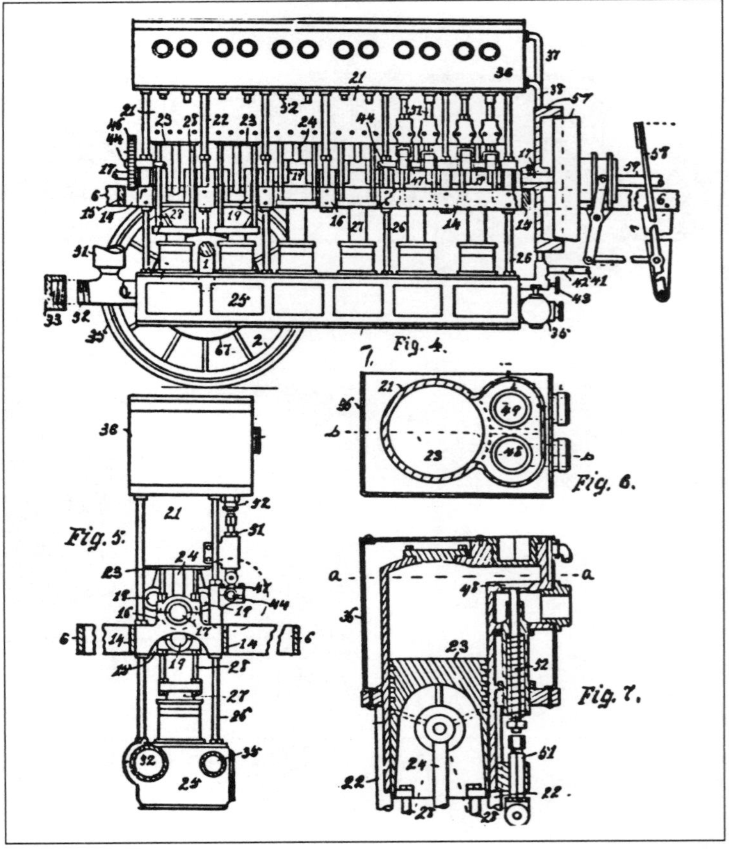

FIGURE 235

The William Francis Gibbs Patent

Also from this era before there were any fully-trained and educated automotive engineers was a 1909 patent, number 918,140, granted to William Francis Gibbs. During subsequent years Mr. Gibbs, the inventor of the New York Fire Department's Super Pumper System, achieved prominence in his fields of marine engineering and marine architecture. The patent drawings for Gibbs' "Gasolene (his spelling) Fire Engine" are reproduced as FIGURES 236 and 237. This invention included not one, but a pair of rotary pumps, with discharge lines so valved that water could be ejected from either one or from both simultaneously, and from either side of the fire engine. Later on, of course, the first one of these capabilities came to be refined as "parallel" and "series" pumping.

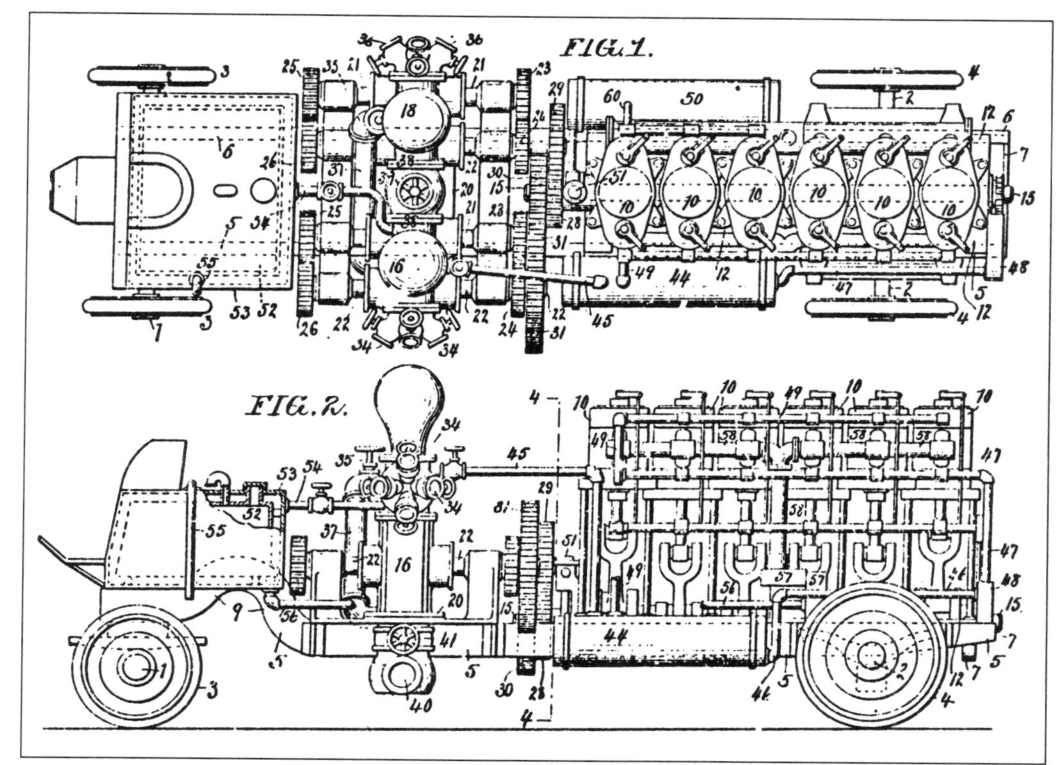

FIGURE 236

Staged Pumping

The first-ever statement of staged pumping is in patent number 932,282, inventors Frank Barrington Hunter and Harry Bettis Hunter, both from Memphis, Tennessee. They specified that their fire engine be equipped with two centrifugal pumps with piping and valves arranged so that, "water may be directed through the pumps separately or in parallel, thereby giving a maximum capacity or, if desired, the water may be caused to first flow through one pump, then return through the passage and through the other pump, thereby giving maximum pressure." To this time no more clearly succinct explanation of pump staging has ever been formulated. FIGURES 238 and 239 were the drawings for this patent.

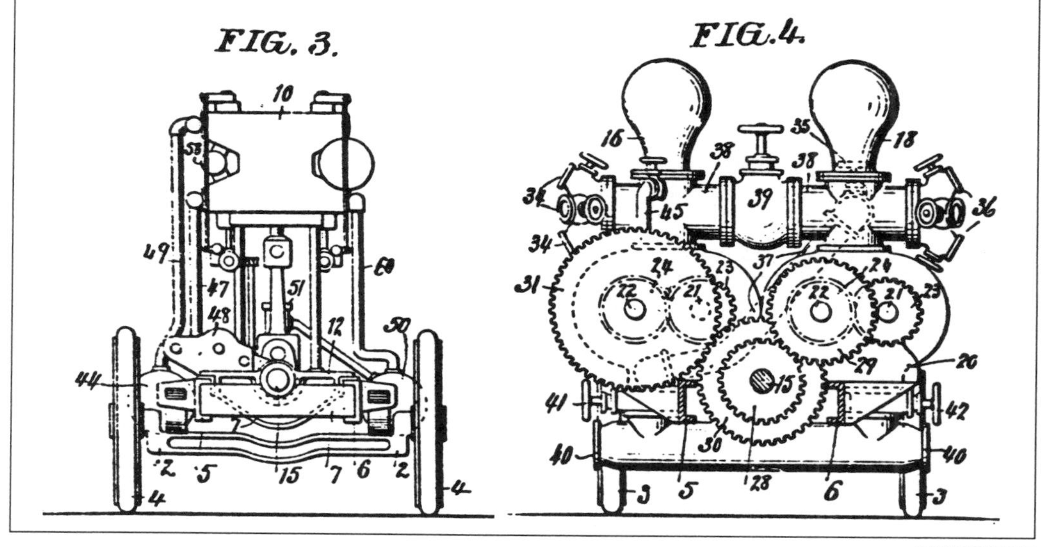

FIGURE 237

U.S. Patent number 1,000,000 was issued during the first week of August, 1911, or just about 12 years after Commissioner of Patents Charles H. Duell had pontificated that since "everything that can be invented has been invented" the patent office ought to be shut down! Patent number 1,004,816 dated October 3, 1911 was received by Haverford, Pennsylvania resident John Livingston Poultney, who assigned his

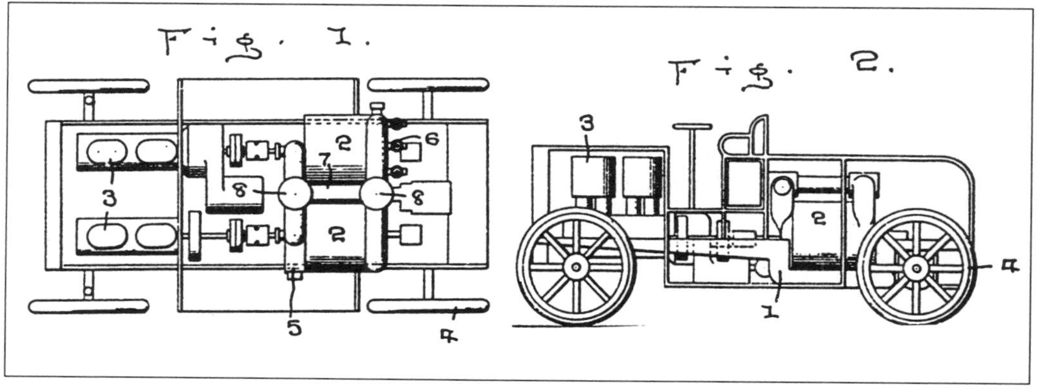

FIGURE 238

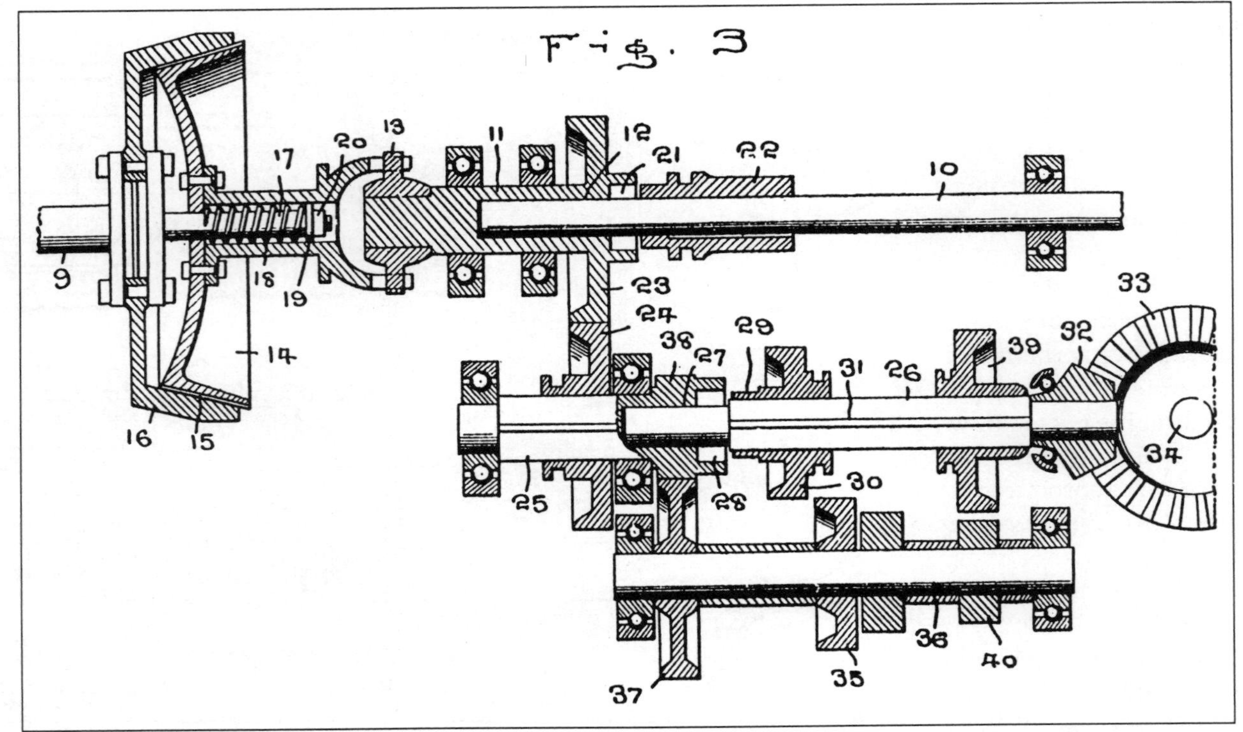

FIGURE 239

rights to the Philadelphia concern of James Boyd and Brother, Incorporated. The drawing is shown in FIGURE 240. This invention featured a method for operation of a rotary pump, through appropriate gearing, from the propelling motor of the vehicle.

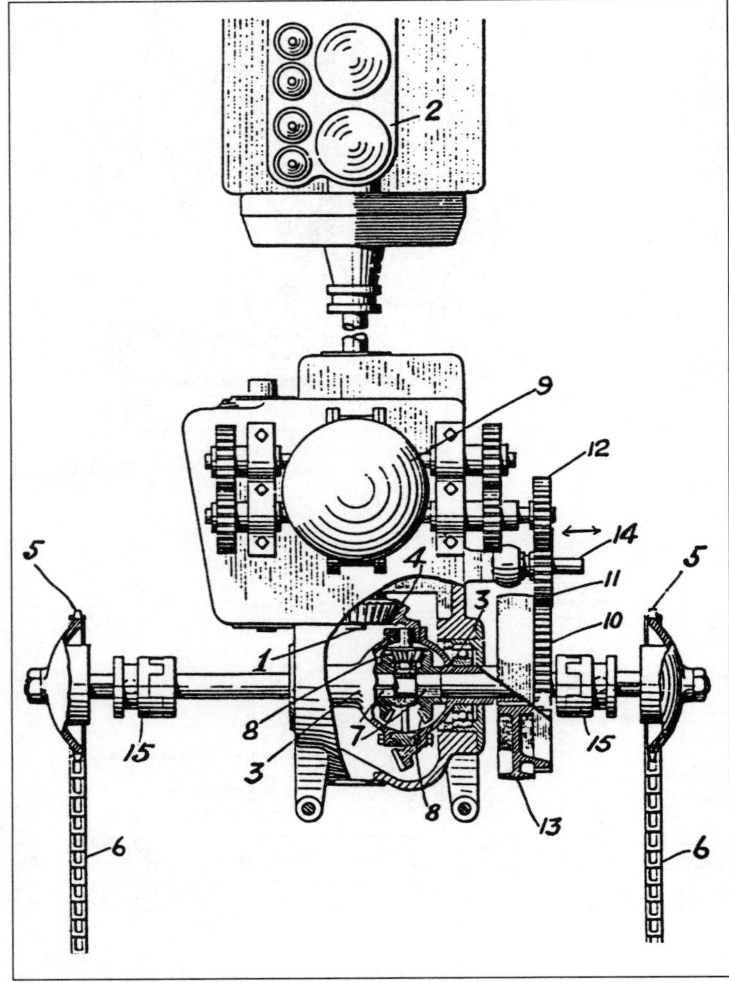

FIGURE 240

Albert C. Webb

Inventor Albert C. Webb was an auto race driver and a gentleman who must have personified the successful engineer–entrepreneur of the earlier years of the twentieth century. Webb moved his operations about from time to time, locating his factory first in Vincennes, Indiana, then in Allentown, Pennsylvania, and finally in St. Louis, Missouri. By 1914 over 100 communities had purchased Webb fire apparatus but, just like the Knox, Luitweiler, Martin, Nott, Robinson, and other concerns, the company did not survive. Back in those good old days the bottom line for most commercial operations was as much the quality of product as it was the return on investment. It is not clear whether the Webb Company failed by reason of faulty product or insufficient funds, or both. Mr. Webb filed two patent applications on the tenth day of July, 1911. One described his rotary pump and the other a piston pump which he had invented (patent number 1,005,586, FIGURE 241, and patent number 1,018,795, FIGURES 242 and 243).

Webb always claimed that he was the original supplier of motorized fire pumpers, an allegation quickly disputed by other fire engine manufacturers. Historian Walter McCall, by the way, has credited this honor to the Waterous Engine Works because, in 1906, the Radnor Fire Company of Wayne, Pennsylvania was operating a Waterous pumper and hose wagon, a combined outfit which looked as if it had been designed by Harry Eisenbise (FIGURE 244). The oldest available catalogue photograph of a Webb pumper, FIGURE 245, illustrated an apparatus which was delivered in 1907.

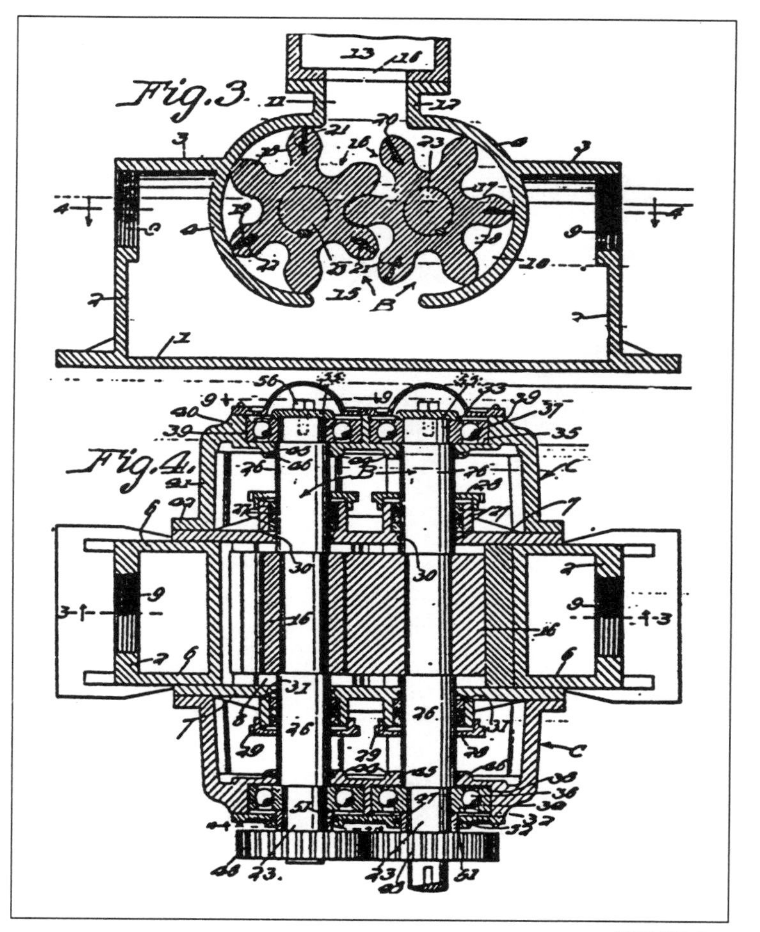

FIGURE 241

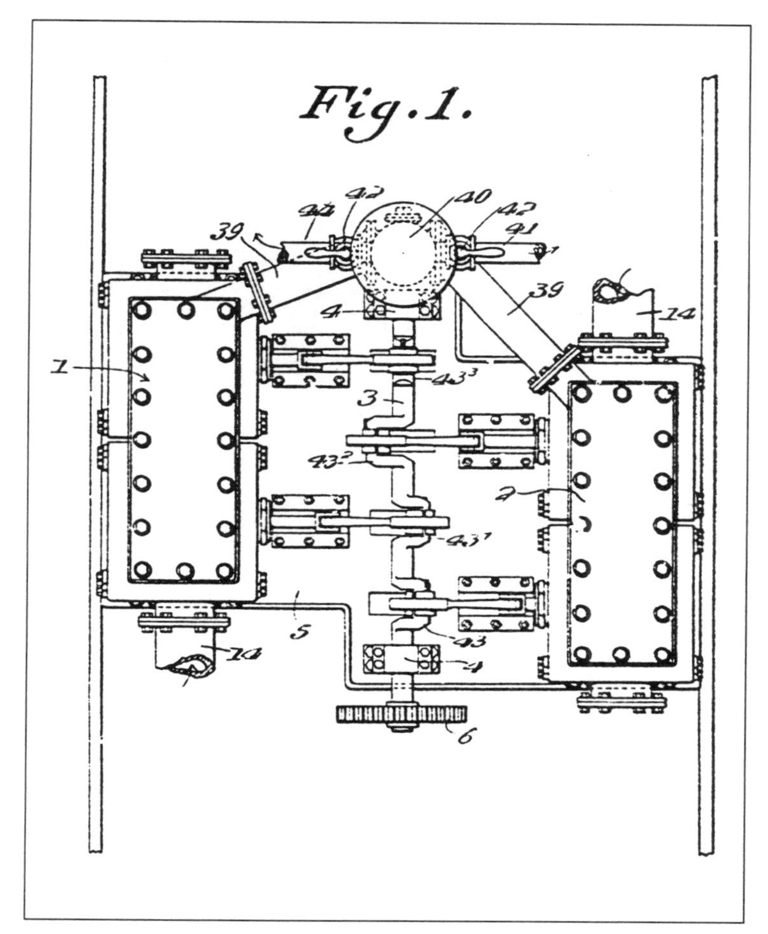

FIGURE 243

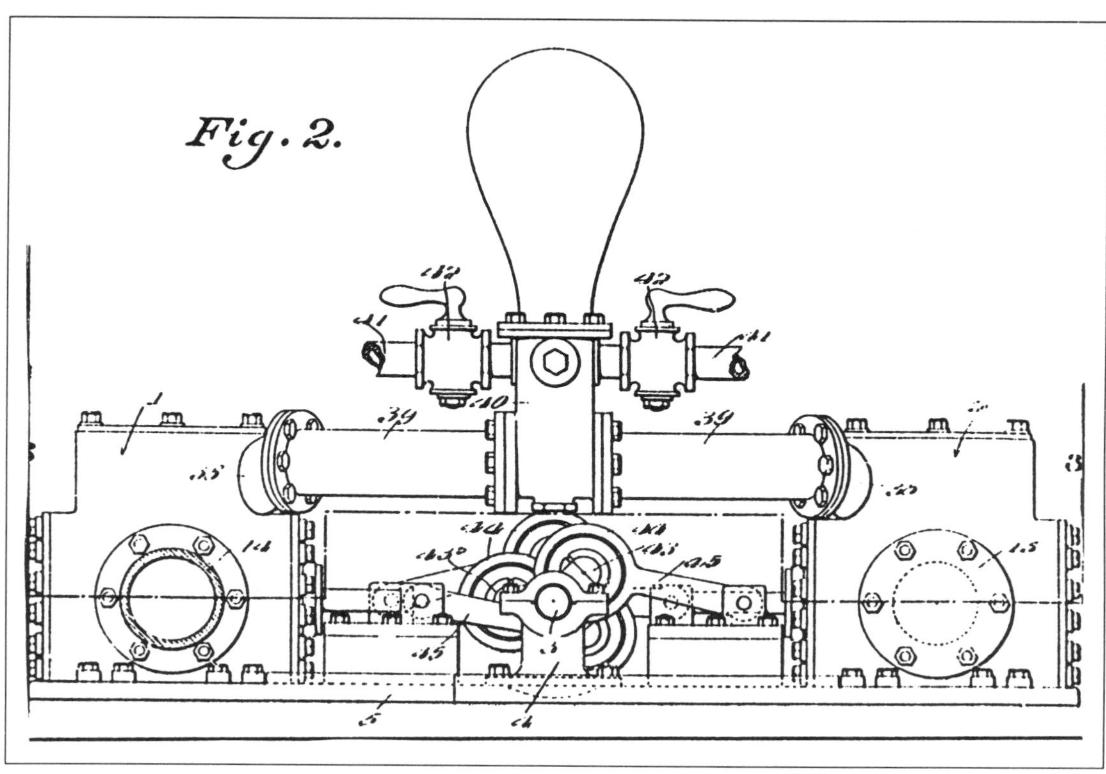

FIGURE 242

FIGURE 244

FIGURE 245

In parallel with the explosive expansion of the automobile industry during the years 1910-1920 in the United States, and the opening of paved streets, gasoline stations, and auto repair and replacement parts concerns, the level of activity among inventors, designers and manufacturers of motor fire engines also accelerated. Several of the fire vehicle patents of that era deserve brief notation.

Fuchs, Lent, Schoonmaker and Taurman Patents

Leonhard Fuchs, Rochester, New York received patent number 1,007,551 on October 31, 1911 (FIGURE 246) for his invention of a means of correlating engine speed and pump output. He had designed a single shifter-rod method for regulating the pump as well as for selecting the engine transmission positions, and he built a massive mechanism for vibration-free pump operation.

Leon B. Lent of Brewster, New York, a small town just off Interstate Route 84, and near the Croton Falls Reservoir received patent number 1,010,158 dated November 28, 1911 (FIGURE 247) for his invention in which the drive shaft of the motor passed axially through a pair of centrifugal pumps. Thus, the pump impellors would revolve whenever the motor was operated, and they functioned therefore as fly wheels for the motor. In a curious omission Mr. Lent did not discuss the implications of his having fitted two pumps to a fire engine.

James M. Schoonmaker of Pittsburgh, Pennsylvania received patent number 1,013,622 on the second day of the year 1912 (FIGURES 248 and 249 are his first and fourth patent drawings). This inventor's primary novelty claims related to control devices and a complex clutch–transmission mechanism (notice the two foot pedals and three lever handles), but his most significant contribution was a new method for engine cooling while pumping water. He provided for a pipe, designated by the number 85, to carry water from the pump to the engine jacket.

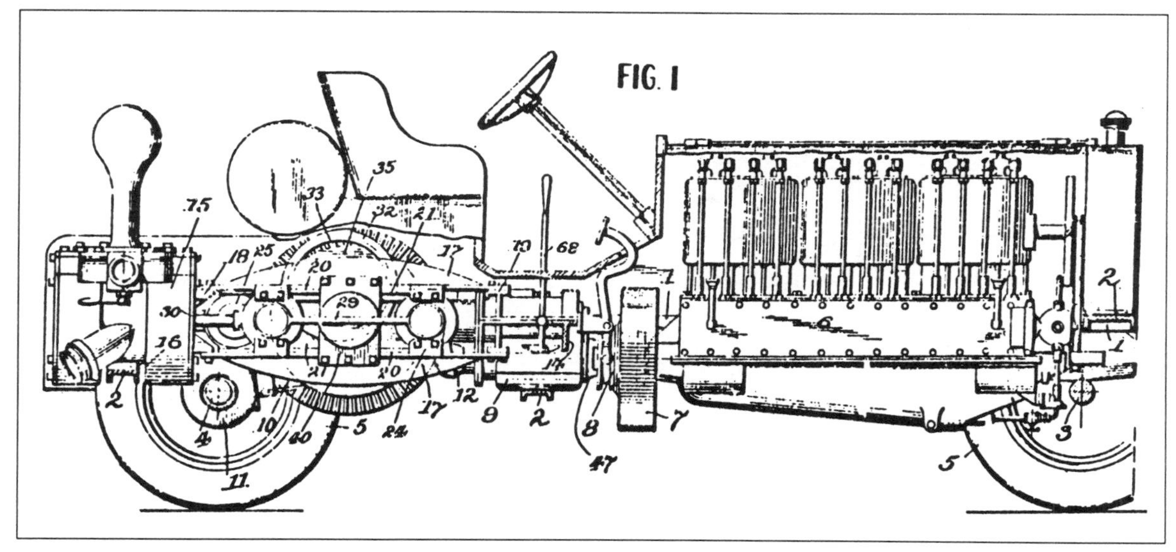

FIGURE 246

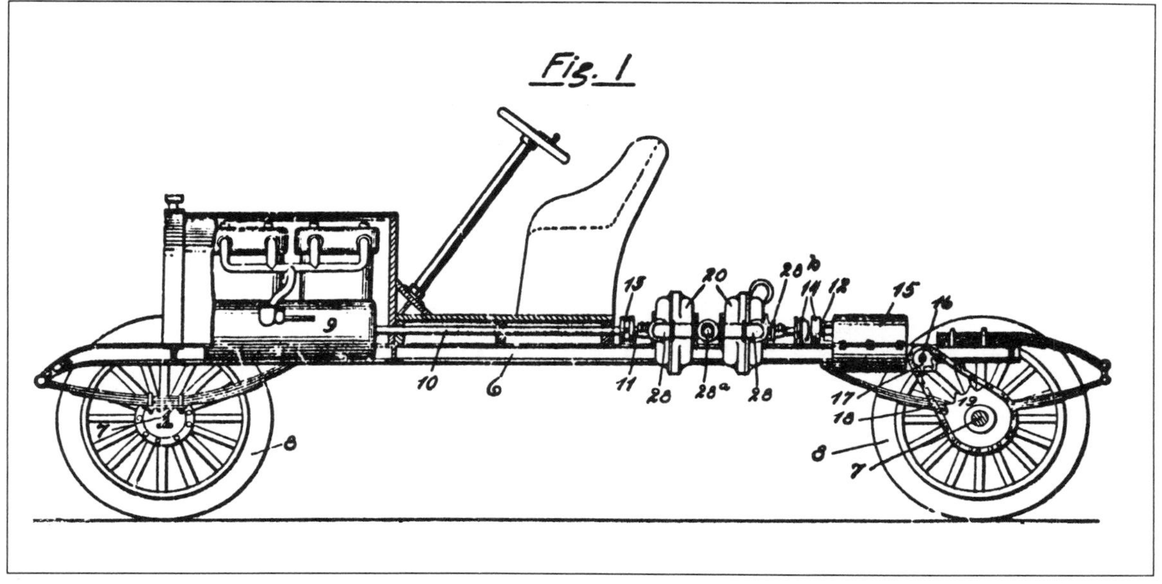

FIGURE 247

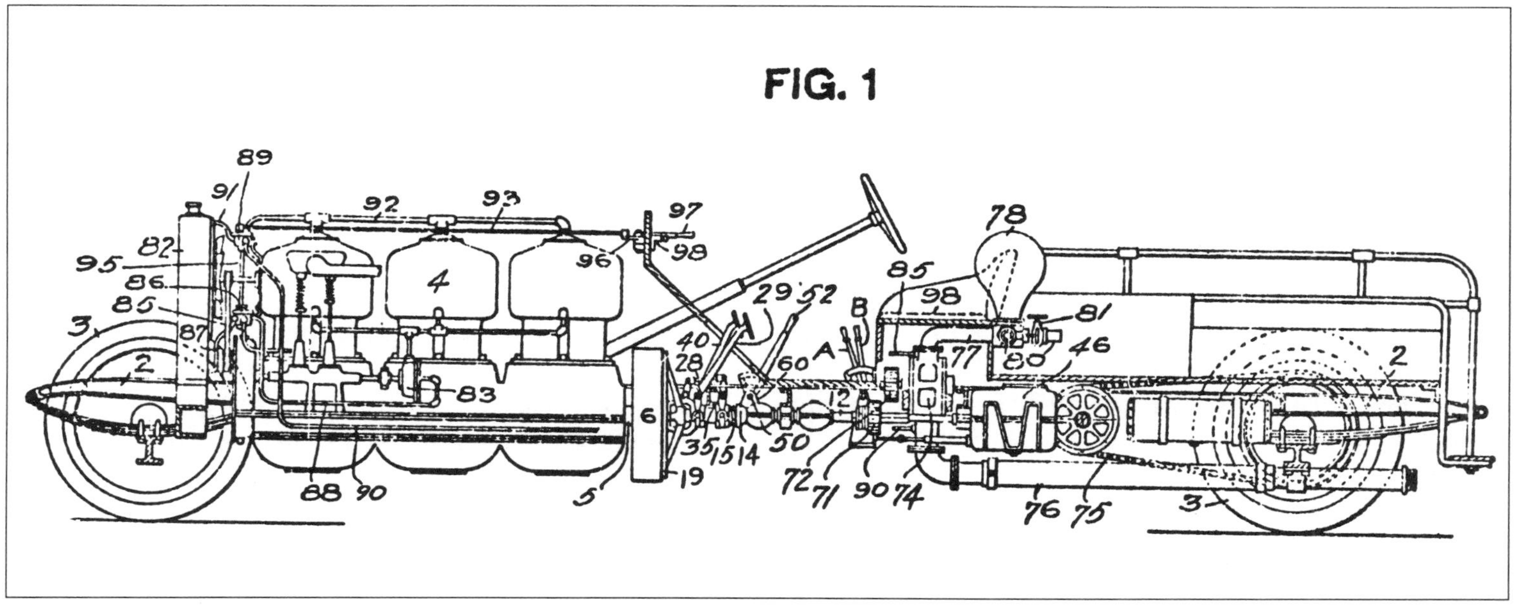

FIGURE 248

A Cincinnatian who was not affiliated with the Ahrens-Fox Fire Engine Company, Chastain Taurman received patent number 1,071,275 dated August 26, 1913 for the unusual fire engine design seen in FIGURE 250. Taurman specified paired reciprocating piston pumps, placed horizontally and at right angles to the main drive shaft. Worm gears were to transmit motion from the drive shaft to each pump shaft. He also specified that the air dome for the pump system was to be close to the driver's seat and adjacent to the foot board of the vehicle, although no logic was presented for this location.

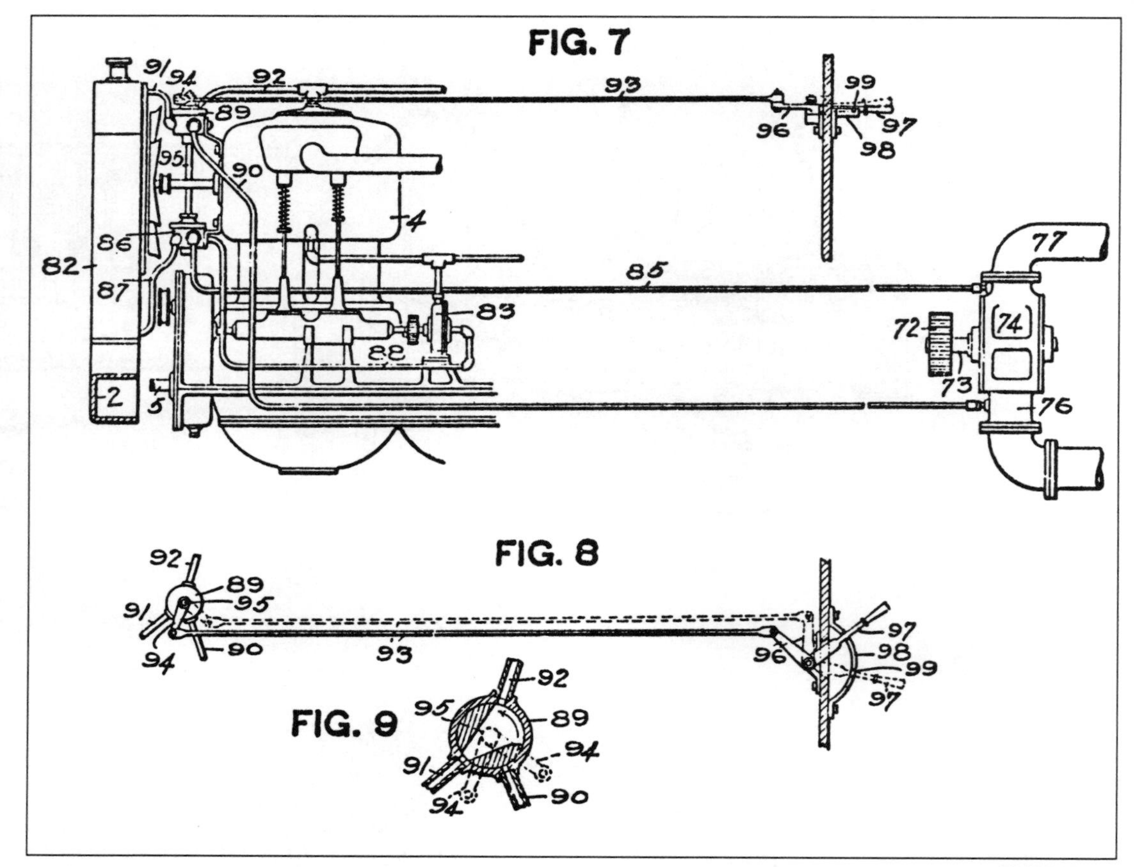

FIGURE 249

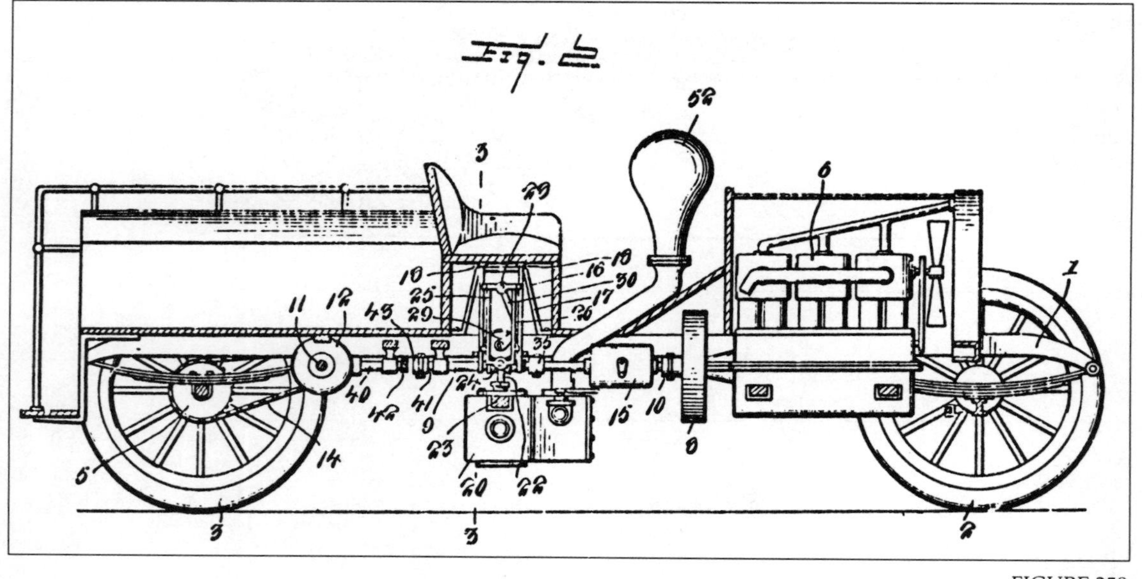

FIGURE 250

The Luitwieler Fire Engine

A fascinating one-of-a-kind fire engine from these old days was the Luitwieler. In his commentary about fire apparatus built during the year 1909 Walter McCall, in "American Fire Engines Since 1900" noted (page 13) that "L.W. Luitwieler of Rochester, New York, advertised a new type of automobile fire engine. The Luitwieler featured a full-length hose basket mounted above an open pump compartment. It is believed that only one of these was ever built by the Luitwieler Pump Engineering Company." And although McCall had no photograph of this fire engine to publish on that page, careful readers might remember that a pair of small line drawings of this apparatus were used earlier in the book, on page five. FIGURE 251, however, is a photograph of this unique fire rig (from "Motorized Fire Apparatus of the West 1900-1960, by Wayne Sorensen and Donald F. Wood). And FIGURE 252 is a drawing from Samuel W. Luitwieler's patent number 1,152,113, filed in November, 1911 and received on August 31, 1915. This, surely, is the fire engine mentioned by McCall and illustrated in Sorensen and Wood. The pump was placed underneath the driver's seat, as shown, and the motor was placed at the rear, a novel location currently a feature only of Emergency One "Hush" apparatus.

FIGURE 251

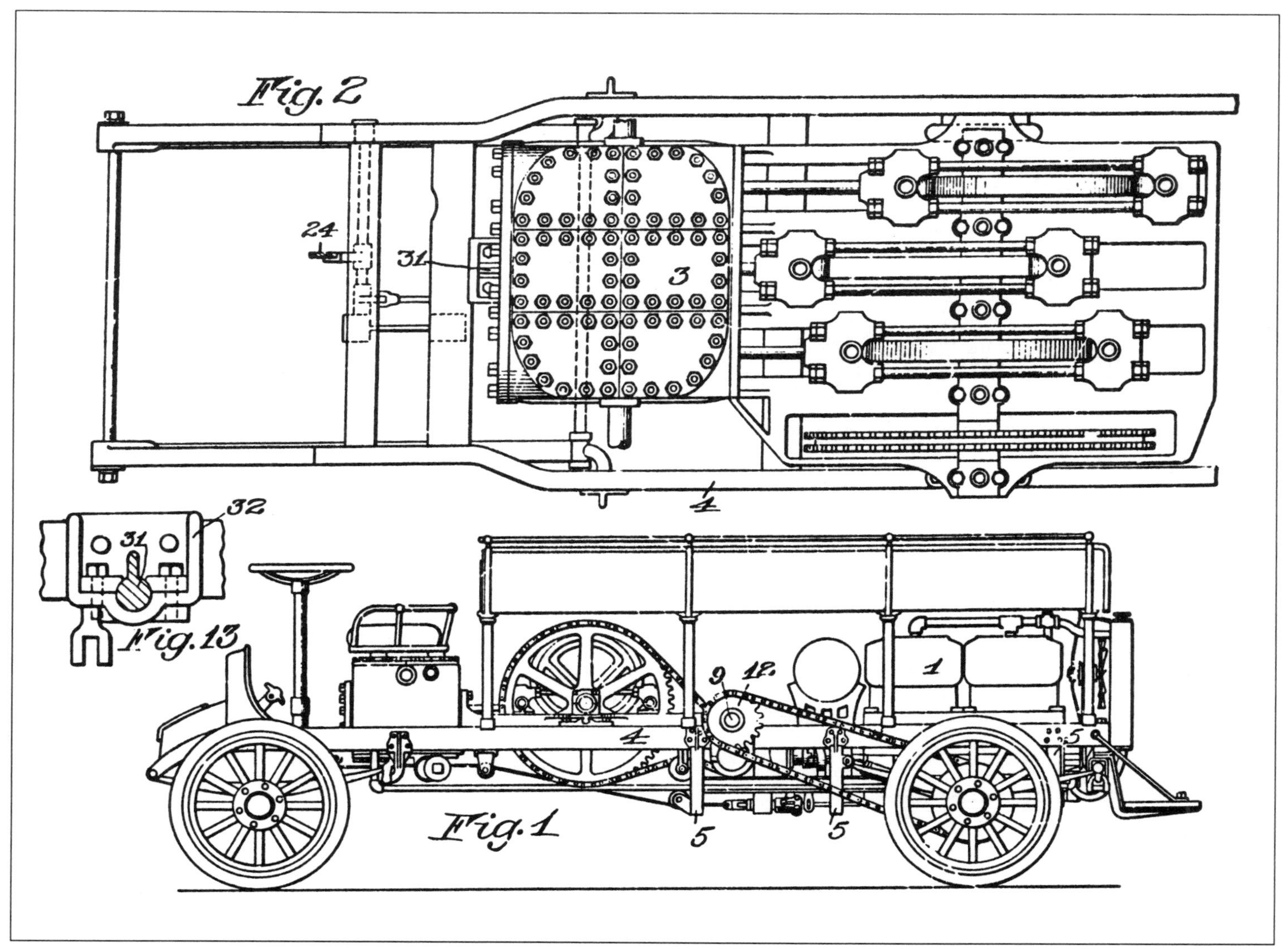

FIGURE 252

H.G. Farr

Looking rather neat and small in FIGURES 253 and 254 was the fire pumper for which patent 1,191,293 was received by Herman G. Farr of Springfield, Massachusetts on July 18, 1916. The inventor started his detailed description with the words, "The vehicle as actually manufactured ...," leading one to believe that a working model was built. Farr then went on to point out that his rig had a short wheel base, about the equal of a steam fire engine minus its horses. Farr was quite sure that the degree of traffic hazard risk was determined by wheel-base length. The motor which powered Farr's rig was located above the rear axle, with its crankshaft connected to a front-mounted transmission which distributed the motion to a transversely-oriented drive shaft. Sprockets which were mounted on either end of the drive shaft turned the rear wheels via a chain mechanism. Notice in these patent drawings that a front-located hose connection was designated by letter w, the pump by letter y, and the pump shaft by letter n. Other components included the clutch foot lever, f; the gear shift lever, z; a lever for control of the clutch, e; the clutch, k; the engine drive shaft, c; the transmission, l; an intermediate shaft for adjusting drive chain tension and catenary, b; the rear axle, a; and the hood, m.

Also evident in the drawings, although not bearing any label, were a hand crank for the motor, a fuel tank behind the driver's seat and a water tank underneath it, a hand-brake lever, a bell, a searchlight, and a kerosene lamp.

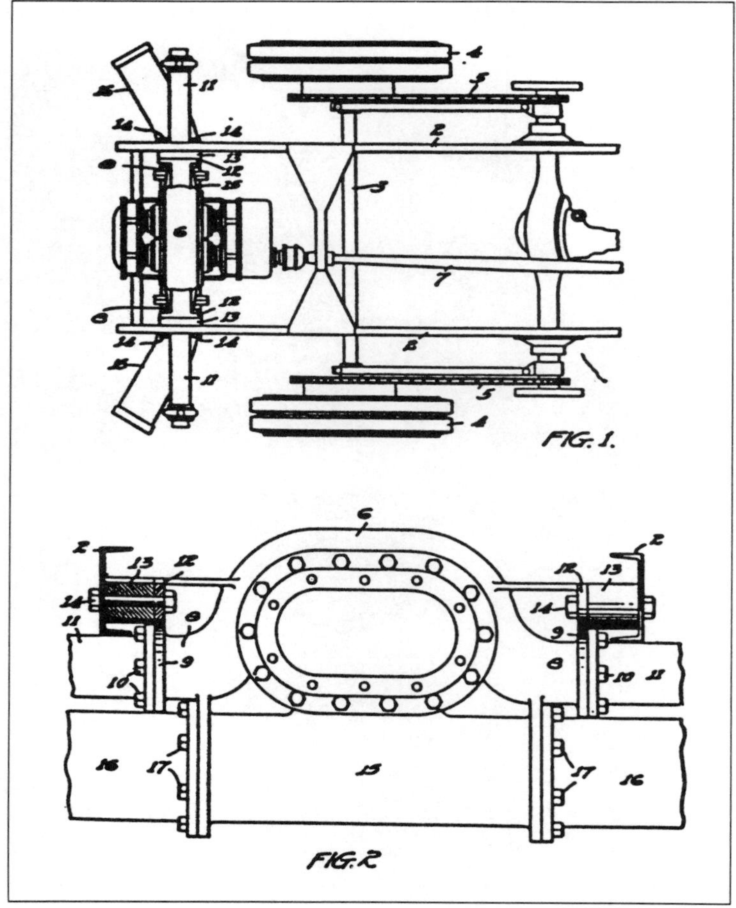

FIGURE 255

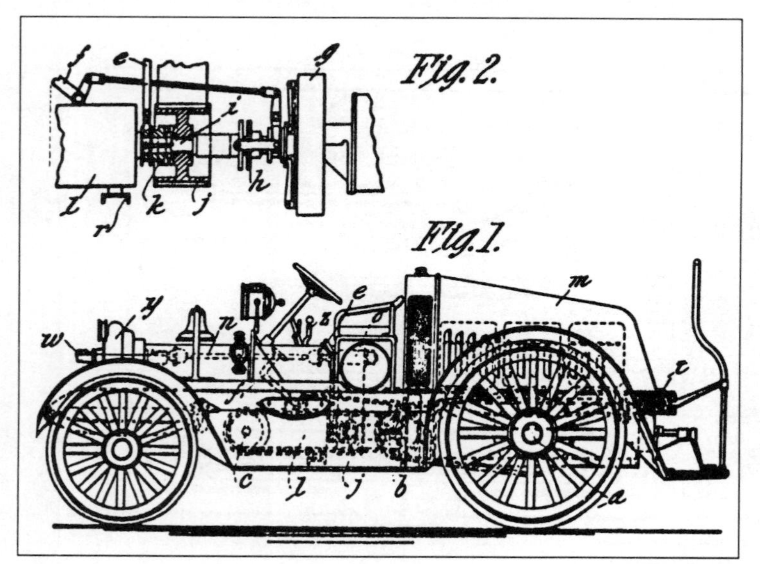

FIGURE 253

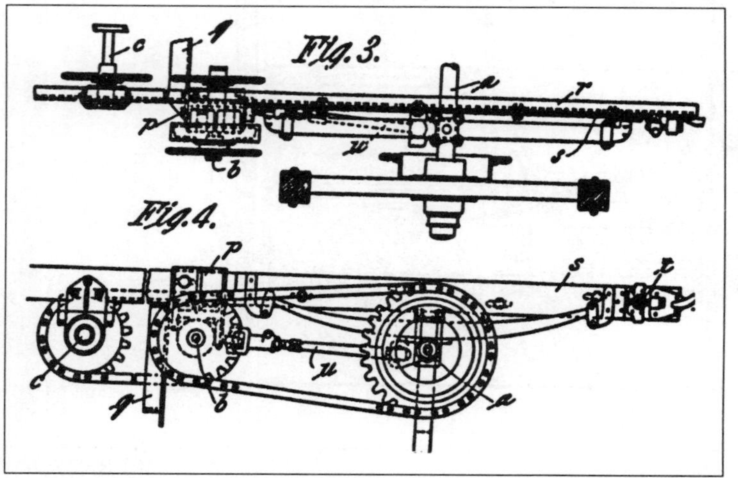

FIGURE 254

C.R. Waterous

Long an honored name in fire apparatus history, and still in operation today, is the Waterous Company, formerly the Waterous Engine Works Company of St. Paul, Minnesota. On April 30, 1912, Mr. Charles P. Waterous received a patent for his improved piston pump, and during 1917 he received patents number 1,217,671 and 1,238,853. The first one of these inventions was a new manner for hanging a pump from a fire engine chassis (FIGURE 255). By simply removing four bolts (number 14 in the illustration) the pump could be quickly detached from the vehicle. In the second patent Waterous laid out his method and mechanism for engine block temperature moderation while pumping (FIGURES 256 and 257). A small-diameter pipe, designated as number 19, carried cool hydrant water from the suction line, 17, to the engine block-radiator system. Jacketed around this pipe was a conduit of larger diameter, 22, which carried engine exhaust gases. This line was controlled by an appropriate valve and was used to increase water temperature in cold weather situations.

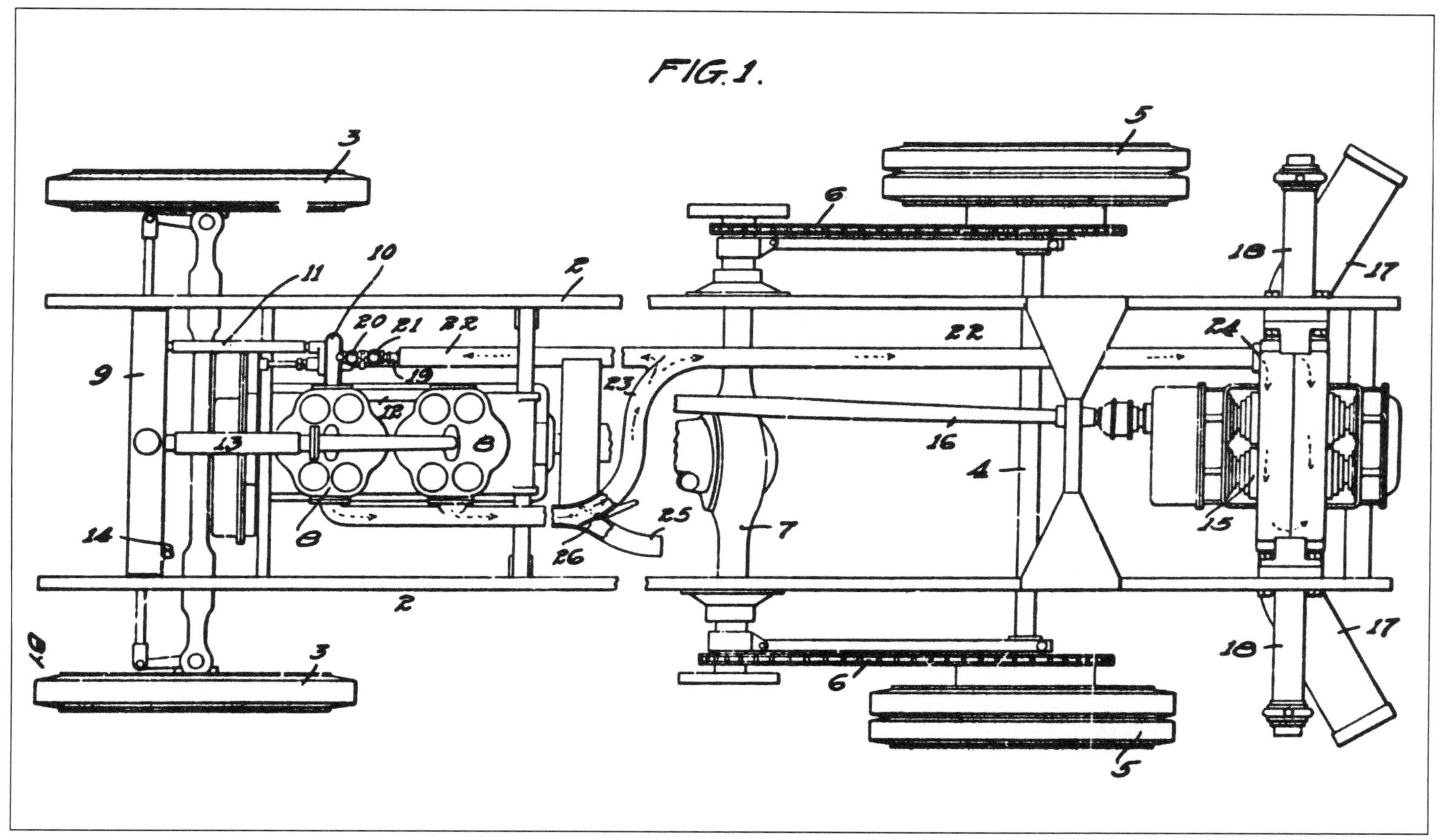

FIGURE 256

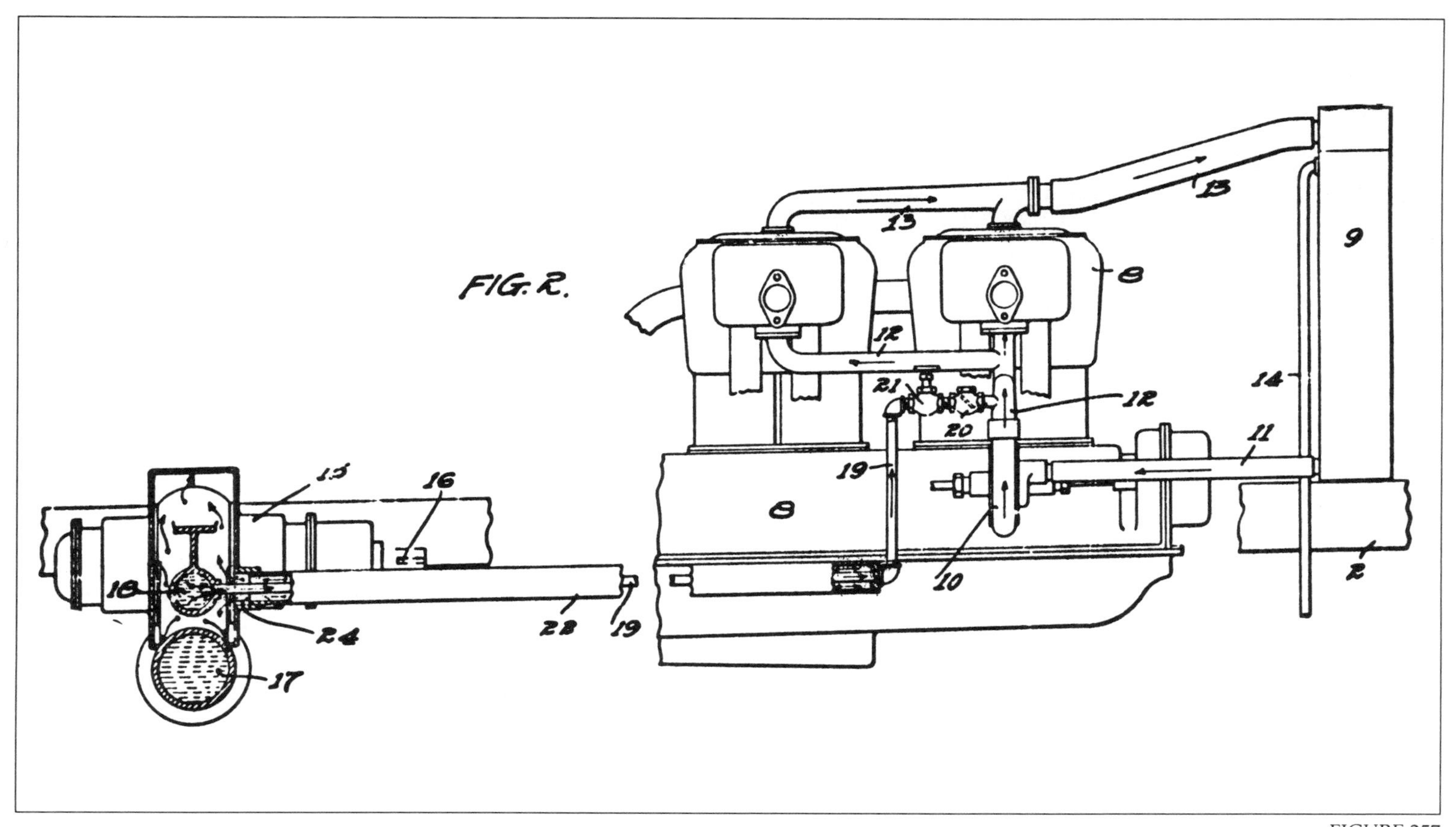

FIGURE 257

FIGURE 258

FIGURE 259

The Robinson Fire Engine

Thus by the middle teen years of our century it was becoming increasingly clear that the motor fire engine, while still a bit avant garde, was no longer a novelty, although still a luxury if your department had just invested in new steamers and horses. The motor fire engine was about to become a necessary item in the baggage of city life, and the inventors and affectionados were already working to modify and to alter components, and to add the latest gadgets and mechanisms. With reference to a Robinson apparatus the following material, exerpted from Scientific American Supplement Number 1862, page 165 of September 9, 1911, described the current state of the art of motor fire apparatus construction and function (FIGURES 258 and 259).

"The triple automobile combination chemical engine, hose wagon, and fire engine of the Robinson 'Jumbo' type is equipped with an engine of 110 horse power capacity, which drives the pump at full capacity at a speed of 900 revolutions per minute, the same motor giving abundant powerful high speed travel for hill climbing and pulling the car through mud or snow. A vertical motor having six cylinders cast in pairs with a stroke of six and one-half inches, and a bore of five and five-eighth inches is utilized on this triple auto fire engine. This engine was designed to operate at any speed up to 1,400 revolutions per minute, both intake and exhaust valves being mechanically operated and located in the heads of the cylinders. The cooling of the engine is effected by a gear pump forcing the water through a large square tube radiator, a bypass from the main pump direct to the motor cylinders being provided with regulating valves enabling the operator to keep the motor cylinders at any desired temperature when pumping."

"The car is geared for a normal speed of 30 miles per hour, but can be put on double this speed if desired. The hose body has a capacity for 1,000 feet of two and one-half inch rubber lined cotton hose, and the chemical equipment consists of a 35 gallon chemical tank of six and one-half pound hammered copper heavily tinned on the inside and provided with a fan agitator, the entire length of the tank being revolved by a brass scroll wheel to effect thorough dissolution of the soda and to prevent its settling at the bottom of the tank. An independent acid pump is located at the end of the tank. The fire pump is of the triplex piston type with cylinders five and three-fourth inches in diameter and of eight inch stroke. This pump is back geared and driven by a spur gear on the main driving shaft, the pump being such that when operating, the spur gear is engaged by a positive jaw clutch. The air dome is of 100 pound hammered copper tested to 350 pounds per square inch. In actual service the fire pump on this novel triple automobile fire fighting equipment delivers no less than 700 to 800 gallons of water per minute and will supply as many as three fire streams from one and one-eighth inch ring nozzles. It will deliver 750 gallons per minute through two lines of two and one-half inch hose 100 feet in length, each provided with a one and one-half inch smooth bore nozzle drawing from dead suction against a net pump pressure of 110 pounds. It is claimed that the engine will run many hours continuously without overheating or showing loss of power." (Additional information about the Robinson Jumbo, and a description of one fully restored at a cost of $40,000, as well as particulars and data on the Robinson Fire Apparatus Manufacturing Company can be found in Enjine!-Enjine!, a periodical published by the Society for the Preservation and Appreciation of Antique Motor Fire Apparatus in America, or, SPAAMFAA, issue 1984-3).

Tractors For Old Fire Engines

While on the other hand, as Cincinnatian Henry M. Gabel pointed out (patent number 1,047,197, December 17, 1912), "There is at present in use a large number of horse-drawn steam fire engines, and their performance is so eminently satisfactory that long pause is to be given to the thought of abandoning them in favor of new automobile types." A large number of inventors attempted to devise methods and equipment for maintaining these fire engines, and the Christie system (over 600 units sold, 1911-1918) can probably be judged as the most successful one. Gabel (FIGURE 260) devised a motor and transmission located at the rear of the steamer and operated by the driver. There were a couple of problem areas the solution of which eluded him, such as the tendency for his rear-wheel chain drives to revolve at unequal rates, and for the intrinsic speed of the motor to be insufficient for down-hill coasting.

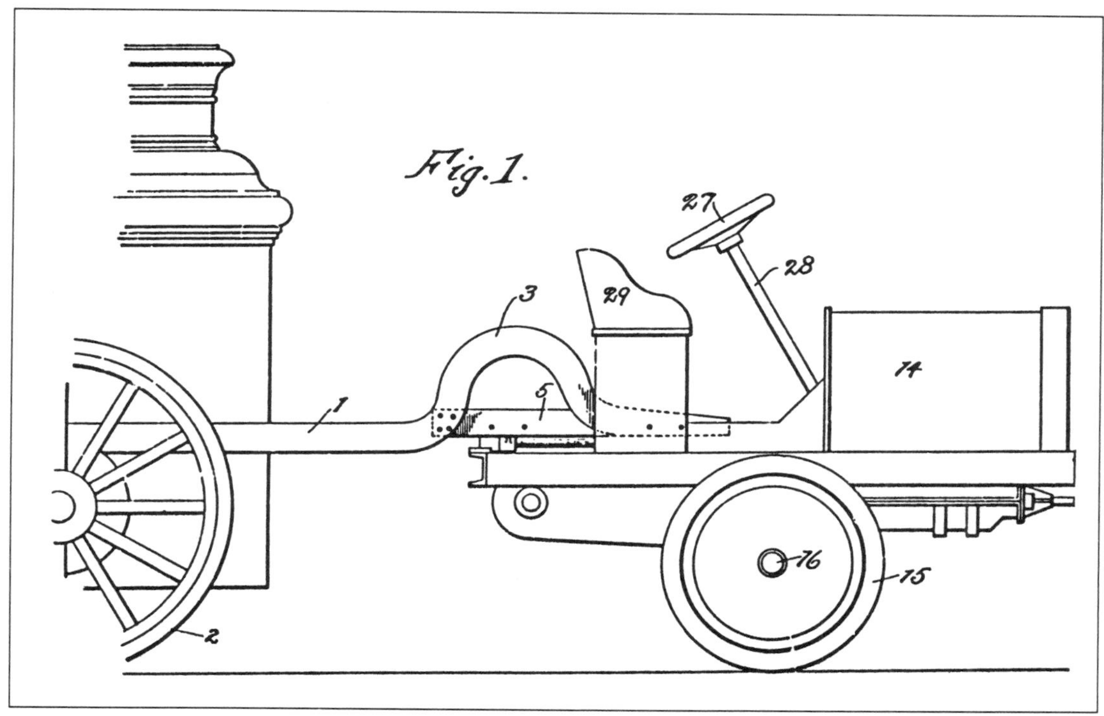

FIGURE 261

An invention of a related kind was patented in 1916 by William Guy Hawley, of Elmira, New York, who assigned the rights (patent number 1,195,216) to the American LaFrance Fire Engine Company of that city. Hawley devised an improvement in the way that old steam fire engines were attached to new front-located tractors. His patent drawings are illustrated in FIGURES 261 and 262. The most important component of this apparatus was a longitudinally-oriented shaft, identified as number 13, which extended through bearings to permit a degree of lateral oscillation between the tractor frame and the main body frame of the steamer.

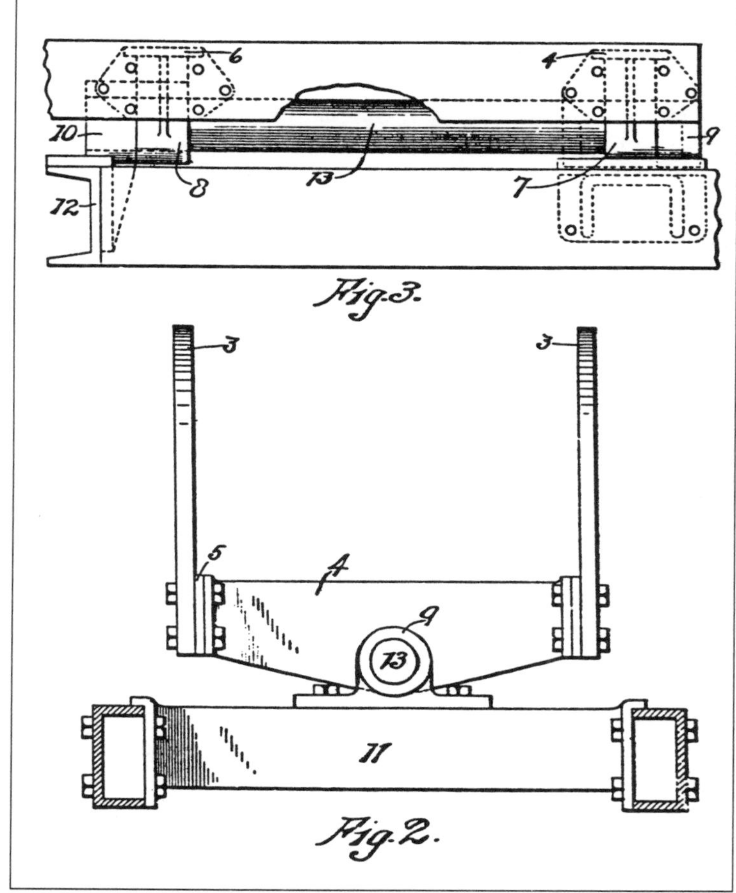

FIGURE 262

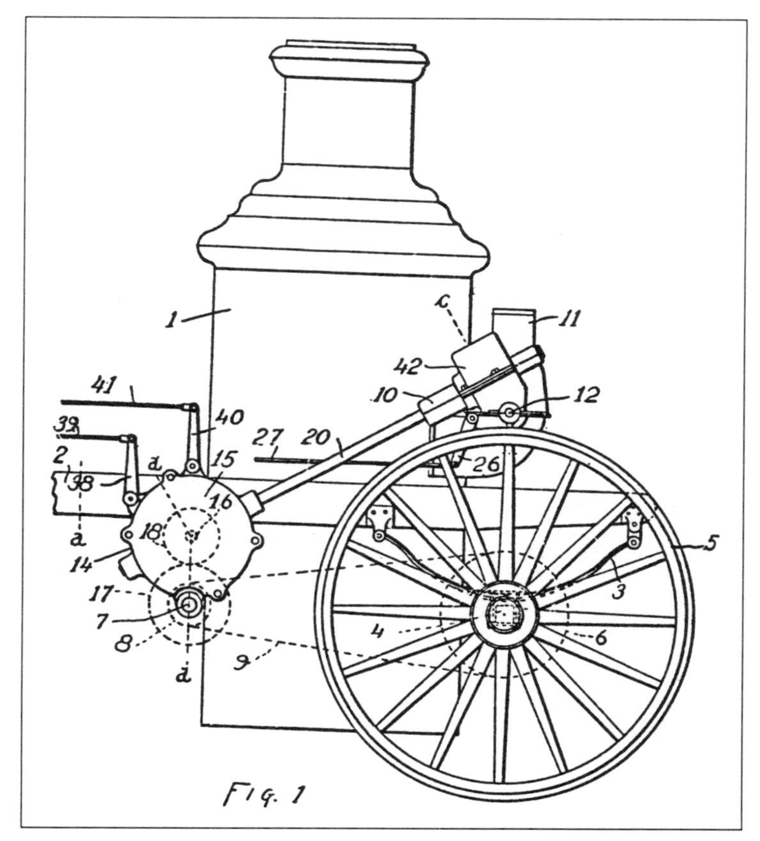

FIGURE 260

Charles Hurst Fox

It was about in 1913 to 1915 that the Ahrens-Fox Fire Engine Company introduced the booster system, and it was in 1914 that the company announced its plan to phase out the twin-dome piston pumpers of its Continental series. The first Ahrens-Fox fire pumpers with the large spherical air chamber, the single most distinctive structural feature ever to appear on a fire vehicle, were introduced that same year. These were the "K" models. Charles H. Fox had filed a patent application covering this piston pump in October of 1914, and he received this patent, number 1,189,085, on June 27, 1916. In the drawing for this patent the spherical ball chamber was, unfortunately, not well illustrated (FIGURE 263). Fox configured this pump so that each piston crank was set one-eighth of a turn in advance of the corresponding one on the other shaft of the pair. Thus, by successive piston action in this order, every one-eighth turn of the crank shafts, produced a nearly steady water stream, a distinctly Ahrens-Fox characteristic, and a marked improvement from the typically intermittent-type flow from the other, "standard" piston pumps.

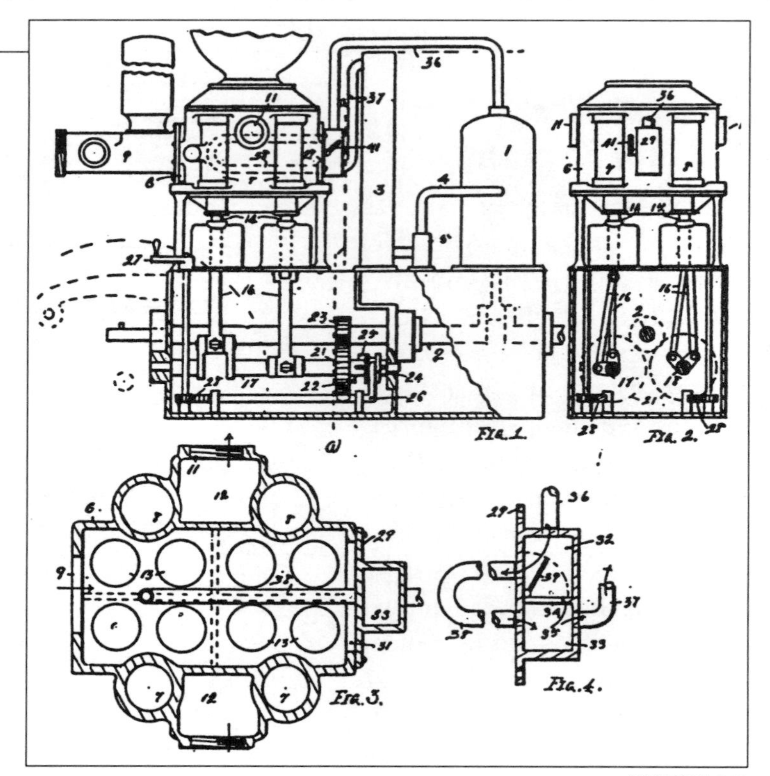

FIGURE 263

Charles Hurst Fox was born in 1860 and died in 1950. He was a son-in-law of Chris Ahrens. In 1910 he became president of the newly-named Ahrens-Fox Fire Engine Company. During his lifetime he was a fireman, an assistant fire chief, an engineer, a business executive, and a distinguished inventor, the holder of 29 patents. Fox can easily be considered as one of the most important men and one of the best minds ever to grace the American fire engine industry. On the occasion of the 150th anniversary of the United States Patent Office, in 1940, he was one of the 15 "Modern Pioneers of Research and Invention" singled out for honors. We will shortly resume the discussion of a number of his creations.

FIGURE 264

But over the years, or so it had been said, some Ahrens-Fox connoisseurs and aesthetes had declaimed and even embellished a legend that these spherical ball air chambers (FIGURES 264 and 265) were fabricated with but a single sheet of metal, hammered and shaped by the hand of a single, elderly craftsman. In his book, "Ahrens-Fox: The Rolls-Royce of Fire Engines," author Ed Hass commented about this, stating that, "This colorful tale had one fatal flaw: it would be virtually impossible to machine a single piece of metal into a perfect hollow globe, much less beat one out by hand. Recent restorations of several pre-1923 Foxes have revealed that underneath the nickel plating that gives the illusion of one piece, early air chambers were really 3 pieces. A long, narrow band of copper was cut along the top and bottom edges in a square gear-tooth pattern, and bent into a ring to form the main ball section (about 2/3 of the total ball); the ring's joint was then welded. Copper caps with matching square-tooth edges were welded to top and bottom to complete the sphere. When plated, the 3 welded seams became invisible, hence the widespread belief that the air chamber was one piece."

FIGURE 265

Some of Charles H. Fox's pump inventions will receive attention presently, but his areas of expertise extended well beyond the pump panel. FIGURES 266, 267, 268, and 269 illustrate Fox patents on methods for suspending ladders and rigid suction hoses upon the side panels of fire engines. Notice in the drawings for the first one of these patents, number 1,890,940, which Fox received in December of 1932, that paired brackets, designated as numbers 22 and 24 held the ladder within their curved ends, 25, by the tension of a spring, 26. The handle-and-spring device for securing suction hose is shown in the drawings for patent number 1,893,166, dated just after the first of the year in 1933. And just six weeks later Fox was awarded patent 1,898,826 for his hose bed invention (FIGURES 270 and 271). The novel features of this invention were the slatted bed bottom and the vertical partitions which fitted between the slats, thus compartmenting the area according to hose size. Patent number 2,037,774 dated April 21, 1936 covered a fire engine design in which the booster water tank formed the forward end of the hose body. In these patent drawings, FIGURES 272, 273, 274 and 275 take note that the wheels were designated as number 1, the motor was number 2, the seat was 3, the pump was 4, the hose bed was 5, and number 6 was the back end of the tank, the structure which was designated by 46. The tank consisted of a fuel section, 47, and a water section, 48, separated by partition, 49. The water tank overflow was 50, with its spout, 51, and cap, 52. The gasoline tank was filled at inlet 53, and the water tank at inlet 54.

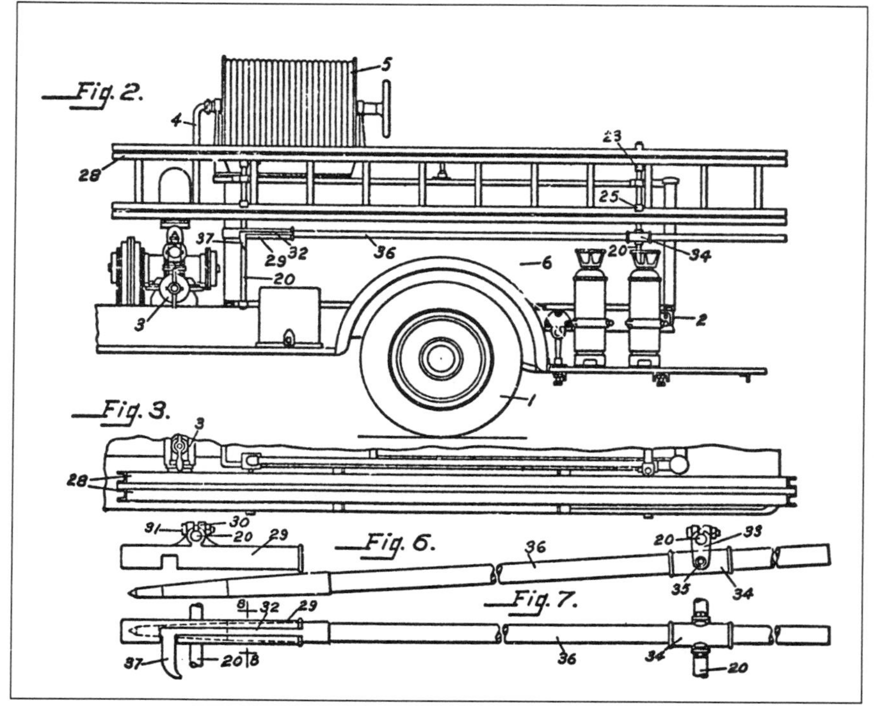

FIGURE 267

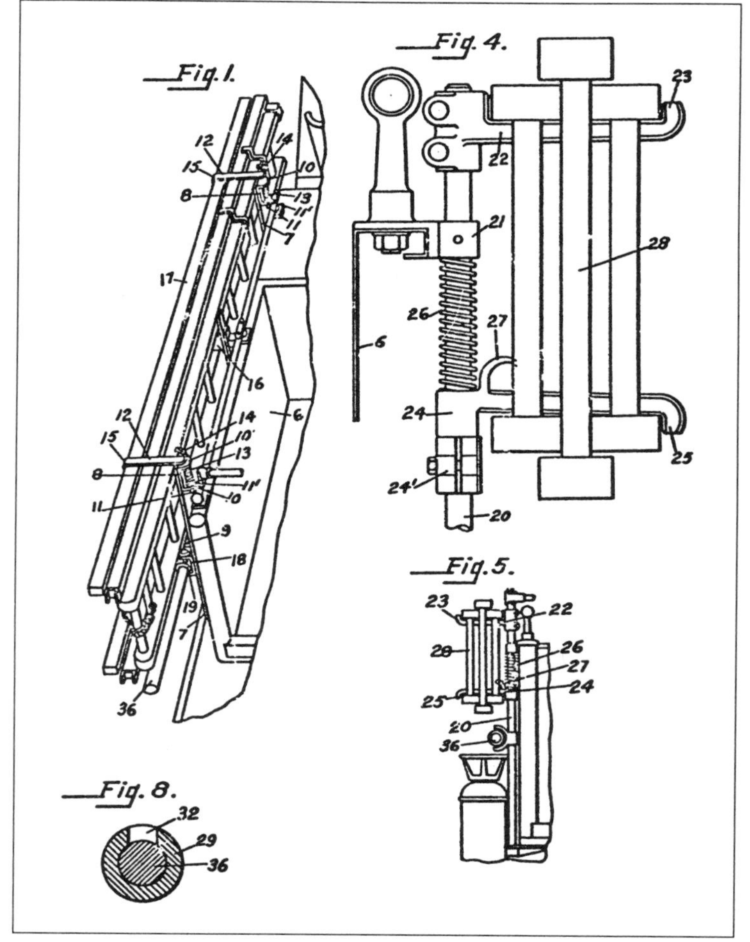

FIGURE 266

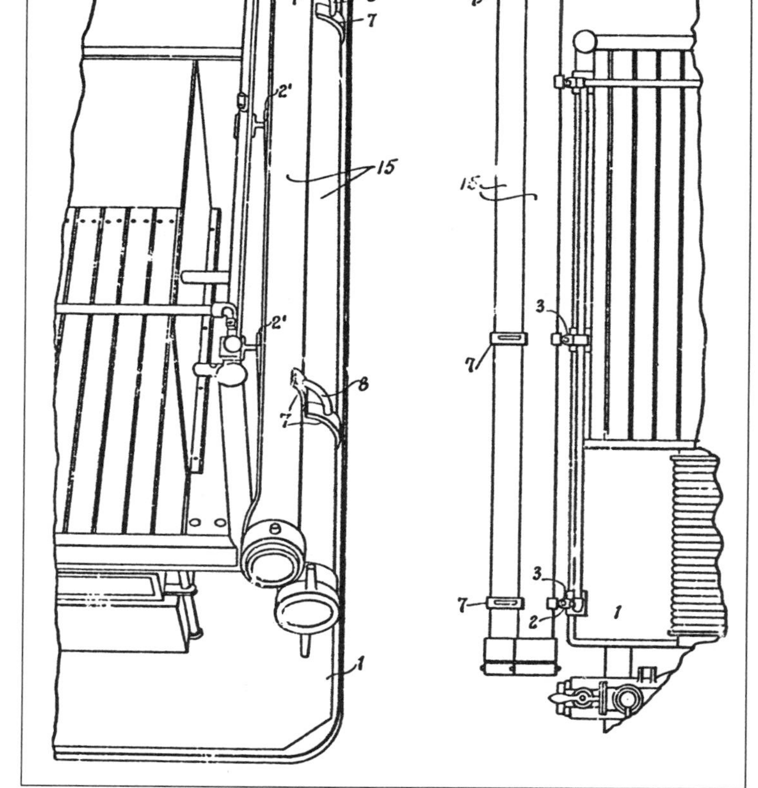

FIGURE 268

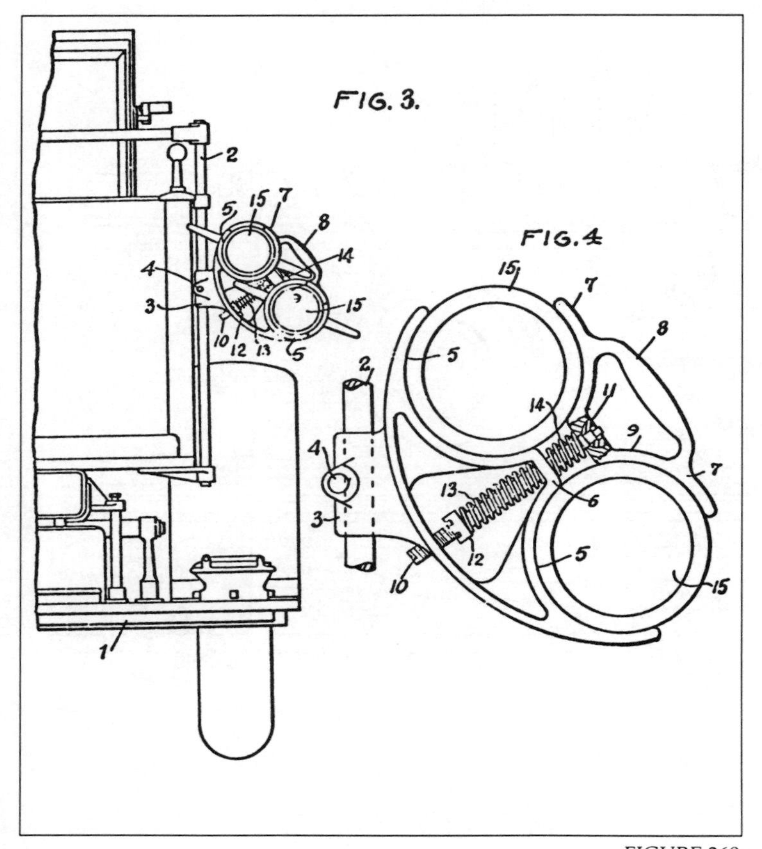

FIGURE 269

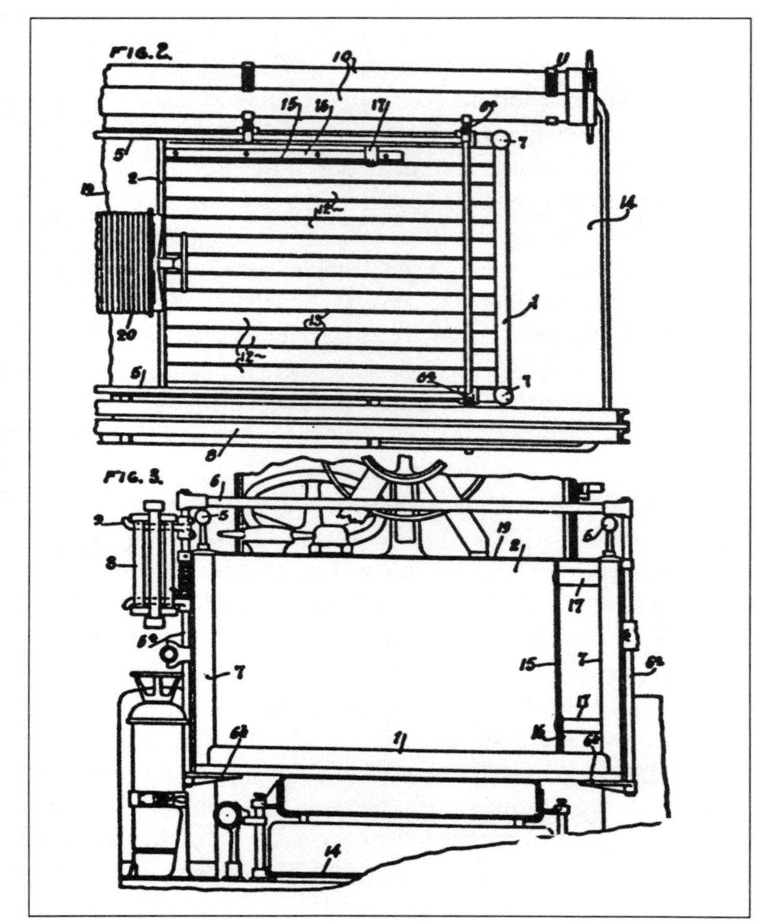

FIGURE 271

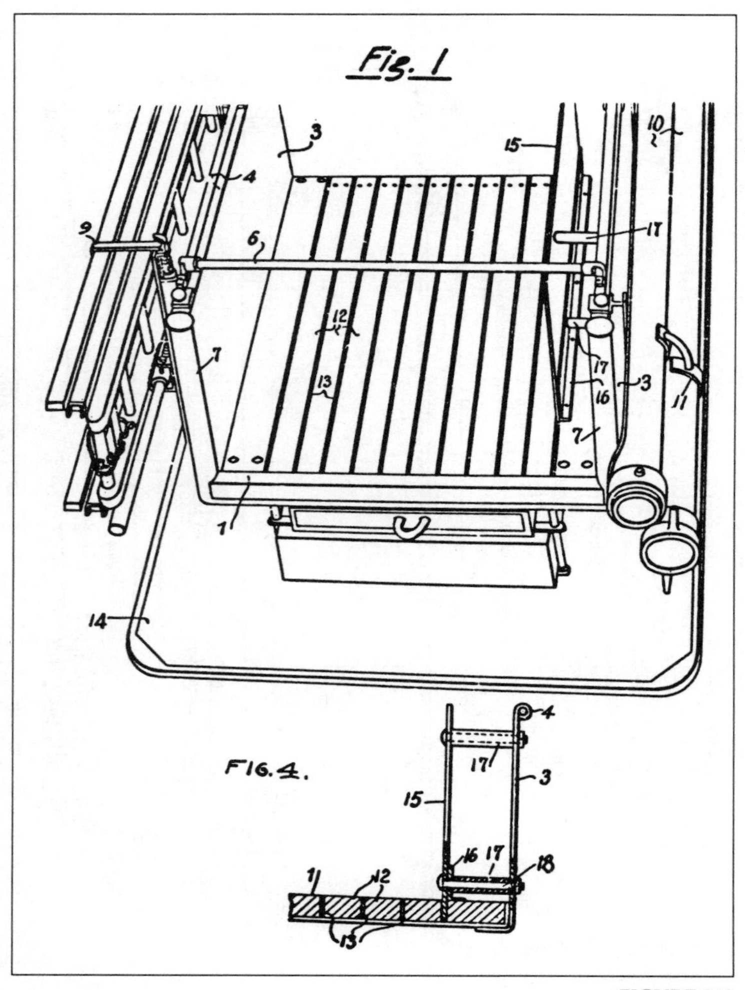

FIGURE 270

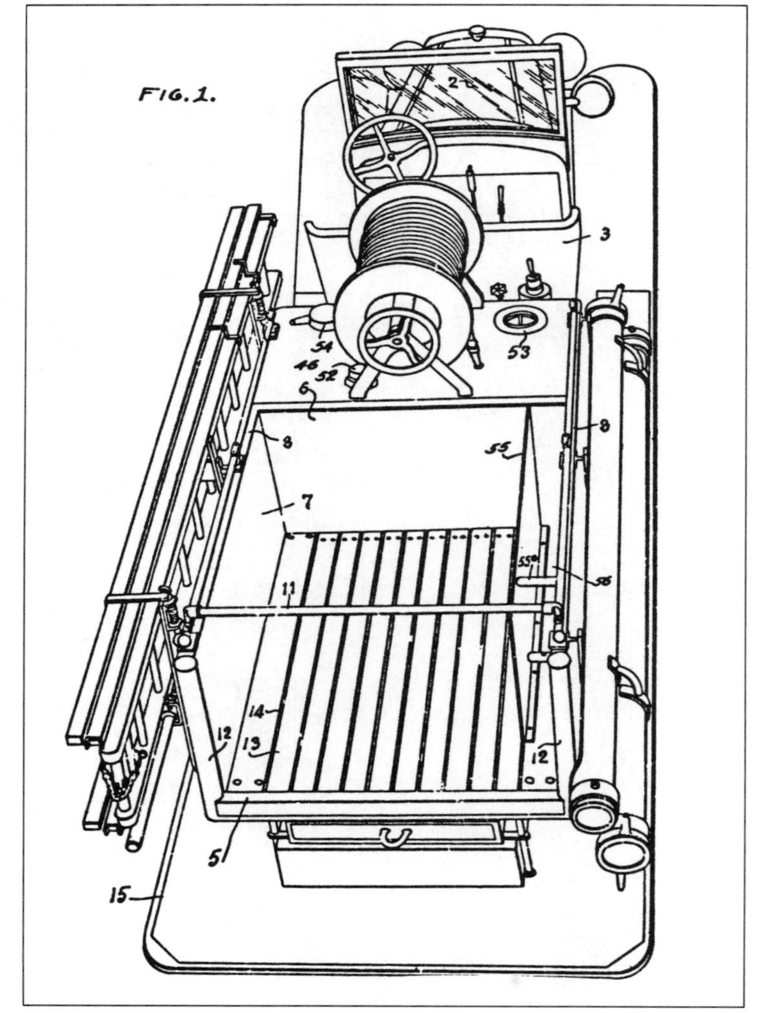

FIGURE 272

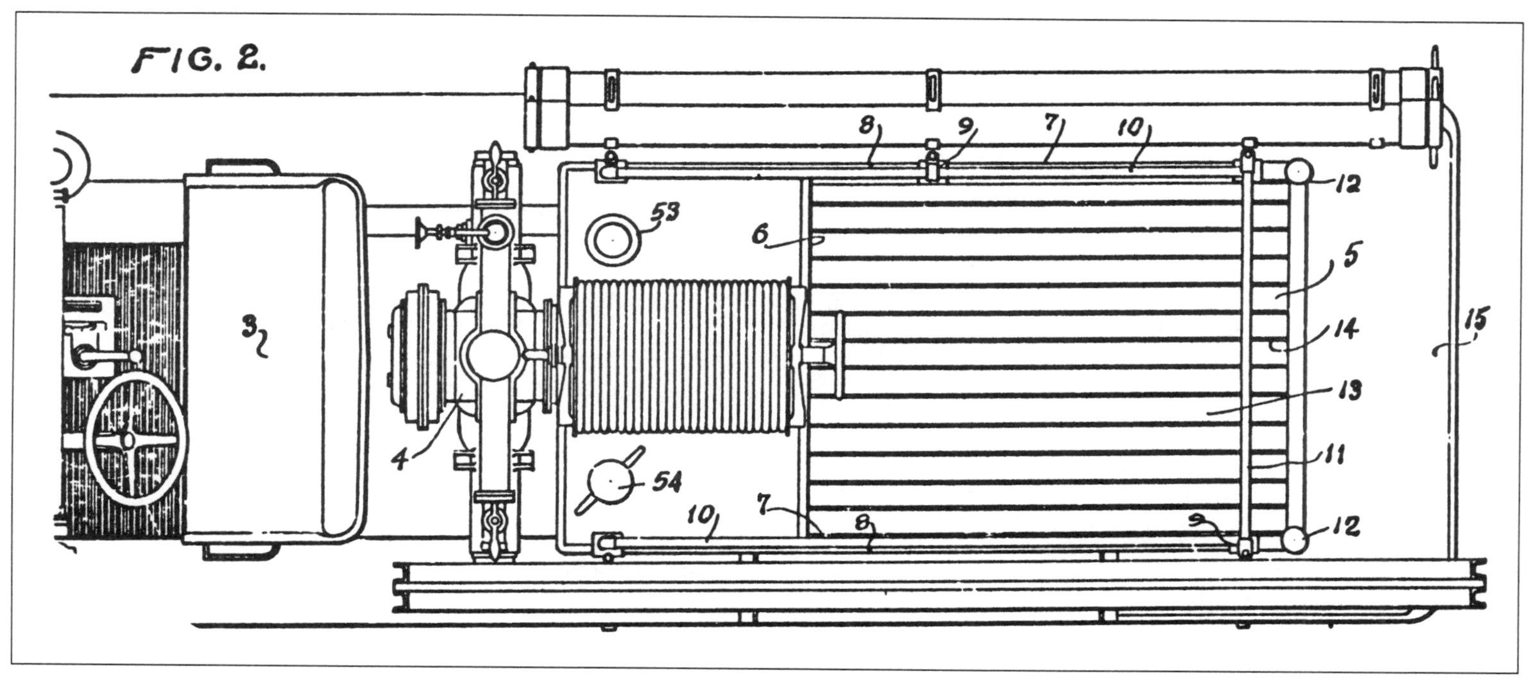

FIGURE 273

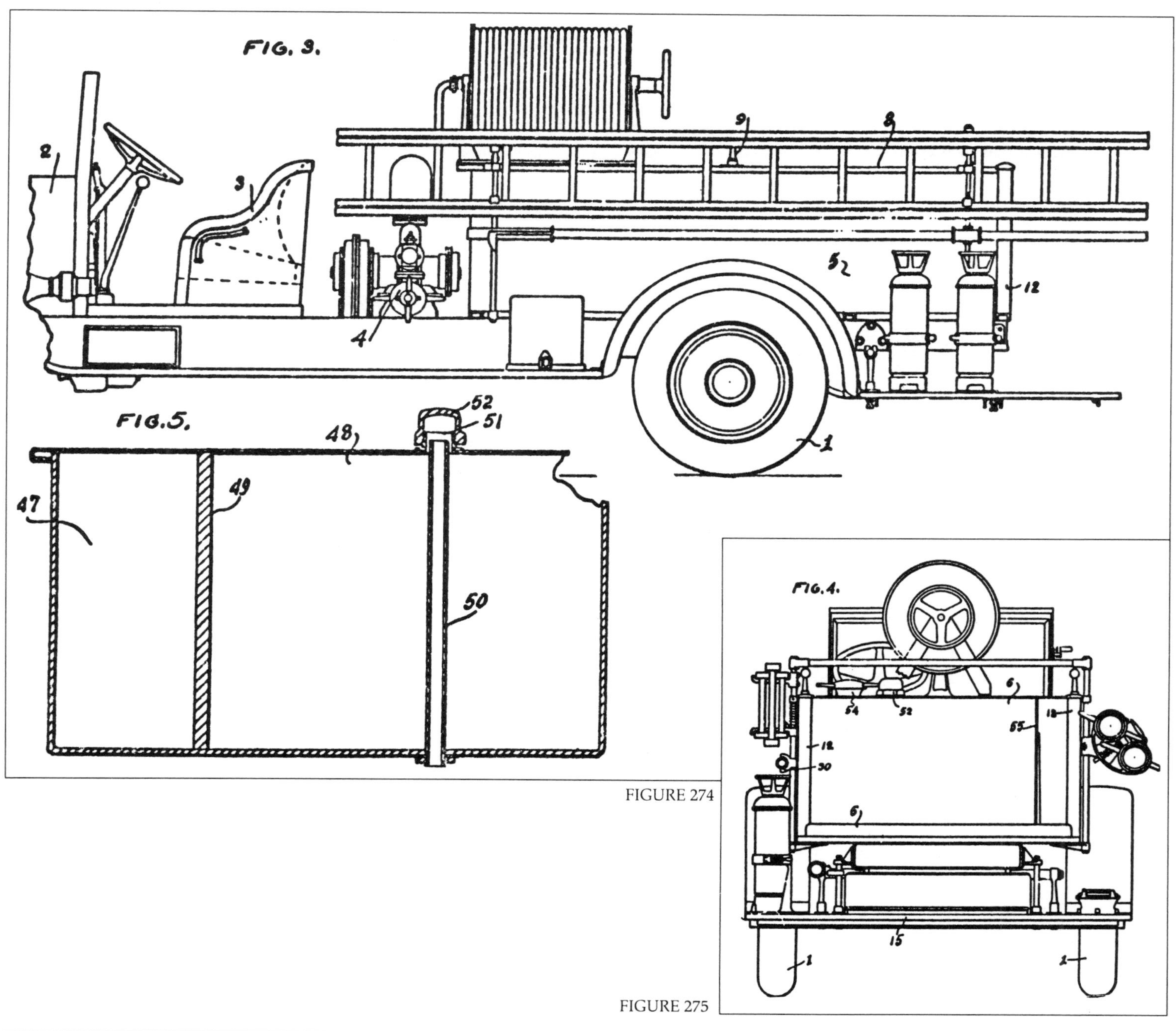

FIGURE 274

FIGURE 275

William S. Darley

W.S. Darley and Company is still today an important fire apparatus manufacturing concern. William S. Darley founded the firm which bears his name in 1908. During the decade of the 1930s he received at least two patents. FIGURE 276 illustrates his invention of a method to convert a pick-up type small truck into a fire ladder carrier (patent 1,967,503 dated July 24, 1934). The drawing shows his design of a simple wrap-around rear platform with securing flanges. This was to have been bolted onto the truck along with posts and braces for mounting ladders. In his patent number 2,005,990 of June 25, 1935 Darley showed how a booster tank ought to be fabricated, particularly citing that perforated baffles should be used to control the motion of water sloshing about within the enclosure. This patent drawing is here shown as FIGURE 277.

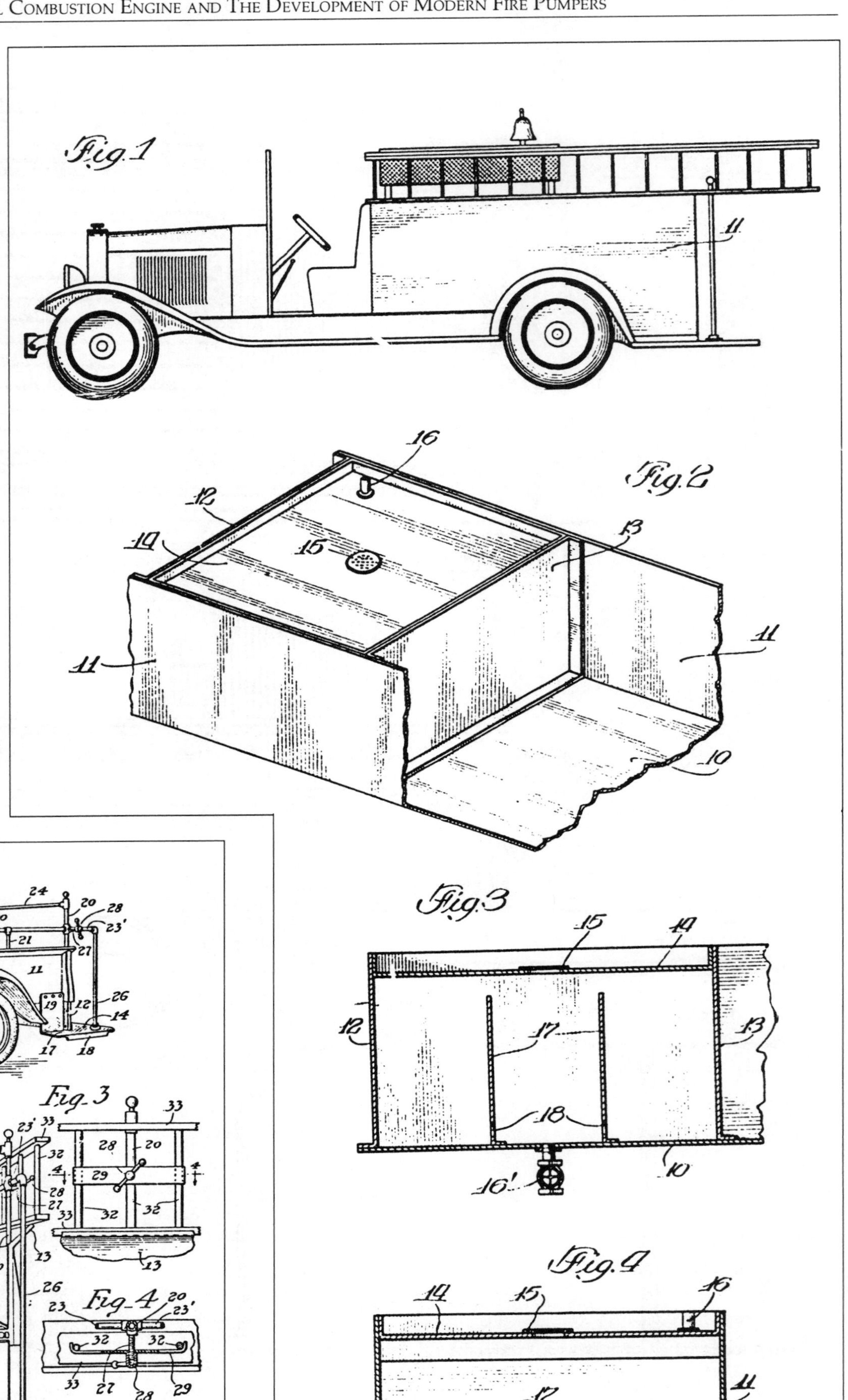

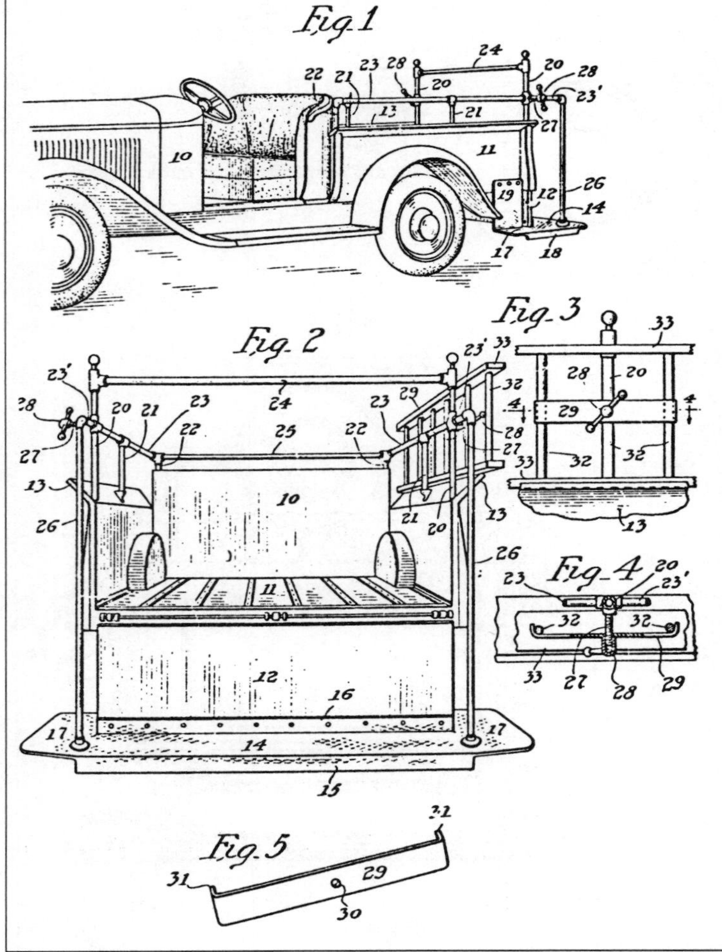

FIGURE 276

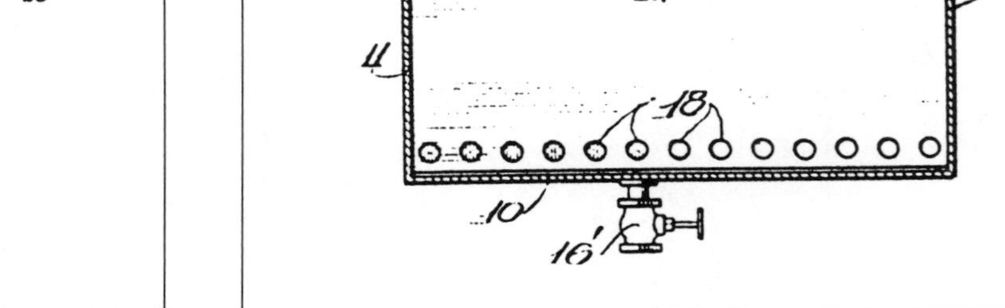

FIGURE 277

American LaFrance

Just as for Ahrens-Fox and Darley the 1930s were also years of great design significance for the American LaFrance Fire Engine Company, known at the time as the American LaFrance Foamite Corporation. ALF introduced their 400 Series fire engines in 1935 and delivered the first Series 500 fire units in 1938. The 400 Series vehicles featured a rather long hood, because the pump was positioned just behind the motor and ahead of the driver's seat. The contour which resulted was aesthetically most pleasing (FIGURE 278). These fire engines were equipped with rotary gear type pumps, usually of 1,000 gallons per minute capacity. By 1937 some closed-cab Series 400 rigs had been produced. And that was the year in which ALF delivered the first two of four "Duplex" pumpers ordered by the Los Angeles Fire Department. These unique vehicles carried two V-12 engines, one aft-mounted to power a 1,000 gallon per minute pump and the other in the conventional forward position where it was the power source for the other pump and for the vehicle itself (FIGURE 279). In the closing years of the decade, just before World War II, ALF was filling orders for their Series 500 fire pumpers (FIGURE 280). All of us who are old enough to remember these vehicles, whether the open cab or the closed cab models, can still see, within our minds and memories, their striking beauty. The pump, now centrifugal, remained mounted under the hood, but the suction intake connections were placed farther to the rear, behind the cab. Overhead was a center-mounted, enclosed ladder rack, faired into the cab roof. ALF was careful in protecting the designs of these fire engines by means of patents. April 18, 1939 was the effective date for patent number 2,154,642 the 400 Series fire rig, Carlisle F. Smith inventor (FIGURE 281). And patent number 2,209,666 dated July 30, 1940, FIGURE 282, was the 500 series pumper, Allan G. Sheppard inventor. Although at first glance one could question that this illustration actually was a 500 Series fire engine, such doubts ought to be eased by looking at the design patent drawings, filed concurrently (Des. 113,183, filed September 30, 1938, and dated January 31, 1939, John J. Grybos inventor) of FIGURES 283 and 284. Mr. Grybos was a designer in the employ of ALF from 1927 until 1948.

During the war years of the 1940s, as might be expected, the engineering talent of the country was directed to tools of death and destruction, and not to refinements of domestic fire engine design. Many of the fire hardware patents registered for that period, 1942-1946 were for such developments as airfield fire apparatus and shipboard fire suppression.

FIGURE 278

FIGURE 279

FIGURE 280

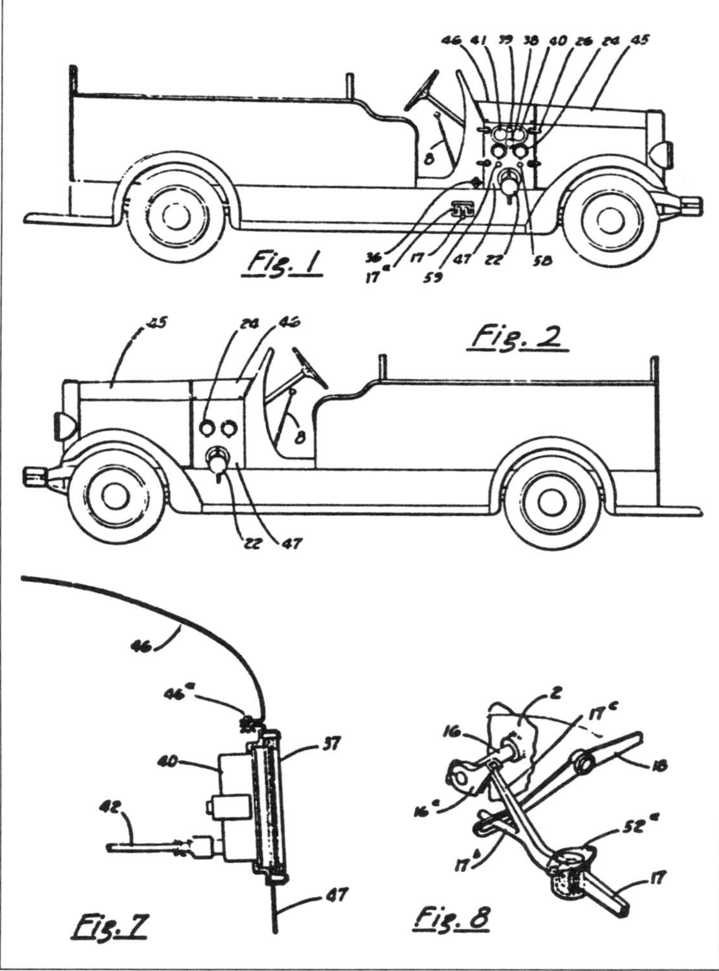

FIGURE 281

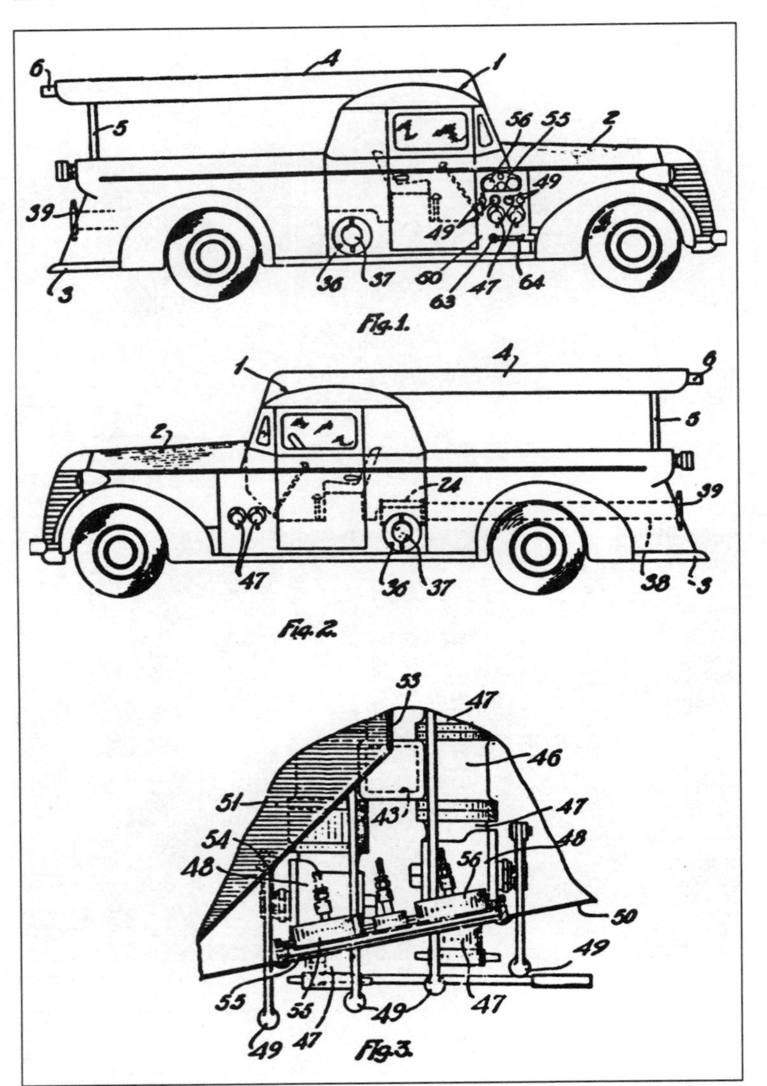

FIGURE 282

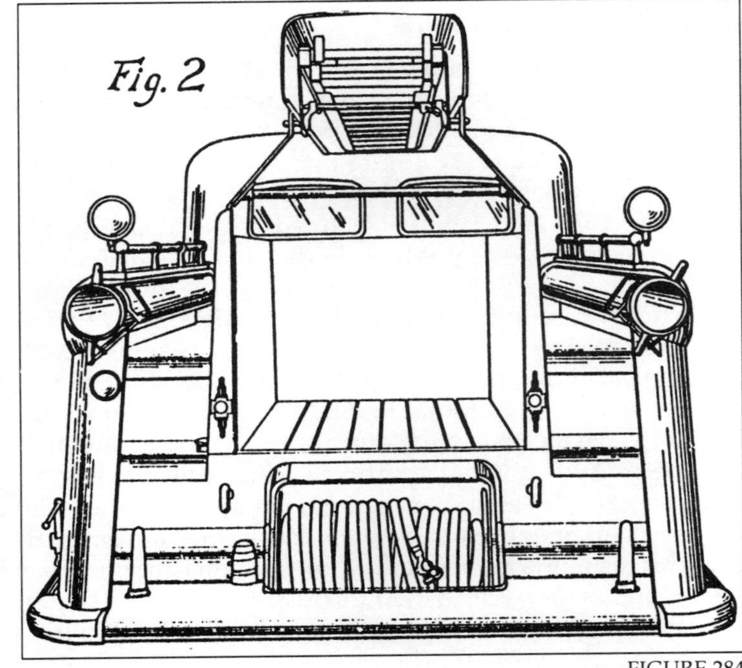

FIGURE 284

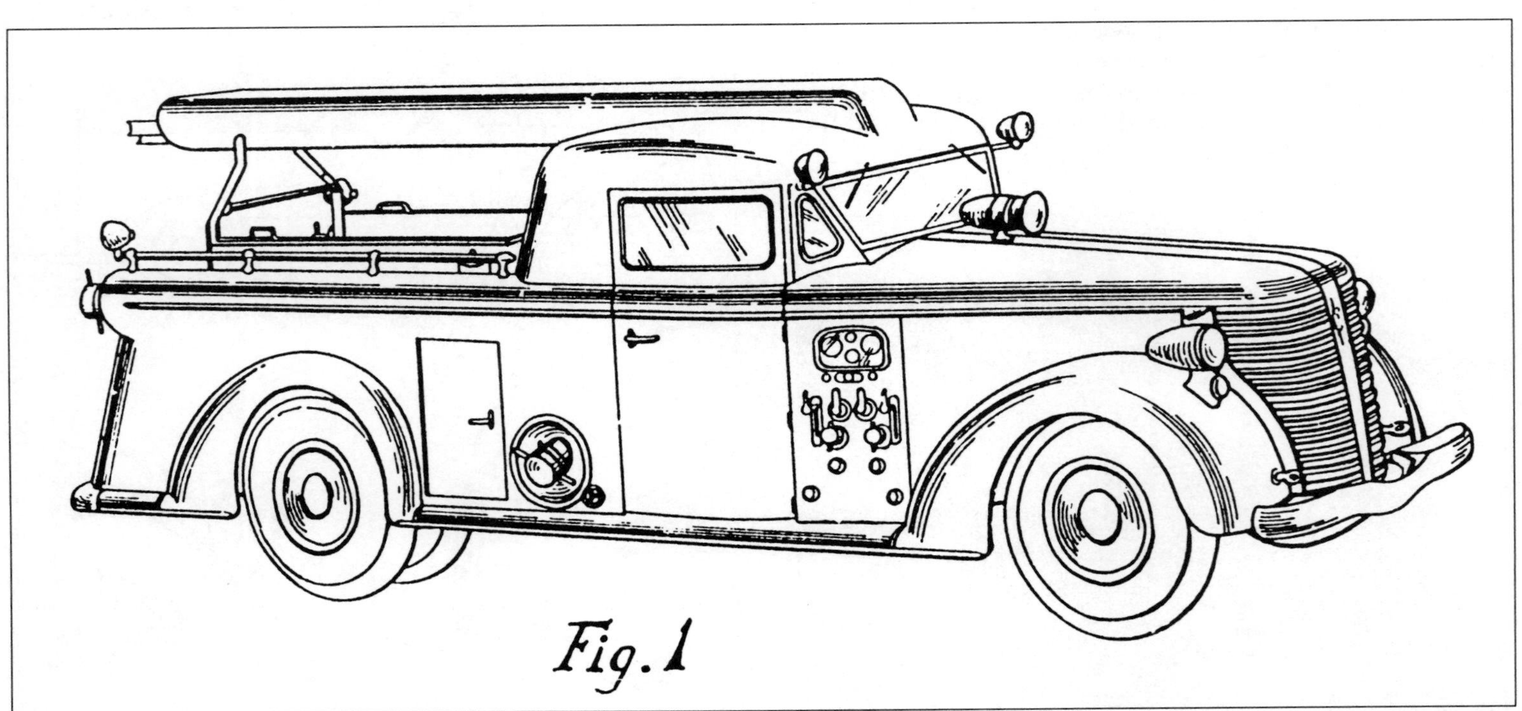

FIGURE 283

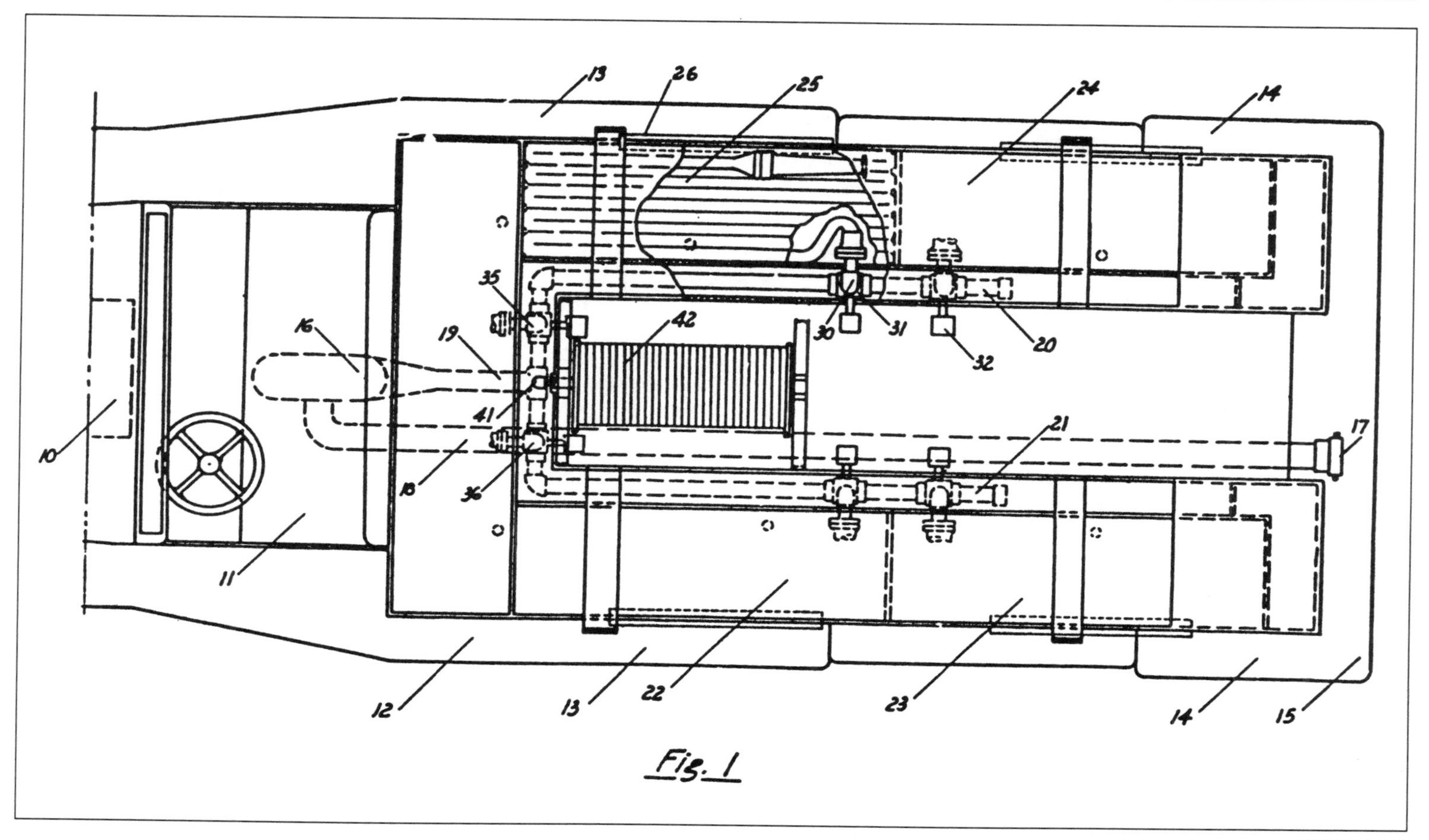

FIGURE 285

Post World War II

In 1950 Hugh N. McNair of Short Hills, New Jersey, received patent number 2,505,055, seen in FIGURE 285. This fire engine had a more-or-less conventional hose bed, but it was also fitted with side compartments containing pre-connected attack lines. There were two of these on each side of the apparatus and two more behind the driver's seat. McNair stipulated the side-mounted lines to be 2 1/2 inches diameter and the rest to be 1 1/2 inches diameter, and all to be at least 200 feet long. Inventor McNair felt that with this hose line configuration the fire attack could be started quite rapidly, faster than was possible with a standard pumper hose load. He also designed electrical switches into each hose connection so that the pump operator could quickly determine when to start flowing water. Referring to the patent drawing notice the following components: the motor, 10; the cab, 11; the body, 12; the running boards, 13 and 14; the pump, 16; rear coupling for a hydrant, 17; pipe from rear steamer fitting to pump, 18; compartments for 2 1/2 inch attack hose lines, 22, 23, 24, and 25.

A fire engine intended to function as a pumper-tanker, and equipped with a large-volume, shallow, U-shaped tank, was the subject of patent 2,533,722. Inventor Joseph H. DeFrees who received his patent during December, 1950, was a resident of Warren, a Pennsylvania town located on the Allegheny River about 20 miles to the south of Jamestown, New York. As his patent drawings, FIGURES 286 and 287, show, the tank was designed to be concave from behind-to-forward at the back end, and concave from above-downward along the top. With such an eccentric tank configuration it would be a matter of complex mathematics to calculate its actual volume capacity. But if we were to assign a set of dimensions a volume estimate can be made. Let us say that the tank length was 15 feet peripherally and 13 feet centrally, and let us divide the tank into six segments each two feet in width and two feet in height. Such a tank would be able to hold about 2,600 gallons of water.

FIGURE 288 illustrates David Wehner's steam jet fire apparatus for use in sub-zero temperature environments. Wehner was from Allison Park, in the metropolitan area of Pittsburgh, Pennsylvania, and he received this patent, number 3,283,828, in November, 1966. This vehicle was supposed to have been plugged in to the house electric supply while garaged, to keep the boiler water hot. Upon leaving the fire station and severing the electric supply an on-board oil burner was to have been fired up. Among the other features of this hypothetical apparatus were a steam generator, water storage tanks, a power plant, and a steam turret.

James P. Triplett from Upperville, a small community in Virginia just to the east of the Appalachian foothills along U.S. Route 50 was the recipient of patent, 3,493,053 in February, 1970. This invention, illustrated in FIGURES 289 and 290 was an interesting system of truck-carried modular pumps, not entirely unlike certain contemporary European fire vehicle systems. This one was obviously intended for use in remote rural areas where a pump-relay method would be required to put water onto a fire.

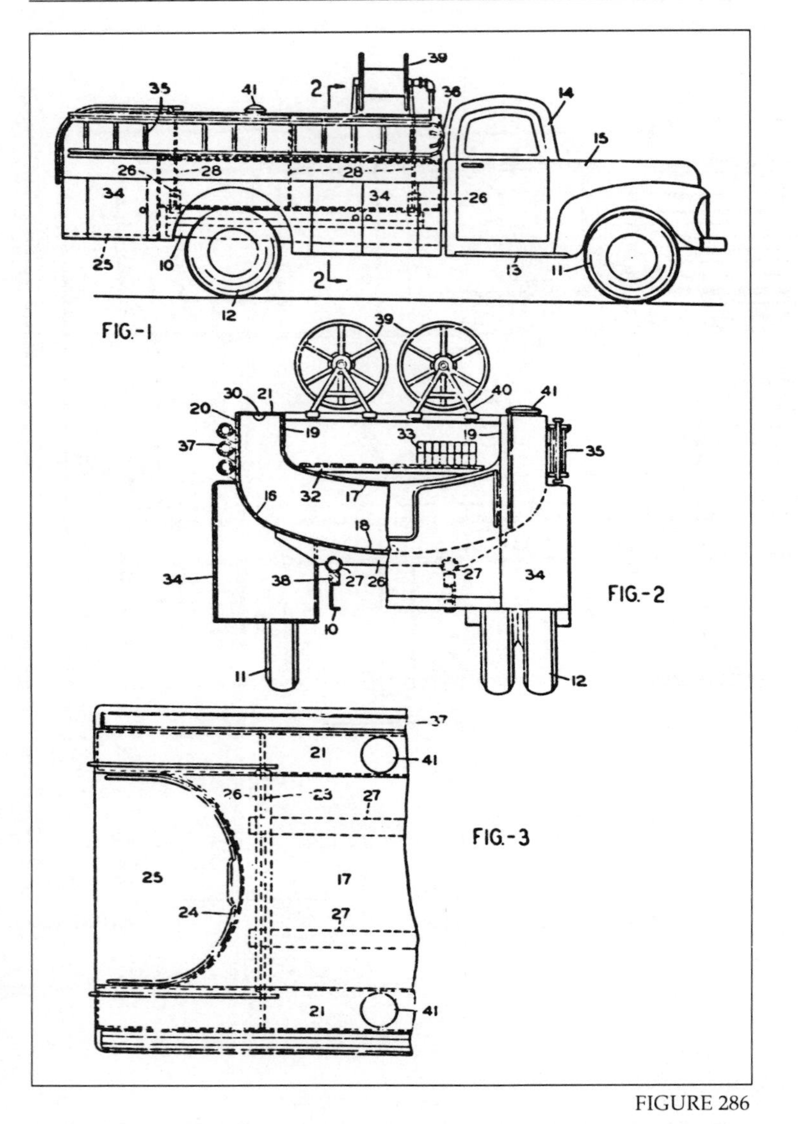

FIGURE 286

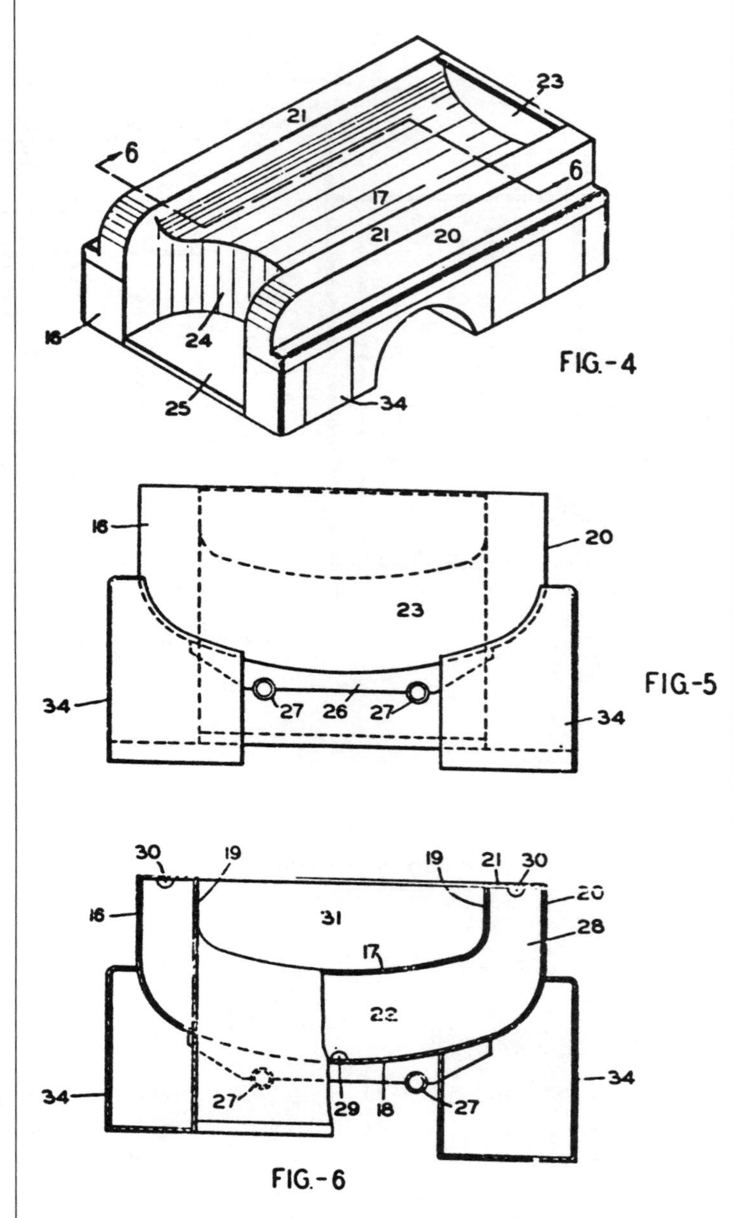

FIGURE 287

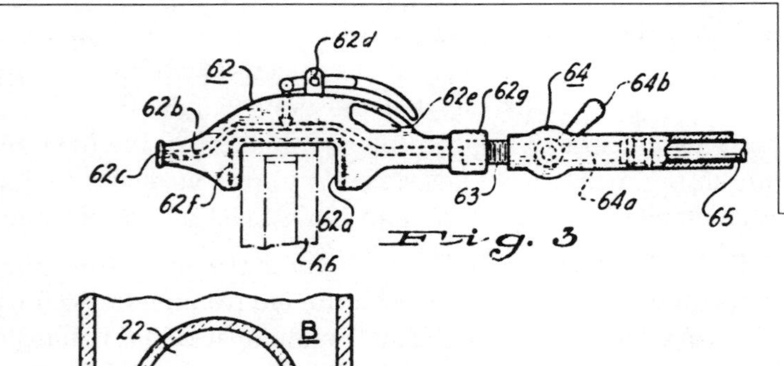

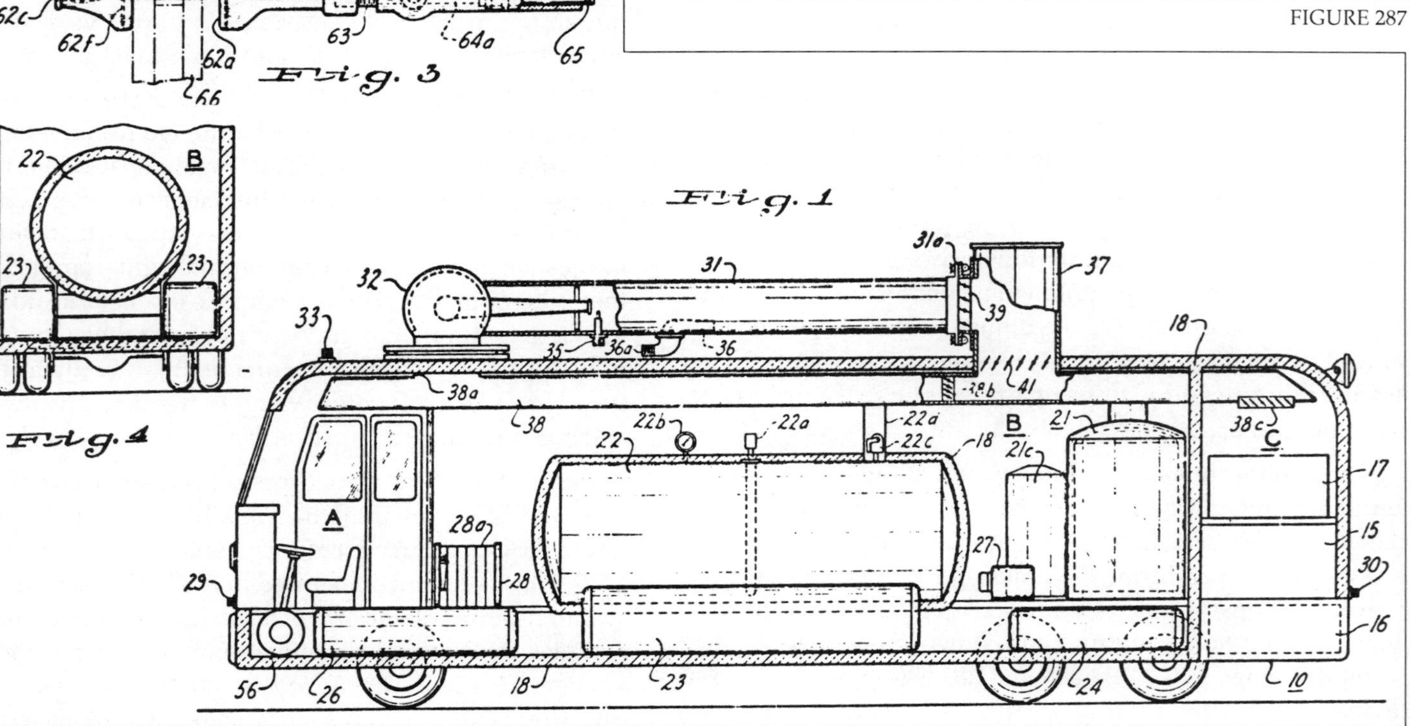

FIGURE 288

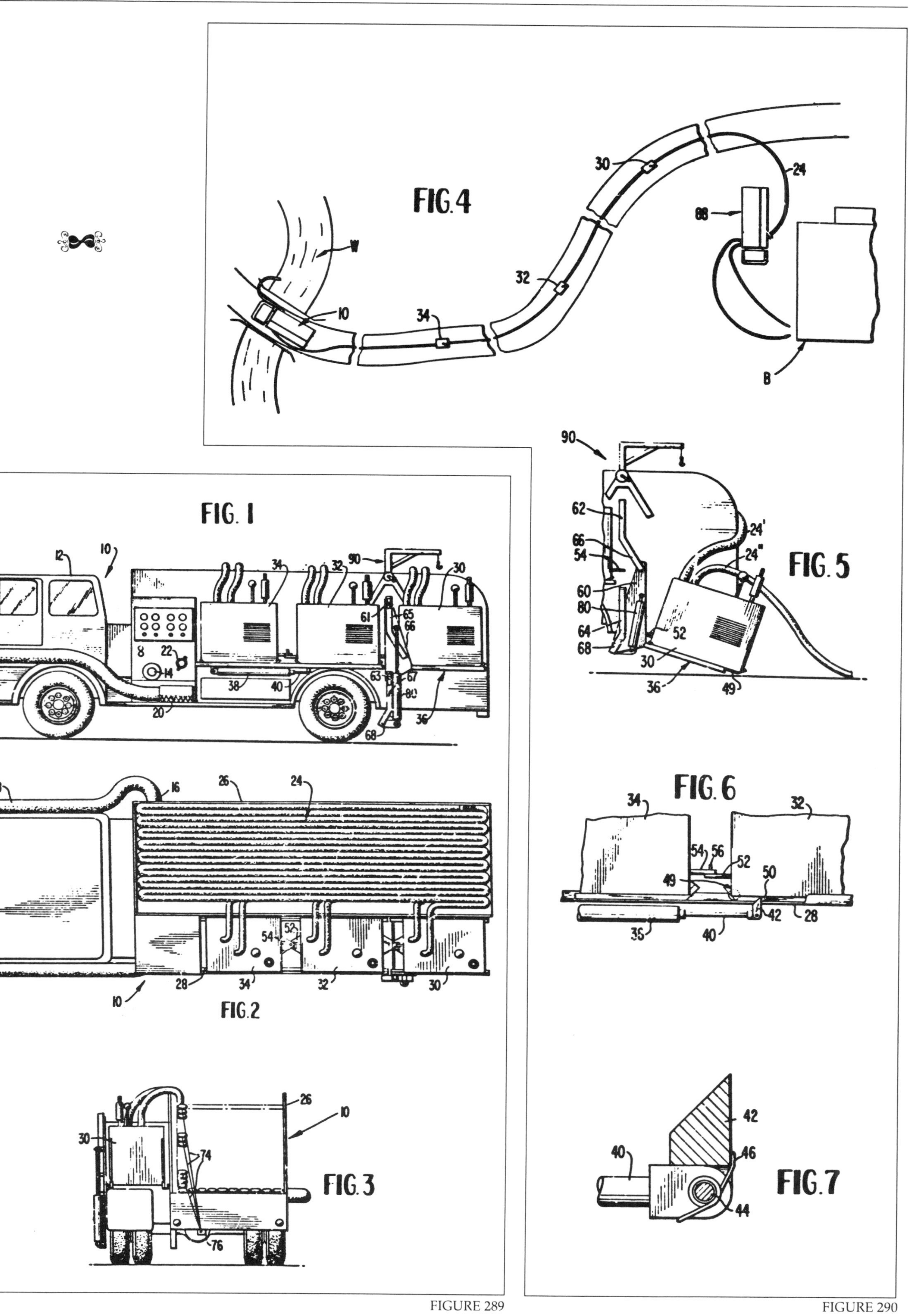

FIGURE 289

FIGURE 290

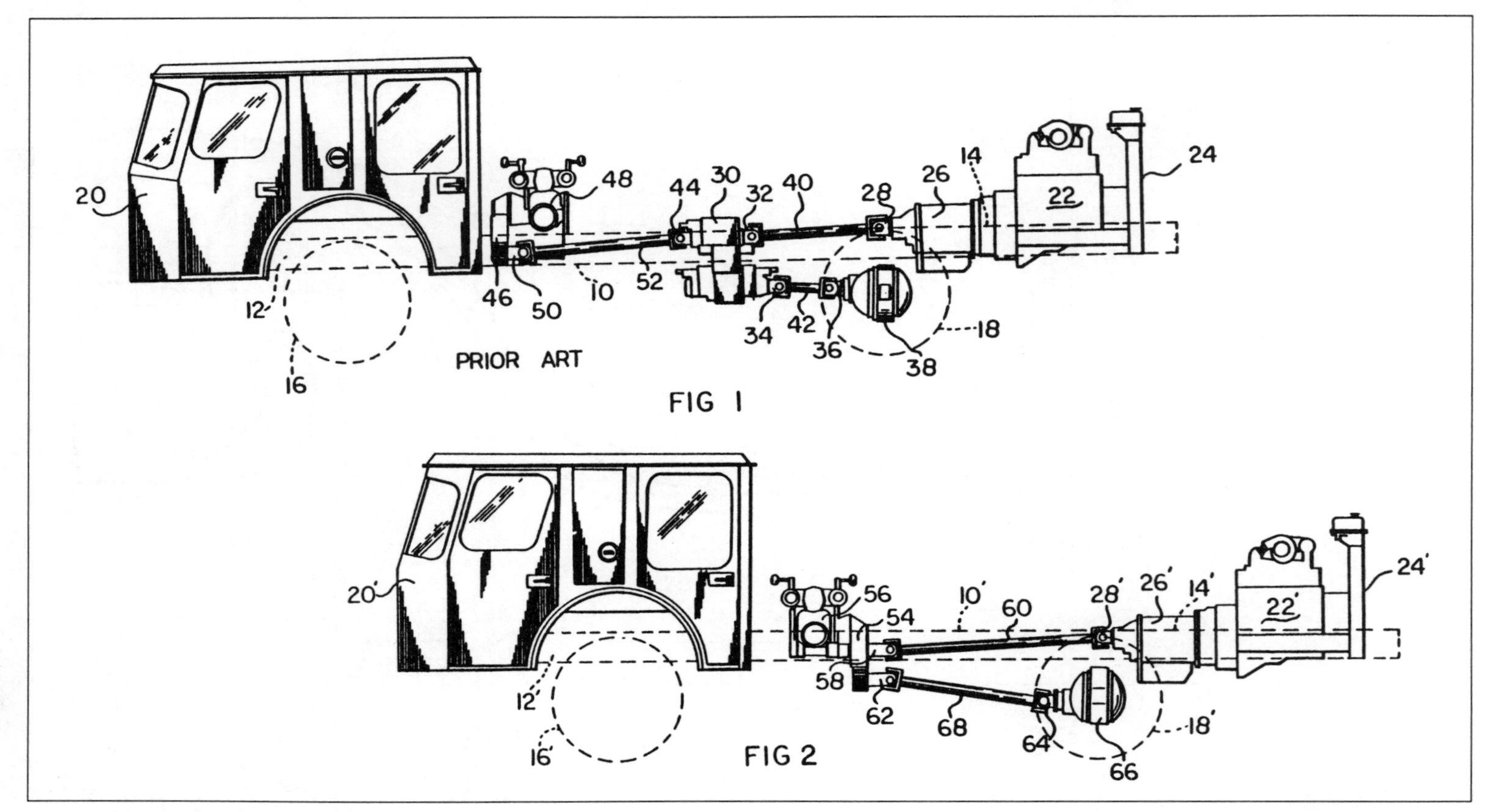

FIGURE 291

The Emergency One "Hush"

All of us who ride fire engines while sitting behind, or above, or beside that big Cummins or Detroit diesel engine knows about and probably have complained about the intensity of the heat and the noise. And while the rest of us were suffering silently and bitching from time to time Ronald L. Ewers and William McCombs, were devising the solution. These gentlemen from Ocala, Florida received patent number 4,811,804 dated March 14, 1989 for their rear engine fire truck. By locating the radiator, the engine, and the transmission behind the rear wheels both the noise and the often-oppressive temperatures within the cab were lowered. The key component of this invention, FIGURES 291 and 292, is the unique fire-pump gear box which is part of the drive train, being a directly-connected intermediate between the transmission and the differential for the rear wheels. Notice the relationship of these key components (first figure): 24 is the rear-most radiator; 22 is the engine; 26 is the transmission; 28 is the transmission output; 60 is the power train-universal joint assembly; 58 is the gear box input; 54 is the fire pump gear box; 56 is the pump; 62 is the gear box output shaft; 68 is the universal joint assembly; 64 is the input shaft of the differential; 66 is the differential.

The drive train of this fire vehicle can be traced out by following these numbered components, with reference to the illustration. The second drawing shows the fire pump and the fire pump gear box. Again note the input and output shafts, numbered 58 and 62. The fire pump drive shaft is 110. The pump is a Hale Model QSMG. Other key parts are the drive gear, 82, the shifter rod, 88, the output shaft gear, 98, and the idler gear, 100.

This invention is the well-respected "Hush" fire vehicle (FIGURES 293 and 294). Ads for these rigs claim that the ambient cab noise is 78 DBA when the speed is 55 miles per hour. This is a remarkable achievement, being about the same level of noise as that associated with late-model passenger cars.

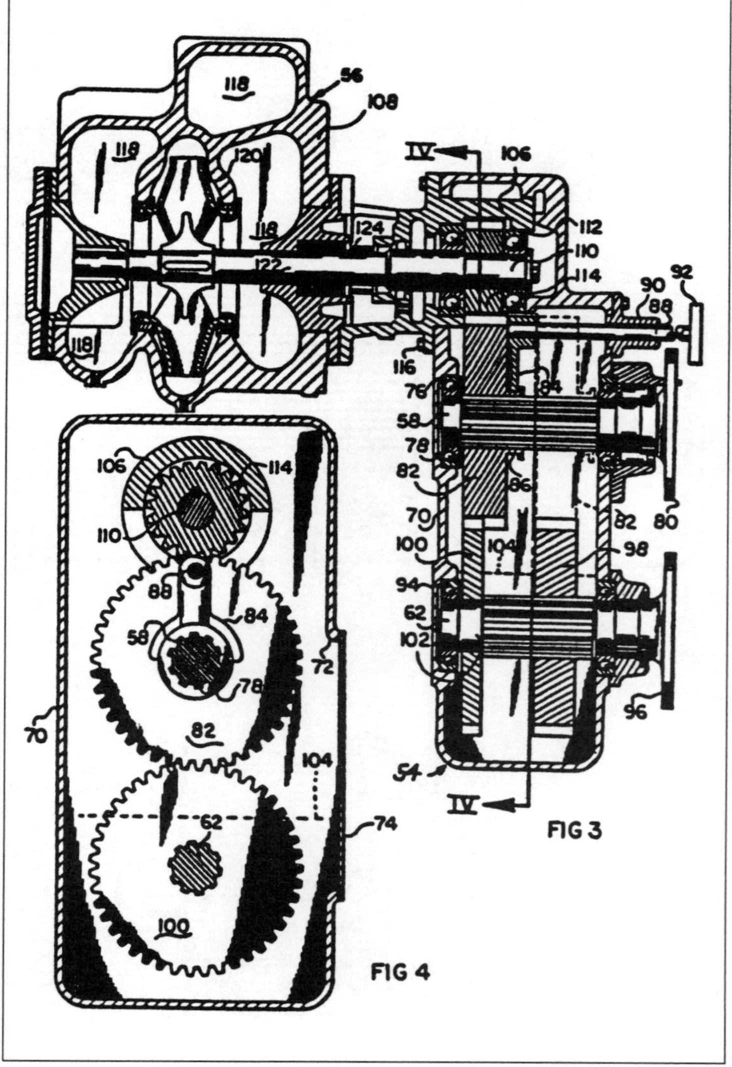

FIGURE 292

FIGURE 293

FIGURE 294

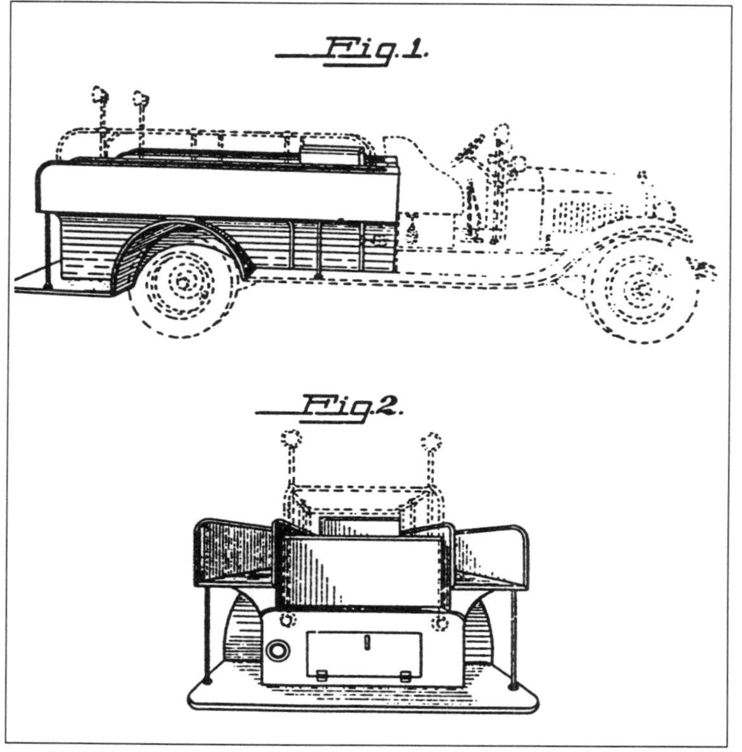

FIGURE 295

Design Patents

These illustrations of handsome Emergency One fire engines prompt a few words about "design patents." Ornamental designs have for many years been patentable, and each design patent is assigned a number, prefixed by the letters "Des." Many design patents are esthetically or artistically quite tasteful. Several motor fire apparatus design patents have been selected as examples. FIGURE 295 is Albert de la Mare's patent number Des. 82,661, dated December 2, 1930. Obviously a much more streamlined vehicular design, FIGURE 296, was patented on October 25, 1938 by John F. Frazier of Detroit, Michigan. This was Des. 111,874, and it was assigned to Gar Wood Industries of that city. The next illustration is FIGURE 297, which shows James W. Fitch's patent Des. 116,841 dated September 26, 1939. Fitch assigned this patent to his employer, Kenworth Motor Truck Corporation of Seattle, Washington. This highly-streamlined vehicle was the first-ever cab-over-engine fire pumper. Kenworth built a total of six fire vehicles, five pumpers and a quad, to this design between 1938 and 1941. The Los Angeles Fire Department purchased the quad and three pumpers, and the fire department of Beverly Hills, California owned one pumper. The very first delivery of one of these fire engines, which by the way carried a "United Fire Engine Corporation" logo, was to the Metro-Goldwyn-Mayer Studio fire department in 1938. The "United Fire Engine Corporation" title was simply advertising hype used for publicity by a Kenworth distributor.

Jack J. Skelley's design patent for a fire engine, Des. 161,060, dated November 28, 1950 is seen in FIGURE 298. And FIGURE 299 is Don L. Johnson's ornamental design for a fire engine cab, Des. 237,370. This 1975 design patent was assigned to FMC Corporation of San Jose, California. One of the founders of Food Machinery Corporation was John Bean. His high-pressure spray pump was patented in 1904, but over 20 years elapsed before it was fitted to a fire engine.

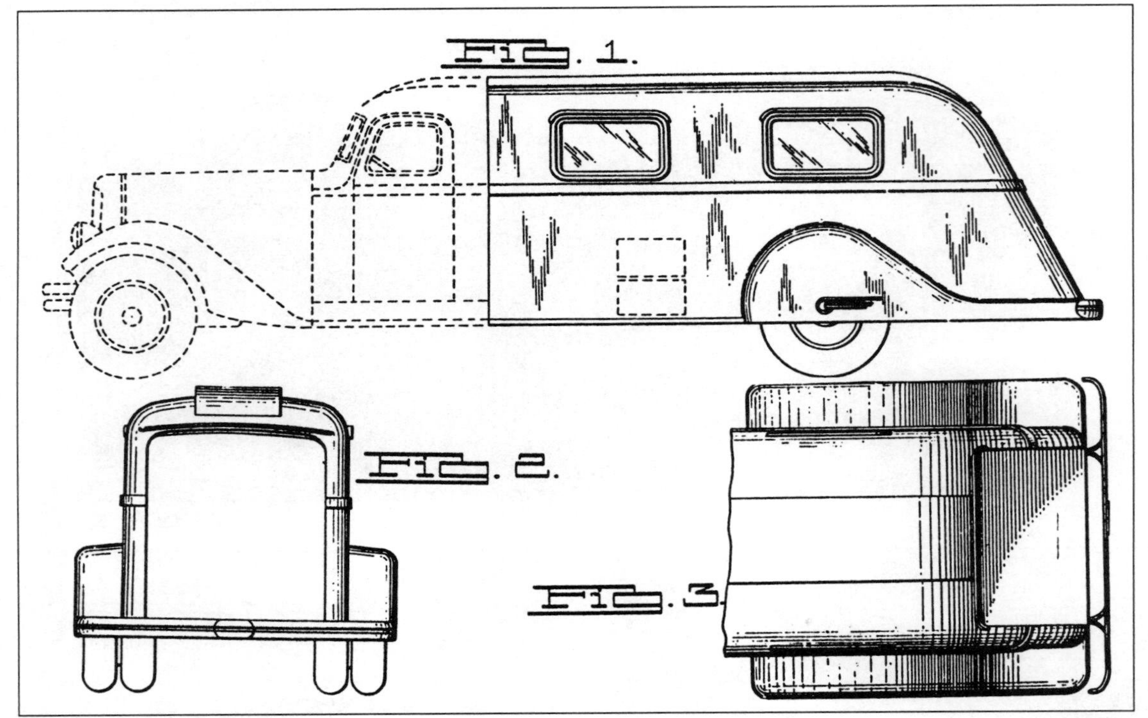

FIGURE 296

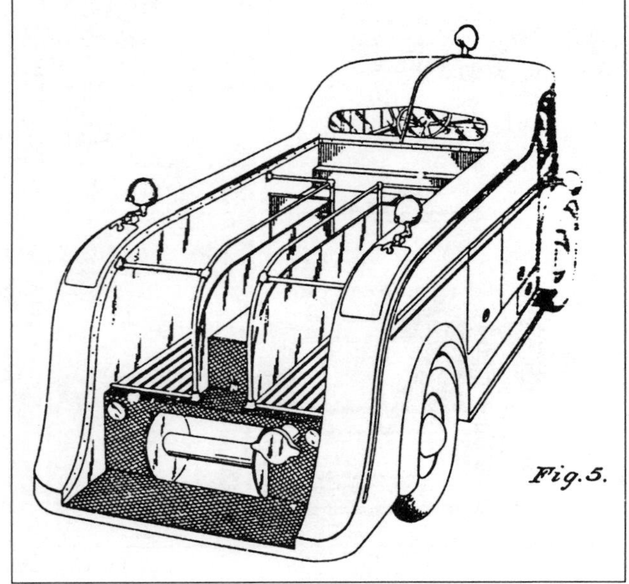

FIGURE 297

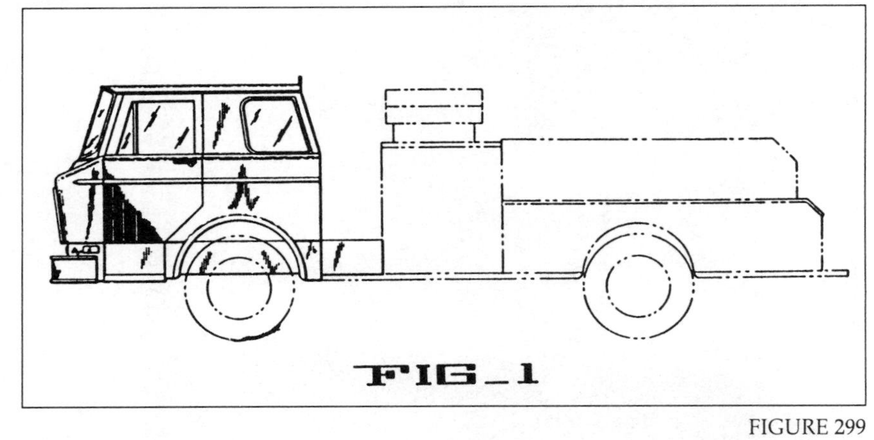

FIGURE 299

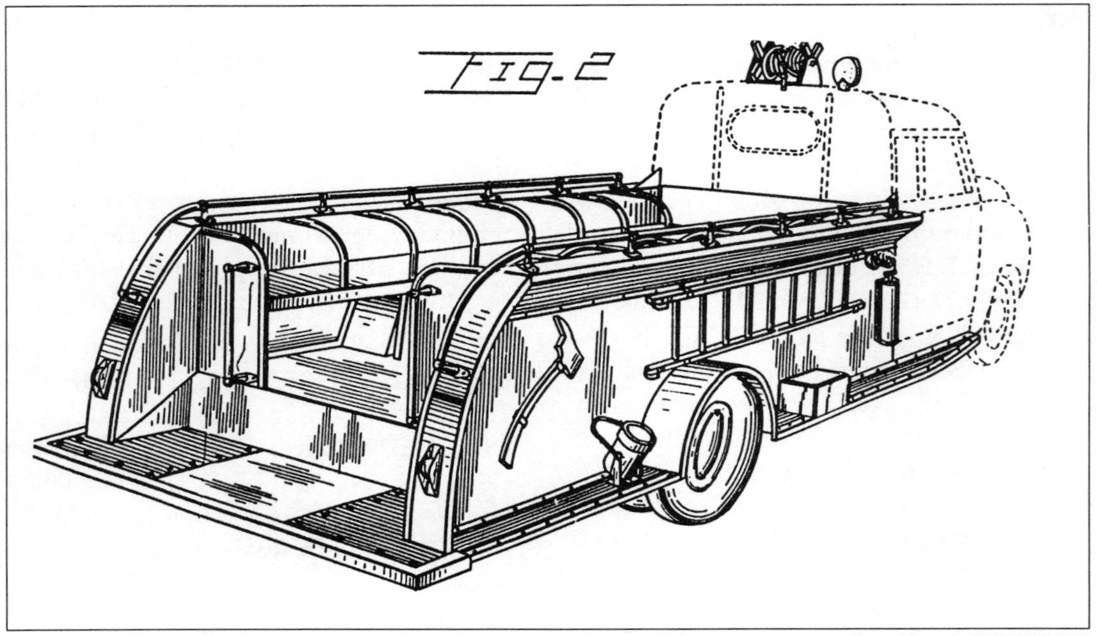

FIGURE 298

Some Accessories and Refinements

Over the years since the internal combustion engine was first assigned responsibility for powering fire engines and their pumps a melange of apparatus accessories, niceties, and refinements had been appearing. Five inventions out of this very broad expanse have been picked for the discussion which follows.

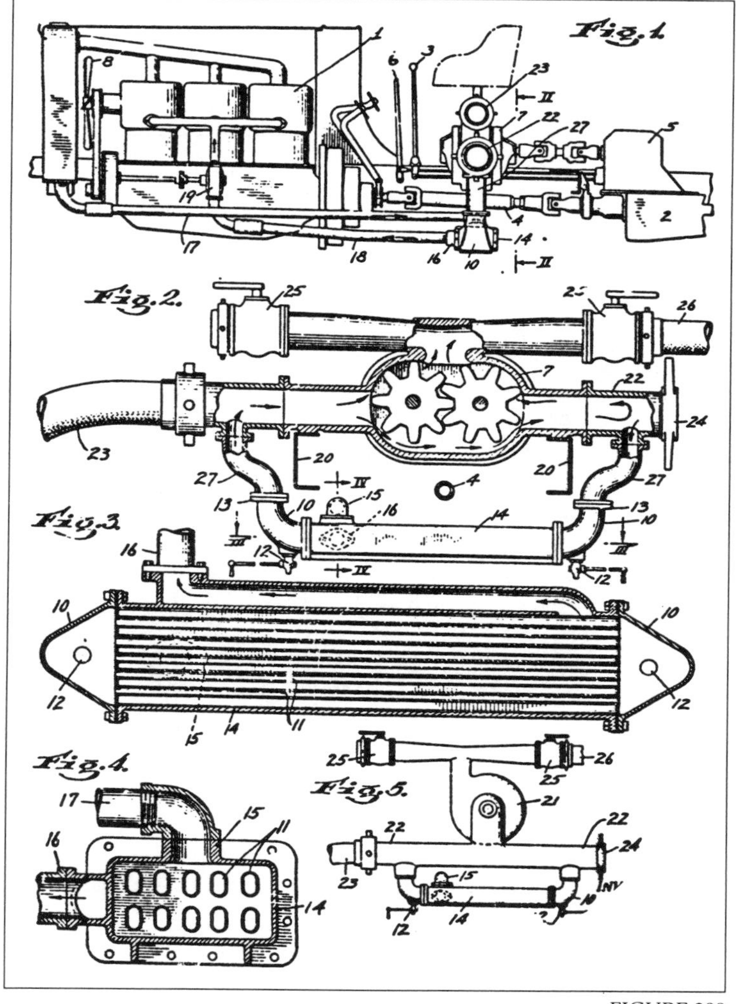

FIGURE 300

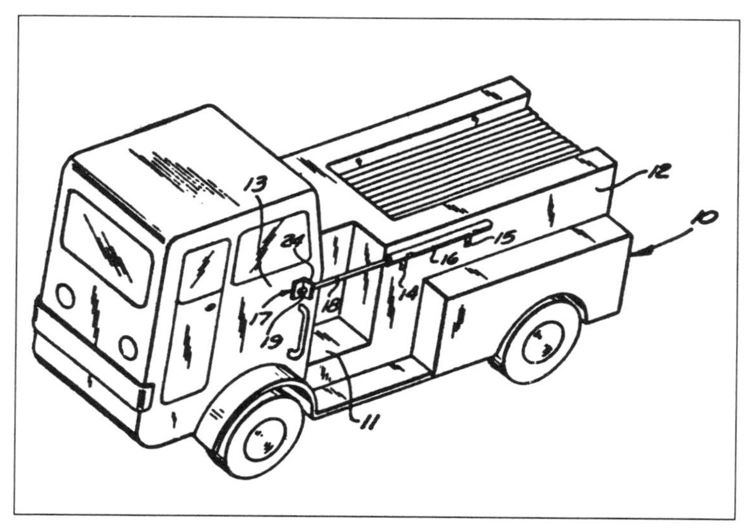

FIGURE 301

Hubert Walker, an American LaFrance and Foamite Corporation employee devised a new way for cooling the engine of an operating fire pumper. His patent was number 1,860,638 dated May 31, 1932 and it is illustrated in FIGURE 300. Walker had designed a heat exchanger in which cool hydrant water was circulated within pipes which were set closely contiguous with piping containing overheated engine water from the radiator. This method was similar, at least in principle, to the way that milk used to be pasteurized.

Leaping ahead to the more recent past there was patent 3,841,409 granted to Syosset, Long Island, New York resident William Rynsky in October, 1974. This was a simple jump-seat safety bar, and it is shown in FIGURE 301.

Inventor Wilber O. Johnson, Jr., of Houston, Texas received a patent in 1975 for his concept of adapting a Pitot tube to a fire engine discharge line, thus making it possible to read water flow directly in units of gallons per minute from a gauge. This was patent 3,876,009 and it is illustrated in FIGURE 302. Fire engine pump panels have classically been equipped with gauges which display line pressure as measured in pounds per square inch. From such numbers the water volume flow discharge rate had either to be guessed, to be estimated, or to be read from a prepared table or graphic display. This conventional method of estimating water flow with pressure gauge readings has been central to fire department hydraulics ever since the publication in 1895 of John R. Freeman's flow tables. The experienced pump operator bypasses this, of course, and sets the pressure which he expects to be adequate for the situation: hose diameter, hose length, nozzle size, and nozzle height above the pump, as well as other factors. Johnson, in contrast, started with the Bernoulli equation and worked out a way to avoid all such estimates.

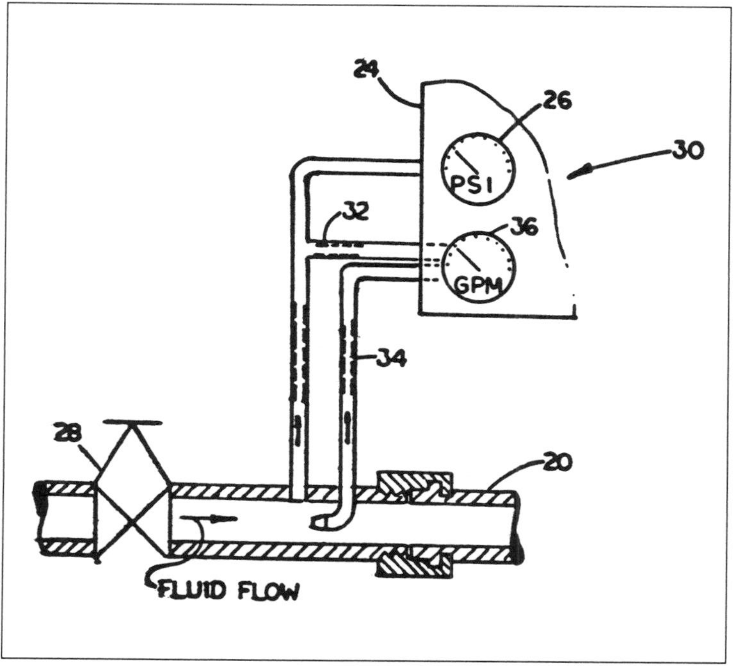

FIGURE 302

A system which was no less than a fully-integrated, computer-controlled, hydrant-to-pump-to-nozzle disposition was set out in 1980 by John McLoughlin of Smithtown, on New York's Long Island. Previously he had invented a nozzle pressure control unit which linked hosemen, the pump operator, and the incident commander by means of hard-wire communications. This new invention was patent number 4,189,005 dated February 19, 1980 (FIGURE 303). In the words of the inventor, "This system is designed to make it simpler for a fire department to hit the fire with water as fast as possible, while giving the nozzleman control over his own pressure and flow. The total evolution requires the pump operator to just engage his pump. The system then starts up, checks all of the important truck parameters, and readies itself for signals from the nozzleman to operate their valves. Upon engaging the pump, engine speed is increased until a pre-set pump pressure is obtained. If the first water flow is from an on-board tank the pump pressure will reach any pre-set point between 70 and 300 pounds per square inch within ten seconds. As the system continues to operate itself it frees the pump operator to pull a supply line from the hose bed and attach it to the pumper's intake. The hydrant valve when attached is opened remotely by radio control. The computer monitors hydrant supply and water tank supply."

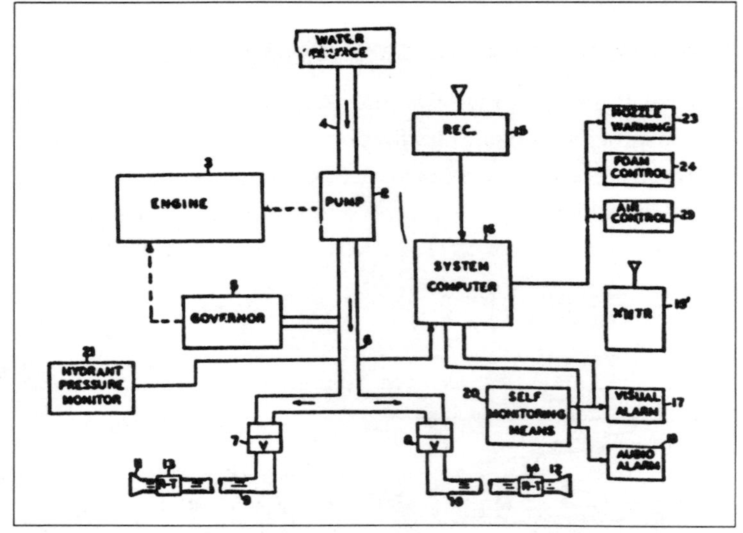

FIGURE 303

FIGURE 304 shows John W. Campbell's patent number 4,512,412 dated April 23, 1985. Campbell, who at the time was a resident of Tonopah, a rather small town situated where U.S. Routes 6 and 95 intersect in southwestern Nevada, designed this panel to give ready access to components behind it in case that repair or maintenance was needed. This panel door initially slides straight outward from the vehicular body, and then turns through 90 degrees using piano-type hinges.

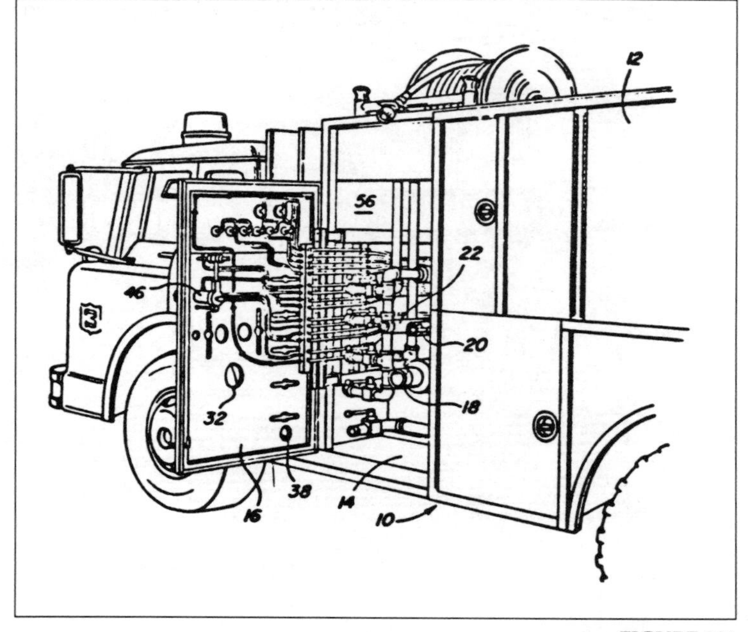

FIGURE 304

Fire Pumps

The subject areas of fire pumps and, especially, of fire department hydraulics are at once fascinating or interminable, subject to mathematical definition or coarsely approximate, and designed by men of genius for operation by semi-cretins. The selection and specifications for fire pumps for the volunteer fire services is generally a function of committee deliberations, although in fact it is the most assertive informed member who finally, as a rule, gets his way. And most often the axiom governing the performance of the selected pump is "bigger is always better." But in the fire service as in life generally there are situations in which bigger is not always better. This was Chief Edwin J. Spahn's thesis when he wrote about "The Art of Pump Selection " in the December, 1982 issue of *Fire Chief* Magazine. Spahn didn't launch an attack upon the large apparatus enthusiast, but he did make a reasoned plea for specifying and designing pumps and apparatus for the environment in which they must function. Such cost-effectiveness concepts can perhaps now be found in the curricula of fire schools, and the International Fire Service Training Association has been the publishing agency for veritable shelves of books devoted to pumps and hydraulics. And such authors as Lawrence W. Erven have been writing on these topics for many years. With such well-written and well-illustrated references so abundantly available any effort at repetition here would be redundant as well as presumptuous.

Thus far, within this chapter in particular, there has been some discussion of pumps, especially piston pumps and rotary pumps. But as we all are aware it is the centrifugal pump which has achieved universal applicability to fire engines, and historically this was predicted many years ago in a National Fire Protection Association Committee on Automobile Fire Apparatus report, dated May 13, 1913: "The multi-stage centrifugal pump has been only lately developed for fire engine service in this country, although used for several years in Europe; its ability to withstand hard usage, its simplicity and ease of repair and its property of automatically regulating discharge and pressure are features tending to make it a most desirable type for fire service."

Sooner or later practically all makers of fire engines were producing units equipped with whatever kind of pump the buyer desired. Of course some purchasers might well have been persuaded by a canny salesman that one kind of pump, his kind, was better suited to his situation than any other type. By and large one would assume, for example, that Ahrens-Fox people would tout their piston pump, while representatives of American LaFrance, Mack, and Stutz would talk only of their own particular rotary pumps. Seagrave introduced the centrifugal pump to the Americans in 1911. Ahrens-Fox began equipping its booster systems with centrifugal pumps in 1913, but of all of the fire engine builders only Seagrave continued to carry the centrifugal pump torch to the exclusion of all others.

True, American LaFrance began to build a few units with centrifugal pumps in 1916 or 1917. But that this was no more than a sideline can be appreciated from the following numbers: from 1910 through 1926 ALF built 4,006 rotary pumpers, 41 piston pumpers, and 5 centrifugal pumpers. By the early 1930s even Mack was turning out centrifugal pumpers, and by the middle of that decade just about the only non-centrifugal units being produced were from Ahrens-Fox, a company just barely holding on and by its teeth. And 1939, incidentally, was the year which saw another vision of the fire engines of the future: the Hartford, Indiana–based New Stutz Fire Engine Company introduced the first ever diesel-powered rig. Mack soon followed, and, of course, surpassed this small and fading company.

In sampling the developments and advances in pump technology of the last 75 years it is striking to see that nearly all pump patents were assigned by inventors to one or another of the major manufacturers. Obviously it was the "in house" engineer-inventors who were filing these patent applications on behalf of their employers, not surprising when you think about the extensive supporting services, such as a machine shop and model maker, which they were prepared to furnish.

All of those many and necessary equipment items which can be characterized as fire pump accessories will be omitted from the discussion which follows, since these are described in any specific pump manual. These are such devices as priming valves, priming pump lubrication systems, automatic relief valves, pressure-operated governors, pressure-reducing valves, auxiliary cooling systems, bypass proportioners, or details of priming devices such as clutch driven or electric motor driven positive displacement pumps, and engine intake manifold or engine exhaust gas systems.

The Bathrick Rotary Pump

An invention dated July 31, 1923, patent number 1,463,569 was assigned by patentee Charles E. Bathrick to the South Bend Motors Company. This was a rotary pump designed for mounting above and at a right angle to the vehicular transmission mechanism. As can be seen in the illustration, FIGURE 305, this configuration allowed for both suction and discharge ports to be located on each side of the vehicle.

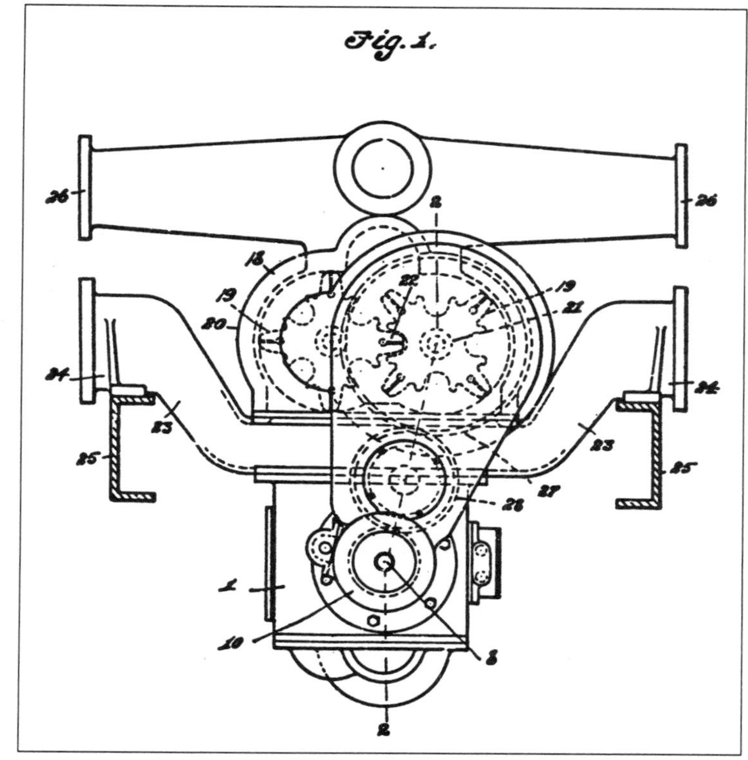

FIGURE 305

Charles Hurst Fox—Again

In addition to his mastery of the piston pump, already described, Charles H. Fox also invented a rotary pump and four variations of centrifugal pumps. Mr. Fox's last pump invention and, in fact, his next-to-last ever patent, as he retired the following year at age 80, was number 2,112,651 dated March 29, 1938. This was his design for a pair of centrifugal pumps which could be worked singly or together and, in the latter case, could be operated in parallel or in series. FIGURE 306 is representative of the patent drawings.

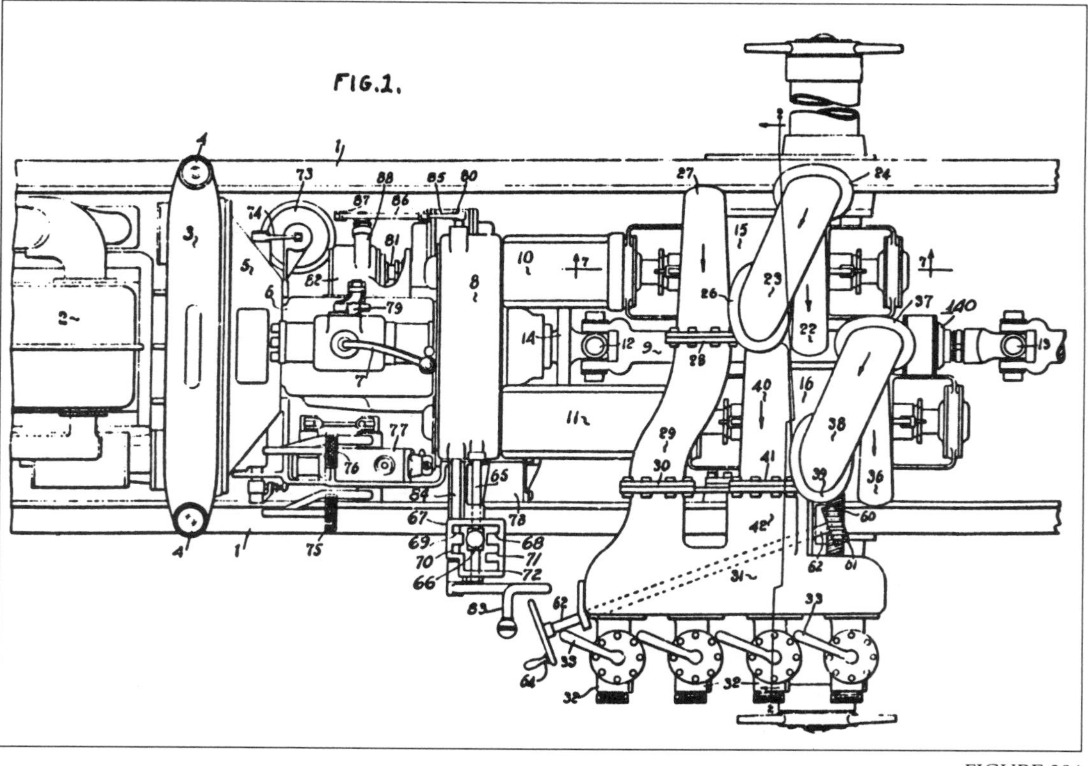

FIGURE 306

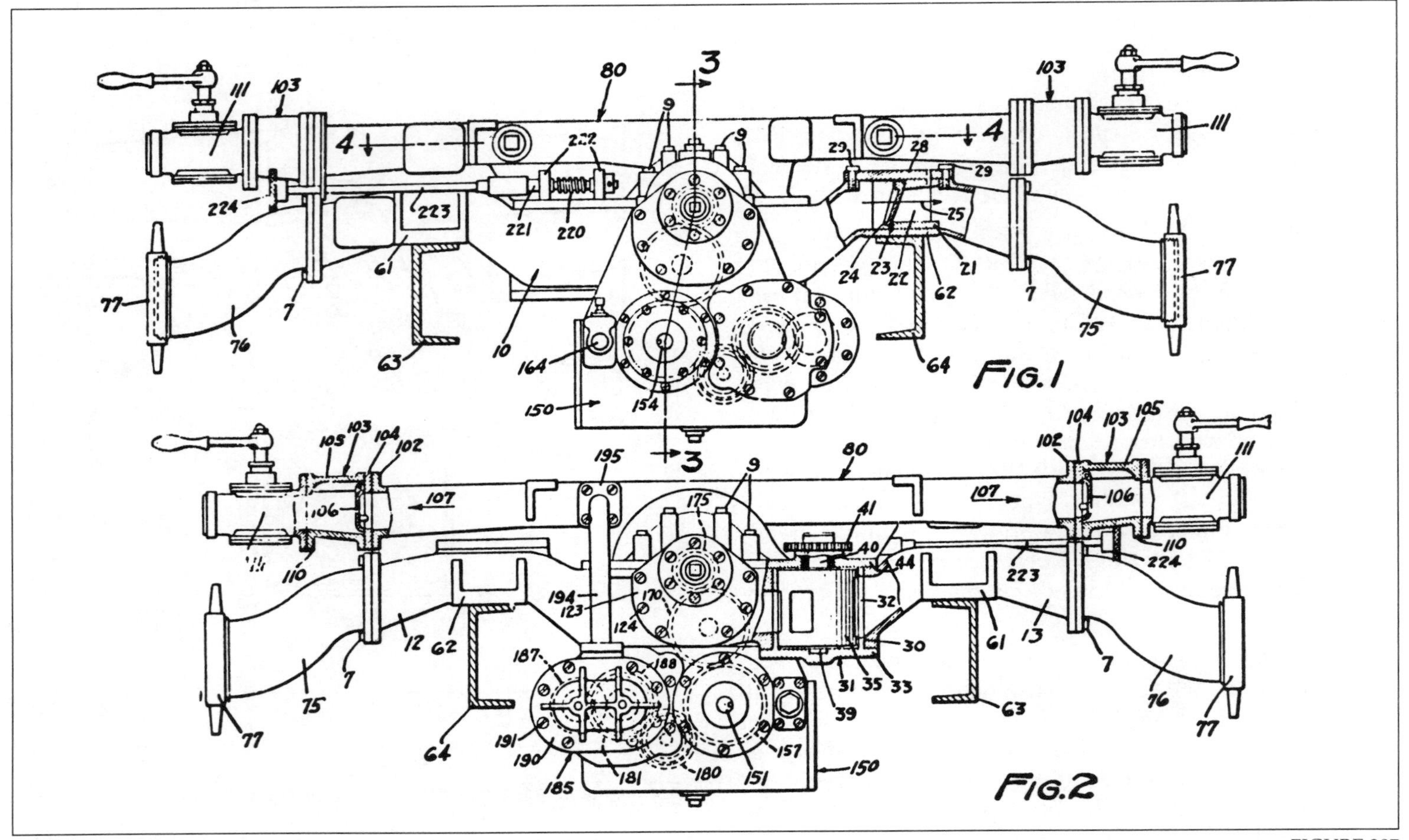

FIGURE 307

Boyles, Wilhelm, Waterous Centrifugal Pump

Ralph R. Boyles and Edward A. Wilhelm of St. Paul, Minnesota, assignors to the Waterous Company, received patent number 2,207,575 in July, 1940 for their invention of a centrifugal pump complete with a built-in rotary primer and a transfer valve. This patent is illustrated in FIGURE 307, which shows the small rotary priming pump assembly.

Barton Pumps

Patent number 2,223,592 dated December 3, 1940 used what the inventors termed a "jetting" principle for automatic priming, and described a duplex pump with stages capable of independent, staged, or series relationships. The inventors were Ben D. Barton, Adrian P. Adney, and Harold A. Bentley, all of Battle Creek, Michigan, and the patent rights were assigned to American-Marsh Pumps, Incorporated, of that city. This patent was one of the forebearers of the "Barton-American" series of pumps. FIGURE 308 is one of the patent drawings, while FIGURES 309, 310 and 311 illustrate later Barton-American products, including a priming valve device. Notice also that by 1962 the corporate name had been changed to the American Fire Apparatus Company, and by the end of that year there was another renaming, this time to the American Fire Pump Company.

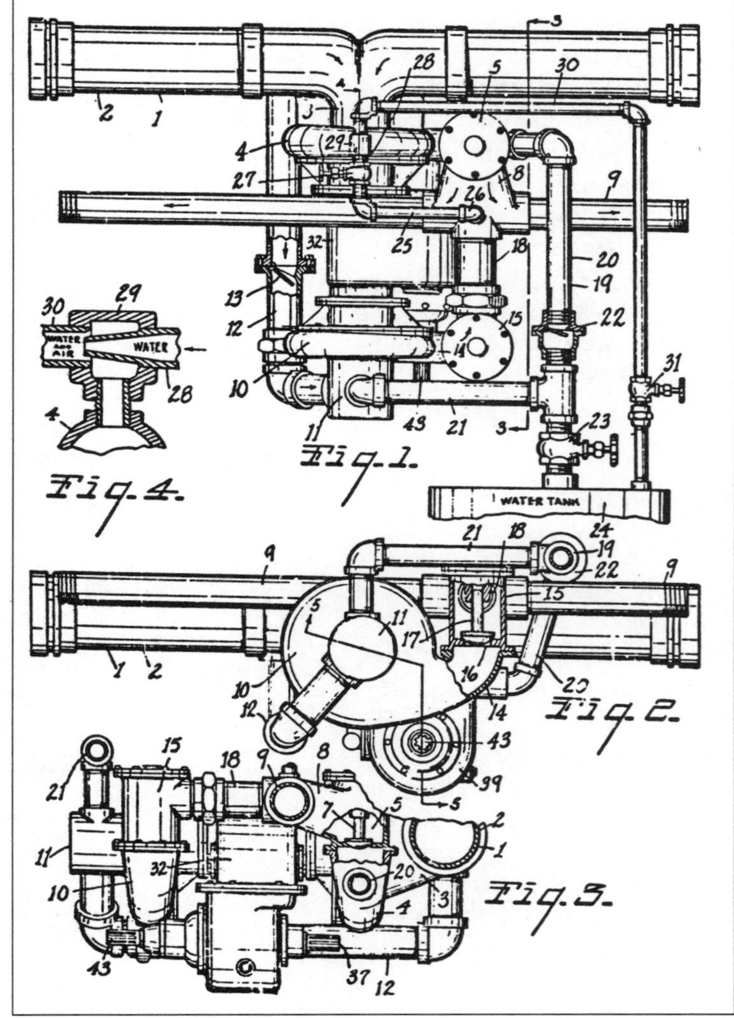

FIGURE 308

GET 3 BIG ADVANTAGES

WITH THE VERSATILE BARTON-AMERICAN DUPLEX-MULTISTAGE CENTRIFUGAL PUMP

You'll get faster control– and higher performance–with a Barton-American Type DMA Pump. These three outstanding design features explain why –and *only* the DMA has them all!

1. In *one* compact pump unit, you have a choice of three pressure and volume selections: High volume at normal pressures, normal high pressures, and extremely high pressures for fog guns. All operate at moderate engine speeds . . . and by merely shifting a *single* lever.

2. Two impellers, *independently* shafted and geared, will operate alone or together in series. Volume pump handles dirty water *without* damage to the high pressure pump.

3. This Barton-American Duplex multistage centrifugal pump has one-point lubrication similar to truck transmission housings. All bearings are bathed in oil inside . . . there's no need for water lubrication.

Yes, these unique, advanced features, plus automatic pressure-operated valves and other advantages provide you with *outstanding* fire-fighting power. For full details, write for . . .

Typical pump panel. One lever gives you instant selection of *Capacity-Pressure-Series* operation—a Duplex Pump feature.

BULLETIN 8733
DMA Duplex Multistage Centrifugal Midship Pumps

BULLETIN 8734
DMB Duplex Four-Stage Midship Pumps

MERICAN FIRE PUMP COMPANY • P. O. BOX 219-B • 803 MAIN STREET ROAD • BATTLE CREEK, MICHIGAN

FIGURE 309

Barton-American ENGINE VACUUM PRIMER

. . . another superior feature you get in American FIRE TRUCKS

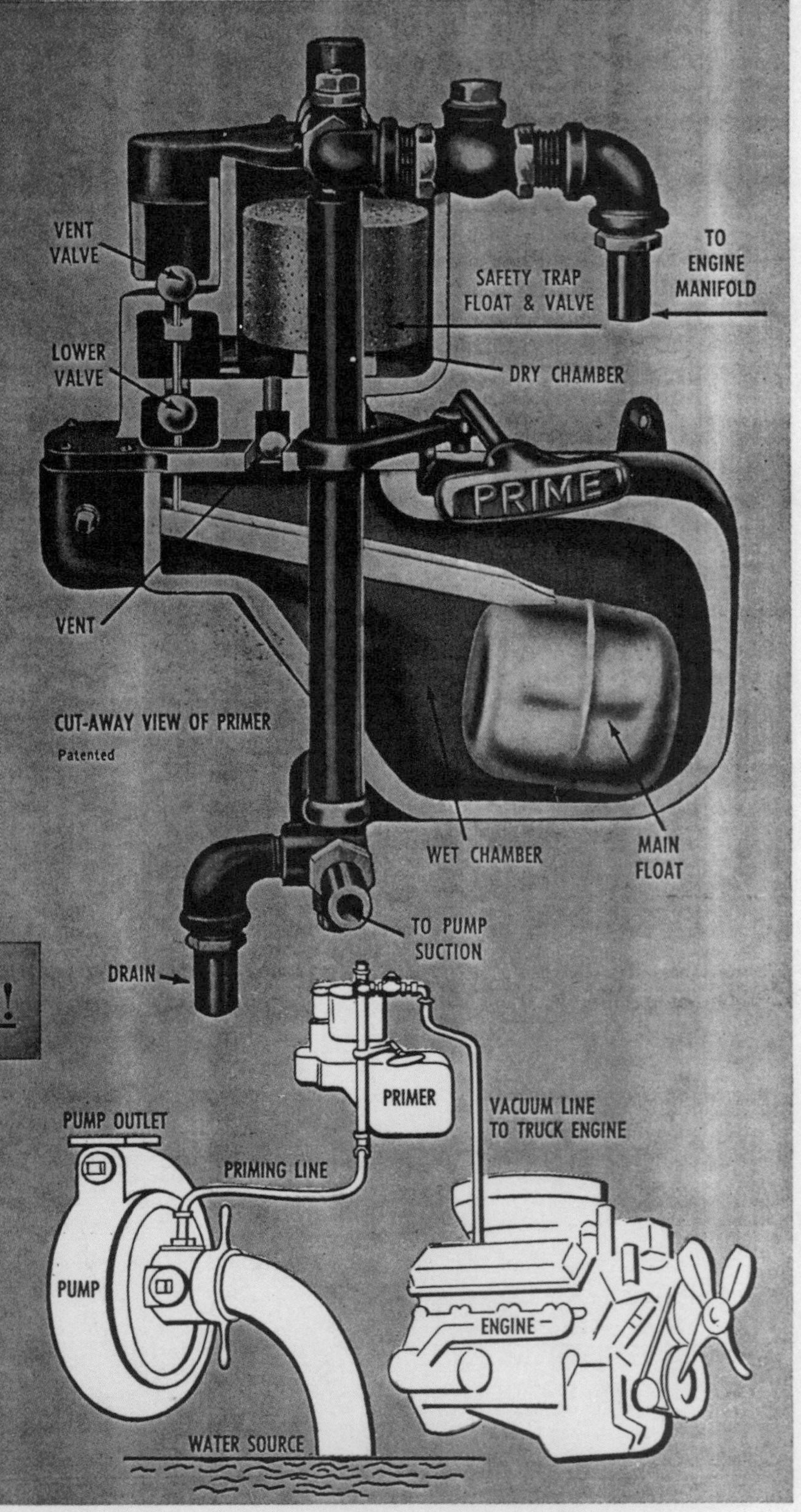

Always operates with like-new efficiency — not affected by engine wear

Employs a simple, trouble-proof principle

Requires no oil, no adjusting, no maintenance

Completely frost-proof

Always dependable

This PRIMER IS FAST!

The Barton-American has *none* of the many disadvantages of rotary priming pumps. Superior to exhaust primers, which require engine speeds as high as 3300 RPM, *the Barton-American readily primes at 1000 RPM*. If the truck engine will run, the Barton-American will prime. Patented dual float design makes it impossible for water to pass into the truck engine, eliminating possibility of engine damage.

For more details on this and other superior features you get in American Fire Apparatus — even though you don't specify them — see your nearest American representative.

American

AMERICAN FIRE APPARATUS CO.

FIRE APPARATUS CO. "Always Pioneering with Better Engineering"

Main Office: Main Street Rd., Battle Creek, Mich. Plants at: BATTLE CREEK, MICH., MARSHALLTOWN, IA., STRATFORD, ONT.

FIGURE 310

Famed for dependable fire service . . .

BARTON-AMERICAN PUMPS

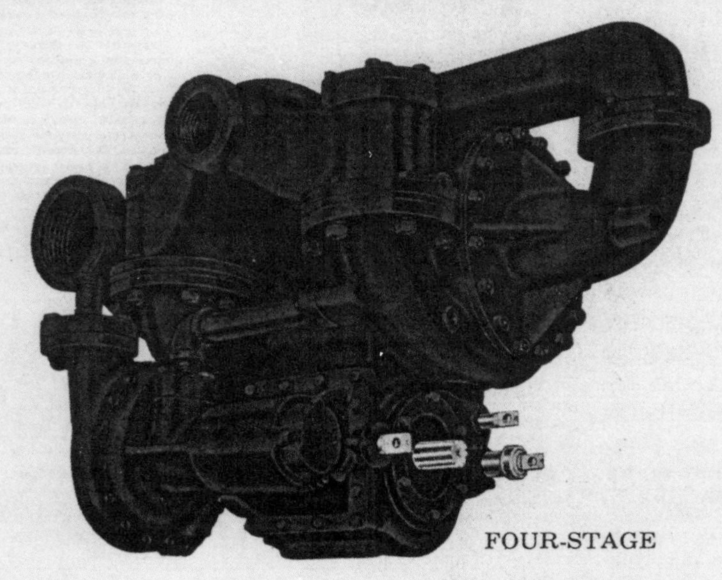

FOUR-STAGE

MIDSHIP

Available in single stage, two stage, three stage and four stage construction. Furnished in 500 to 1250 gpm capacities in pressures to 850 lbs.

Front-mounted and midship pumps are furnished with the famous Barton-American Vacuum Primer.

FRONT-MOUNTED

Capacities 500 to 750 gpm. Pressures to 400 lbs. Furnished with engineered PTO assembly.

500 GPM
TYPE UA-50

750 GPM
TYPE UA-75

PORTABLE

Capacities 200 and 350 gpm. Pressures to 90 and 150 lbs. Other capacities available. Tell us your particular requirements.

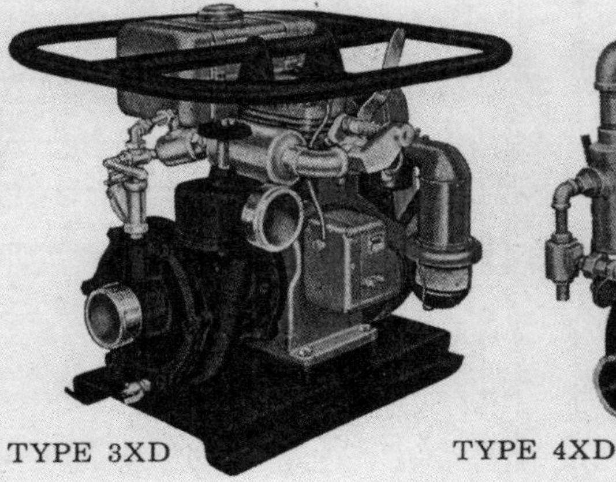

TYPE 3XD

TYPE 4XD

SEE YOUR NEAREST
AMERICAN FIRE PUMP
DEALER

BATTLE CREEK, MICHIGAN, U. S. A.

FIGURE 311

The Hale Fire Pump Company

Harold W. Yocum, who died in 1965, was an engineer with the Hale Fire Pump Company, to which he assigned his patent number 2,468,008, effective April 19, 1949 (FIGURE 312). This was a pressure booster pump adapted for attaching to a centrifugal fire pump adding as it were, a third stage to the system. All of the Hale traditions of quality and ingenuity have been continued to our present day, as some of the more recent inventions of H. Alfred Eberhardt illustrate. His patents numbered 4,209,282 and 4,337,830, received in 1980 and 1982, and assigned to the Hale Company, are shown in FIGURES 313, 314, and 315. Notice that the "Abstract" for each of these patents is identical. The differences between inventions are subtle and deal with retrofitting to an in-service vehicle. What Mr. Eberhardt had developed was a dual pump which used a single rotating shaft. There was a low flow, high pressure booster pump with two stages and a high flow rate, single stage main pump. This pump configuration facilitated rapid fire attack with booster hose while larger diameter lines were being hooked up to it. The capabilities typical of the current two-stage centrifugal pumps, either set for volume operation or for pressure operation, can be visualized readily with the aide of FIGURE 316. This illustration is from Introduction to Fire Apparatus Practices, IFSTA 106, Sixth Edition, and it is also used in current literature as published by the Hale Fire Pump Company.

And finally there is patent number 4,157,733 issued on June 12, 1979 to Ronald L. Ewers and John L. Oakley, and assigned by them to Emergency One, Incorporated of Ocala, Florida. This is a system of paired power take offs driving separate water pumps. With this equipment it became possible to pump water while driving the vehicle, the so-called "pump-and-roll" capability. This of course is essential for airport crash-fire-rescue vehicles. The Ewers Oakley patent is illustrated in FIGURE 317.

[54] PUMP ASSEMBLY
[75] Inventor: H. Alfred Eberhardt, Paoli, Pa.
[73] Assignee: Hale Fire Pump Company, Conshohocken, Pa.
[21] Appl. No.: 902,350
[22] Filed: May 3, 1978
[51] Int. Cl.² F01D 11/04; A62C 25/00
[52] U.S. Cl. .... 417/251; 169/24; 415/110; 415/170 A
[58] Field of Search .... 417/251; 169/13, 16, 169/24; 415/110, 111, 112, 170 A, 170 R; 277/15

[56] References Cited

U.S. PATENT DOCUMENTS

825,964 7/1906 Godfrey 169/13
1,737,870 12/1929 Telfer .... 415/170 A
2,280,626 4/1942 Carpenter .... 417/251 X
2,918,975 12/1959 Conery et al. .... 417/234 X

FOREIGN PATENT DOCUMENTS

1129345 9/1956 France .... 415/170 A

Primary Examiner—Richard E. Gluck
Attorney, Agent, or Firm—Harding, Earley & Follmer

[57] ABSTRACT

A pump assembly for use in fire fighting service is constructed of a single stage main pump and a two stage booster pump connected in series with the discharge of the main pump being connected to a first high flow rate fire fighting application and to the inlet of the booster pump and the discharge of the booster pump being connected to a second low flow high pressure fire fighting application. The impellers for both the main pump and the booster pump are mounted on a common rotating shaft so as to be driven thereby. A flow restriction and conduit means is provided to reduce the pressure on the booster pump seal. A by-pass conduit is arranged to conduct flow from the discharge of the booster pump back to the inlet of the main pump so that whenever the main pump is operated there will be flow through the booster pump to prevent overheating thereof.

7 Claims, 3 Drawing Figures

FIGURE 313

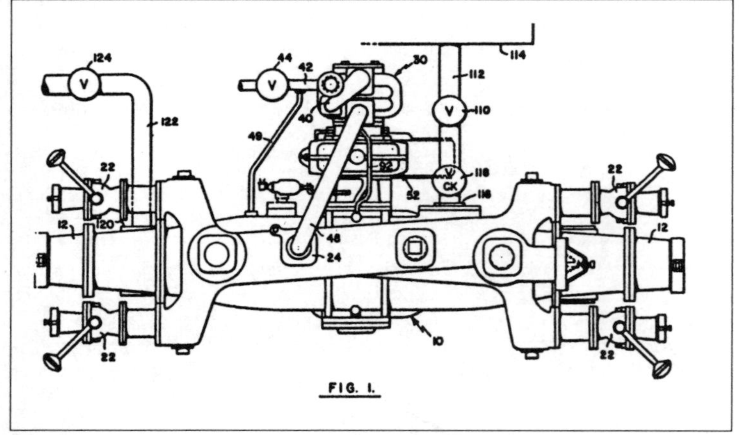

FIGURE 314

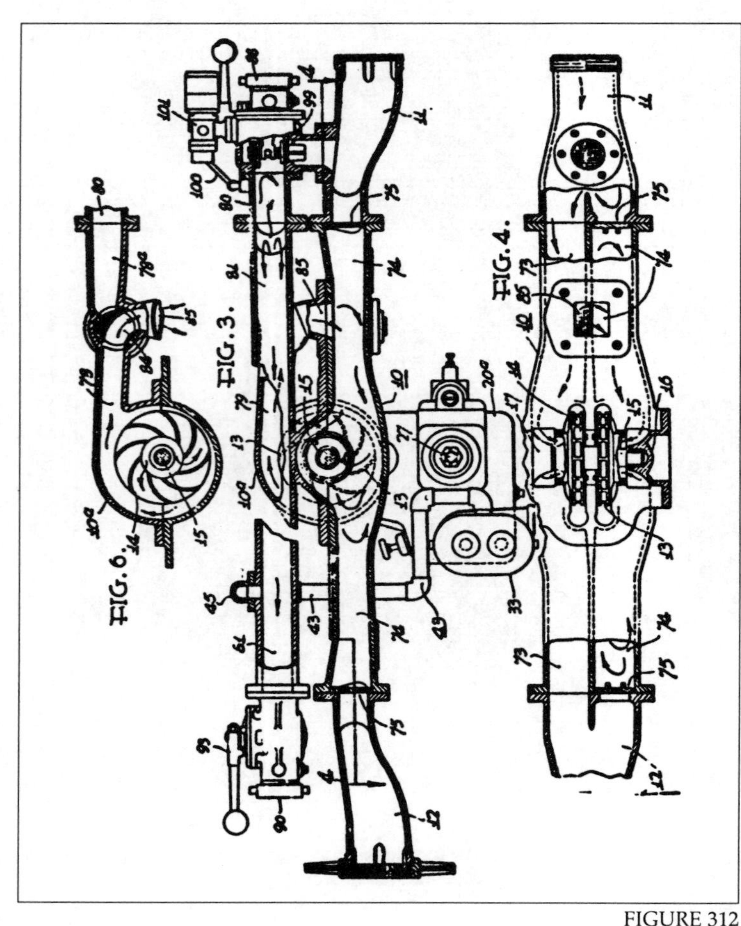

FIGURE 312

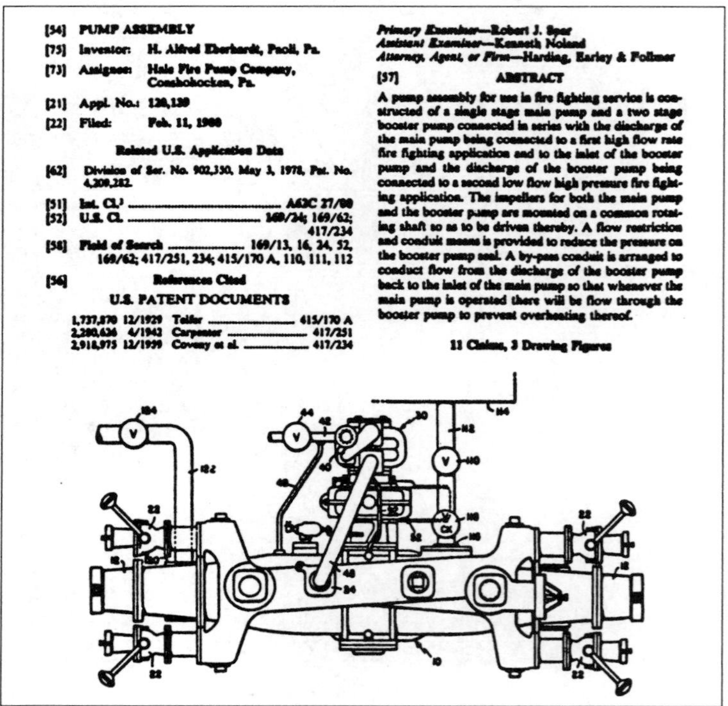

[54] PUMP ASSEMBLY
[75] Inventor: H. Alfred Eberhardt, Paoli, Pa.
[73] Assignee: Hale Fire Pump Company, Conshohocken, Pa.
[21] Appl. No.: 120,139
[22] Filed: Feb. 11, 1980

Related U.S. Application Data

[62] Division of Ser. No. 902,350, May 3, 1978, Pat. No. 4,209,282.
[51] Int. Cl.³ A62C 27/00
[52] U.S. Cl. 169/24; 169/62; 417/234
[58] Field of Search .... 169/13, 16, 24, 52, 169/62; 417/251, 234; 415/170 A, 110, 111, 112

[56] References Cited

U.S. PATENT DOCUMENTS

1,737,870 12/1929 Telfer .... 415/170 A
2,280,626 4/1942 Carpenter 417/251
2,918,975 12/1959 Coveny et al. 417/234

Primary Examiner—Robert J. Spar
Assistant Examiner—Kenneth Noland
Attorney, Agent, or Firm—Harding, Earley & Follmer

[57] ABSTRACT

A pump assembly for use in fire fighting service is constructed of a single stage main pump and a two stage booster pump connected in series with the discharge of the main pump being connected to a first high flow rate fire fighting application and to the inlet of the booster pump and the discharge of the booster pump being connected to a second low flow high pressure fire fighting application. The impellers for both the main pump and the booster pump are mounted on a common rotating shaft so as to be driven thereby. A flow restriction and conduit means is provided to reduce the pressure on the booster pump seal. A by-pass conduit is arranged to conduct flow from the discharge of the booster pump back to the inlet of the main pump so that whenever the main pump is operated there will be flow through the booster pump to prevent overheating thereof.

11 Claims, 3 Drawing Figures

FIGURE 315

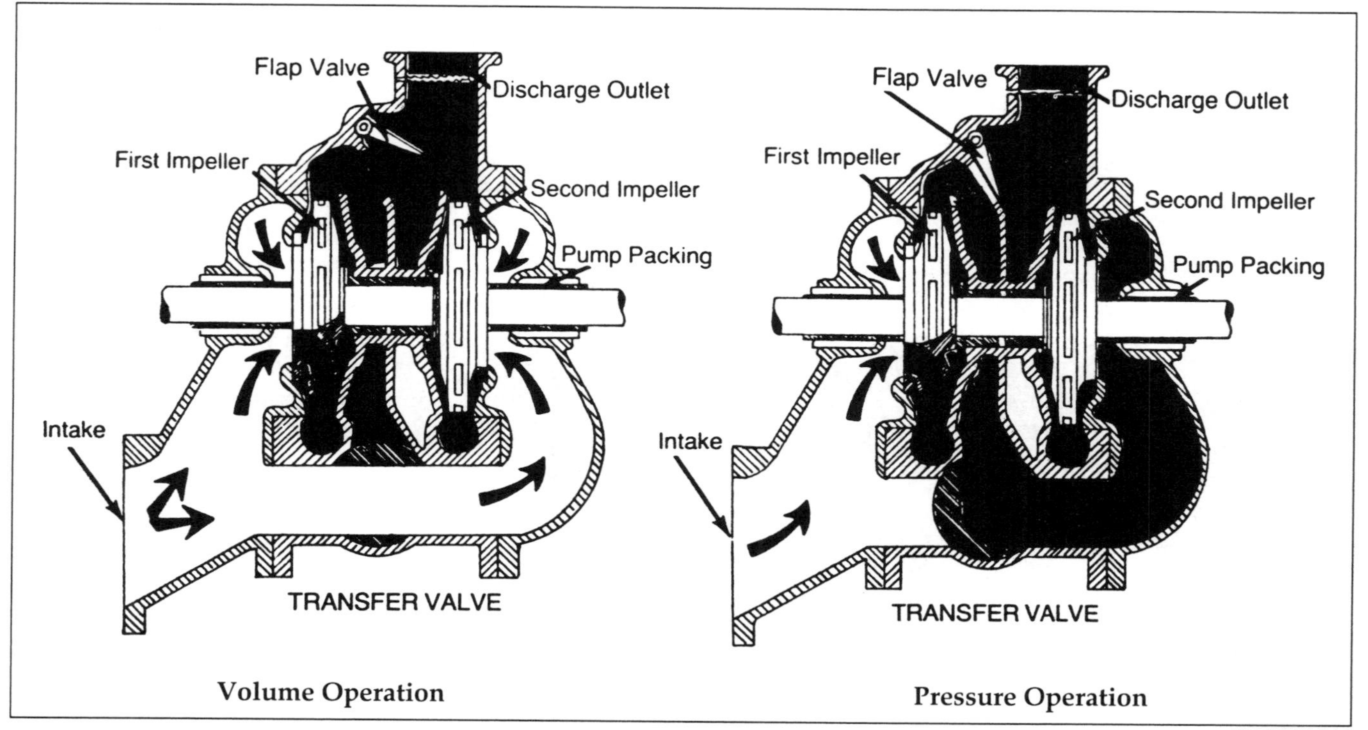

FIGURE 316

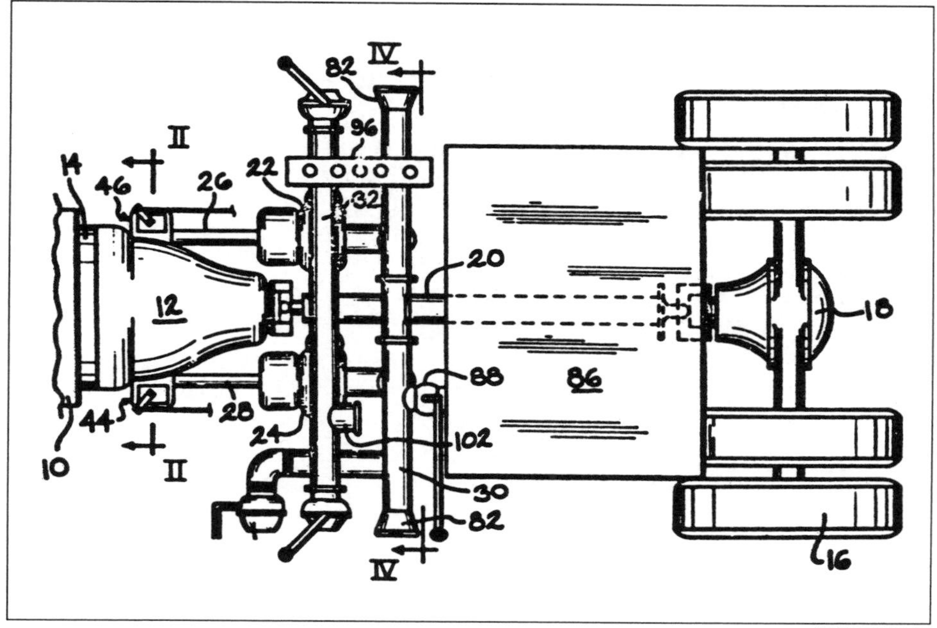

FIGURE 317

NFPA Standards

Many, perhaps most, of the ideas and inventions which have been reviewed have been discarded, being either outdated, undependable or unworkable. Nevertheless most good and workable inventions do survive, at least until some more effective machine is devised. Knowledge of the operational components of modern fire apparatus is now available, in codified format, and no longer scattered about in a variety of references and memories. Publications of the National Fire Protection Association contain such knowledge, the distillation of years of experience, and assembled by experts. These publications include NFPA 414, Aircraft Rescue and Fire Fighting Vehicles; NFPA 1901, Pumper Fire Apparatus, NFPA 1902, Initial Attack Fire Apparatus; NFPA 1903, Mobile Water Supply Fire Apparatus; NFPA 1904, Aerial Ladder and Elevating Platform Fire Apparatus; and NFPA 1911, Service Tests of Pumps on Fire Department Apparatus.

FIGURE 318

FIGURE 319

FIGURE 320

Fire Engine Photo Album

And finally, the last group of illustrations is again a photograph album, this time exhibiting motor fire pumpers, including antique and modern rigs. Generally these photographs were taken at musters or parades, and they represent the apparatus manufacturers which have been favored by fire departments in Pennsylvania, New Jersey and Delaware. Not represented in these illustrations are quads, aerial ladders, elevating platforms and towers, or boom-equipped pumpers, because all such fire apparatus is presented in the second volume of this series.

FIGURES 318 and 319 are Ahrens-Fox Fire Engine Company apparatus. Shown first is a 1929 rotary pump vehicle, model number GW-75-4, Register number 5037, owned by Harold J. Swartz of Pleasantville, New Jersey. The second photograph is a 1933 Ahrens-Fox piston pumper which served the Vigilant Fire Company of York, Pennsylvania. It is a model C-T-4, numbered 4006, and for many years it served as the Vigilant's second piece, sitting behind an ALF 500 Series pumper in the old, and since demolished, West Market Street firehouse.

A series of American LaFrance fire engines are displayed in FIGURES 320, 321, 322, 323, 324, and 325, beginning with a 1921 rig owned by the Shenandoah, Pennsylvania fire department. Also illustrated are 1926 ALF pumpers, from Bethesda, Maryland and Delaware Water Gap, Pennsylvania. The 1949 Ellendale, Delaware rig is a Series 700 pumper. Also shown are a Century Series and a Pioneer fire pumper, serving, respectively the Mountaineer Hose Company of Minersville and the East Lawn Volunteer Fire Company of Upper Nazareth Township, both in Pennsylvania.

FIGURE 321

FIGURE 322

FIGURE 323

FIGURE 324

FIGURE 325

FIGURE 326

Two Autocar fire vehicles are shown. FIGURE 326 is a 1945 pumper which belongs to the Millville Volunteer Fire Company of Delaware. "Old Mom" no longer sees service at fires, but she is quite capable of pumping water. The Hockessin Fire Company is the northernmost fire suppression agency in Delaware, located in the only hilly area of the entire state. Hockessin selected an Autocar, FIGURE 327, for dependability in hill climbing. The rig was built by the Hamburg, Pennsylvania based Hahn Motor Company. It was diesel powered, with a 750 GPM pump.

A 1924 model Brockway is shown in FIGURE 328. This was the first apparatus of the Commonwealth Fire Company of York County, Pennsylvania. From Schuylkill Haven, Pennsylvania, the Liberty Fire Company runs the Car-Mar-Spartan rig of FIGURE 329. FIGURE 330 is a pumper built on a Chevrolet chassis. This rig is the property of the Mary-Del Volunteer Fire Company, located in Delaware. The Crown Firecoach, FIGURE 331, has been uncommon, if not unknown, in the eastern United States. This example formerly served the Covina, California fire department. An example of the many Dodge chassis rigs operated in this area is the small brush truck illustrated in FIGURE 332. The Plainfield, Pennsylvania based West Pennsboro Volunteer Fire Company operates this apparatus. FIGURE 333 is a Duplex-Quality fire vehicle which serves the Bryn Athyn Fire Company.

FIGURE 327

FIGURE 328

FIGURE 329

FIGURE 330

FIGURE 331

FIGURE 332

FIGURE 334 is an Emergency One "Cyclone" pumper which is operated by the Seat Pleasant Volunteer Fire Company. This photograph was taken during an apparatus display just prior to an Ocean City, Maryland parade of the state's volunteer fire organizations.

Fire apparatus by FMC is commonly encountered in this area. Food Machinery Corporation builds both custom and commercial rigs, and FIGURE 335 is an example of the latter. This FMC-Ford Custom Cab pumper belongs to the Mar-Lin Citizens Hose Company. The Winona Fire Company of Bloomsburg, Pennsylvania runs Engine 33, shown in FIGURE 336. John Bean Division of FMC built this pumper, on another Ford chassis. Another fire pumper on a Ford chassis is Engine 66 of the West End Fire Company of Sheridan, Pennsylvania. With its green paint job, FIGURE 337, this rig is uncommonly attractive and eyecatching.

FIGURE 333

FIGURE 334

FIGURE 335

FIGURE 336

FIGURE 337

FIGURE 338

The FWD pumper from Broomall, Pennsylvania, FIGURE 338, appears to be in meticulously restored condition. This fire engine is of about 1960 vintage. Of more recent manufacture is the cab-over-engine pumper of FIGURE 339, an apparatus owned by the Honey Pot Volunteer Hose Company of Nanticoke, Pennsylvania. Otto Zachow's invention of a method to supply power to steerable front wheels was the basis for the "Front Wheel Drive" Company.

Northampton, in the Allentown-Bethlehem metropolitan complex runs the Freightliner-KME top-mount pump panel fire engine seen in FIGURE 340. KME Fire Apparatus Company of Nesquehoning, Pennsylvania, a small town located on U.S. Route 209 about three miles south of Jim Thorpe and 10 miles northward of Tamaqua supplies many area fire companies.

Many fire engines have been built with chassis by GMC, and FIGURE 341 is just such an example. Grumman Emergency Products, Inc., which had absorbed the Howe Fire Apparatus Company of Anderson, Indiana and the Oren-Roanoke Corporation of Vinton, Virginia, was the builder of Engine 422, FIGURE 342, of the Boring Volunteer Fire Company, Baltimore County, Maryland.

FIGURE 339

FIGURE 340

FIGURE 341

The Hahn Motors Company of Hamburg, Pennsylvania had its demise in 1991, leaving a legacy of some old and many recent fire pumpers. The Hahn Company was founded in 1915. A pair of antique Hahn fire pumpers are shown in FIGURES 343 and 344, the first one of late 1920s style, and the second a 1939 model. Three modern, distinctly different Hahn fire engines are shown in FIGURES 345, 346, and 347. Note the five-inch steamer fitting and the racks holding boots and helmets on the apparatus of the Citizens Fire Company. The second photograph is a pumper-tanker from Marydel, while the last photograph of this group, from the Collingdale Fire Company, has a top-mount pump panel.

FIGURE 342

FIGURE 343

FIGURE 344

FIGURE 345

FIGURE 346

FIGURE 347

FIGURE 348

FIGURE 349

FIGURE 350

The Hale Fire Pump Company of Conshohocken, Pennsylvania used to build custom motor fire pumpers. The 1926 model Hale, FIGURE 348, belongs to the Odessa, Delaware, Fire Company. Notice the hand-operated siren and the rack of buckets, as well as the rear-mounted lantern. No Hale fire engines were built after 1940.

Engine 31 of the Brooklyn Hose Company of Lewistown, Pennsylvania is seen in FIGURE 349. This is a Howe-International apparatus with a 750 gallon water tank and a 1,000 gallon per minute pump. Back in 1940 this fire company was equipped with a Buffalo Fire Appliance Corporation sedan pumper, one of only eight such vehicles ever manufactured. The pumper illustrated in FIGURE 350 is also built on an International chassis, but it is a KME "Terminator."

FIGURE 351 is a KME custom fire apparatus of the Scullville Volunteer Fire Company of Egg Harbor Township, New Jersey. This interesting 1988 model is equipped with a 1500 gallon per minute pump.

Justly famous among buffs is the fully-restored Mack 1924 pumper and chemical fire engine belonging to the Berlin, Maryland fire department. This beautiful AB model vehicle is shown in FIGURE 352. The Mack fire engine seen in FIGURE 353, from Engine Company Number 2 of Freehold, New Jersey, a 1930, 750 gallon per minute rotary pumper would be classified by Mack expert Harvey Eckart as an "early" B series, these models being manufactured between the

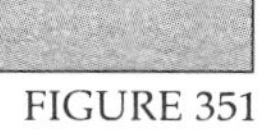
FIGURE 351

FIGURE 352

FIGURE 353

FIGURE 354

FIGURE 355

FIGURE 356

years 1927 and 1941. And in FIGURE 354 there is pictured Eckart's own rig, a 1948 E series Type 75 pumper. This fire engine served New Holland, Pennsylvania for over 32 years. In this particular sedan configuration, entry to the back seat is from the rear of the vehicle. (For additional information, and a multitude of fine illustrations, the following references are suggested: *F.D.N.Y. Mack L-Model Apparatus*, by John A. Calderone; *Mack Fire Apparatus, A Pictorial History*, by Harvey Eckert; *Mack Tilt Cab Fire Apparatus*, by John M. Malecky, and *Mack Fire Apparatus*, by Robert B. Marvin.)

A 1982 Mack MC series model, FIGURE 355, runs with the Wildwood, New Jersey Fire Department. This vehicle is equipped with a 1,500 gallon per minute pump and a 400 gallon water tank. One thousand feet of three inch hose is carried in each of the electrically operated reels located to the rear and over the hose bed. A booster reel is seen above the pump panel. Conestoga Custom Products Company built the body work for Wildwood Engine Number 2. FIGURE 356 is a photograph of Tanker 537 of the Mount Carbon, Pennsylvania, Volunteer Fire Company. This is a 1983 Mack CF series pumper, with bodywork by Pierce. This apparatus carries 2,000 gallons of water and is fitted with a 1,500 gallon per minute pump. The color scheme is, to say the least, intriguing. The pumper tanker of FIGURE 357, operated by the Valley View Fire Company, is a Mack MC model. No descriptive data is at hand for this rig.

FIGURE 358 is a 1985 Mack MS pumper from the Han-Le-Co Volunteer Fire Company. The fire company name is a condensed form meaning "Hanover Township, Lehigh County." This rig, which was built by LTI, carries 500 gallons of water and mounts a 1,000 gallon per minute pump. The Mack CF pumper belonging to Glenside Fire Company, FIGURE 359, shows off the most familiarly-configured rigs of this series. An example of an R Series Mack fire vehicle is seen in FIGURE 360. This tanker with a front-bumper mounted pump runs with a fire company from Anne Arundel County, Maryland, the Riva Volunteer Fire Department.

FIGURE 357

FIGURE 358

FIGURE 359

FIGURE 360

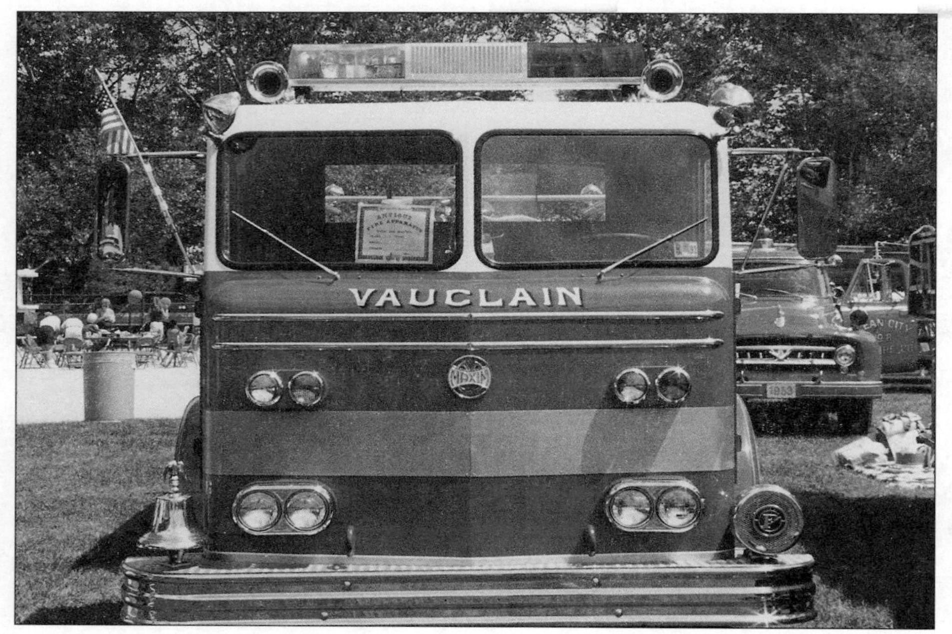

FIGURE 361

FIGURE 362

Photographed in the lineup at a Wilmington, Delaware Fire Department muster is an old Maxim pumper, FIGURE 361. Both Maxim and Seagrave have long been part of the FWD company. The Maxim Motor Company was founded in 1914, in Middletown, Massachusetts, by Charlton Maxim, formally the Middletown fire chief.

FIGURE 362 is an Oren-Ford motor fire pumper owned by the Sudlersville Volunteer Fire Company of Queen Anne's County, Maryland. Oren-Roanoke and Howe affiliated with Grumman Emergency Products.

In 1986 the Upper Frederick Volunteer Fire Company of Perkomenville, Pennsylvania took delivery of the rig seen in FIGURE 363. This combination brush fire unit and community pumper was built by Young Fire Equipment on an Ottawa Brimont "Commando" chassis. This chassis is powered by a Midliner 175 horsepower turbo diesel engine, via a six-speed manual transmission and two-speed transfer case. Other features include all-wheel steering and all-wheel drive. Fire suppression equipment includes a 750 gallon stainless steel water tank, an American Godiva PSD 750 gallon per minute transfer case PTO driven pump, a Hale gasoline-powered auxiliary system giving a pump and roll capability, and paired Hannay booster reels which either system can supply. A forward fire suppression spray system is incorporated with the front brush guard.

Bowers Fire Company, Station 40, of Kent County, Delaware, was proudly exhibiting their pumper-tanker unit, FIGURE 364, in a parade. Pemfab built this apparatus.

Pierce Manufacturing of Appleton, Wisconsin, located just to the north of Lake Winnebago, between Oshkosh and Green Bay, has supplied apparatus to many area fire companies for many years. Three examples of Pierce fire pumpers are shown. FIGURE 365 is a photo of Engine 304 of the New Castle County, Delaware-based Belvedere Volunteer Fire Company. This unit has an air-conditioned crew cab and includes hydraulic ladder and hard suction hose racks.

FIGURE 363

FIGURE 364

FIGURE 365

FIGURE 366

FIGURE 367

FIGURE 368

Engine 48 of the Hershey, Pennsylvania Fire Department is illustrated in FIGURE 366. Notice the pump panel steamer fittings for large-diameter hose, and the set of four cross-lays above. A Pierce "Suburban-750" of the Humane Fire Company of Pottsville, Pennsylvania is seen in FIGURE 367. This well-outfitted attack pumper is powered by a diesel engine, has an extended front bumper with a small-diameter hose tray, cross-lays over the pump panel, and a large-diameter steamer valve.

The late and lamented Peter Pirsch and Sons Company is here represented by FIGURES 368 and 369. The aging open-cab vehicle all dressed in yellow is from Company Number 7 of the fire department of Chevy Chase, Maryland. The second illustration is a Pirsch fire engine of the Good Will Fire Company of the Port Carbon, Pennsylvania Fire Department.

FIGURE 370 is a 1935 Reo pumper which is now the property of the Millsboro, Delaware Fire Company. According to local lore this old fire engine was the first ever apparatus to be run by the Bethany Beach Volunteer Fire Company which, sad to say, junked it when it was no longer serviceable. Rescue and restoration by Millsboro followed. The clean and meticulously painted Reo of FIGURE 371, which looks to be about of 1940 vintage, is owned by the Stroud Township Volunteer Fire Department, a Pennsylvania organization.

FIGURE 369

FIGURE 370

FIGURE 371

FIGURE 372

FIGURE 373

FIGURE 374

The fire company of McSherrystown, Pennsylvania formerly used a 1933 Seagrave Suburbanite pumper, and this was followed by the Seagrave 70th Anniversary Series engine shown in FIGURE 372. Rigs of this series were built by Seagrave from 1951 to 1970. The first cab forward Seagrave was built in 1959, while 1965 was the year of their last sedan pumper. The New Haven, Connecticut Fire Department received the last ever Seagrave engine ahead of cab apparatus in 1970, the same year which saw Jackson, Michigan receive the final Seagrave canopy cab fire engine. Contemporary Seagrave fire apparatus is illustrated in FIGURES 373 and 374. The massive pumper-tanker runs from the Leonardtown Volunteer Fire Department as St. Mary's County Engine 14. The modern rig lettered as Engine 54 of the Freeland Fire Department was photographed at a Fire Expo show in Harrisburg, Pennsylvania a couple of years back. The colorful old Seagrave pumper of FIGURE 375 falls into the group which expert Matthew Lee has characterized as representing the "Full Grille Era." This fire engine looks to be of 1936-1940 style, and is owned by the Citizens Fire Company of Mt. Holly Springs, Pennsylvania.

Each of the following two fire engine photographs are also from Harrisburg, both being snapped during a SPAAMFAA (Society for the Preservation and Appreciation of Antique Motor Fire Apparatus in America) convention and apparatus exhibition along the Susquehanna River waterfront. FIGURE 376 is a 1923 Stoughton equipped with a rotary pump. FIGURE 377 is a superbly maintained and beautifully groomed 1937 Studebaker pumper, lettered for Yardley, Pennsylvania, a small town directly across the Delaware River from Trenton, New Jersey.

FIGURE 375

FIGURE 376

FIGURE 377

FIGURE 378

FIGURE 379

FIGURE 380

Custom pumpers and aerial towers built by the Sutphen Company continue to be readily recognized by their headlight and turn signal configuration. The Pleasant Hill Volunteer Fire Company runs the Sutphen pumper seen in FIGURE 378.

Three photographs represent fire pumpers produced by the defunct Ward LaFrance Company (FIGURES 379, 380, and 381). The open-cab apparatus lettered for Trenton Psychiatric Hospital was photographed at Mays Landing, New Jersey. A reasonable estimate of its year of production is 1950, plus-or-minus two years. The second open-cab model, from the New Minersville Fire Company, probably was built in 1955, again give-or-take two years. Ward LaFrance introduced its Ambassador line of double-angled windshield fire vehicles about 1964, and therefore the orange-painted pumper shown last in this group of illustrations is probably a product from 1963-1967. This rig is from the West End Fire Company of Mt. Carmel, Pennsylvania.

FIGURE 382 is a pumper-tanker belonging to the Swansboro, North Carolina, Fire Department. This vehicle was built by the White Motor Company, with fire suppression gear by the Welch Fire Equipment Company of Marion, Wisconsin.

Tanker 2 of the New Jersey-based Villas Fire Company, FIGURE 383, is a product of Young Fire Equipment Corporation, with the chassis by Ford. Villas is located in the southern end of the state, on the Delaware Bay shore, about 10 miles due west of Wildwood, New Jersey, where this picture was taken during a volunteer firemen's parade.

FIGURE 381

FIGURE 382

FIGURE 383

FIGURE 384

The Young Crusader seen in FIGURE 384 is from Chincoteague, Virginia, on the eastern shore of the Old Dominion State.

A photograph of Zabeck pumper would have completed this album from "A" through "Z." But having none, we close without it.

INDEX

— D —

— E —

—F—

—G—

—H—

—I—

—J—

—K—

— T —

— U —

— V —

— W —

—Y—

—Z—

BIBLIOGRAPHY

Ahrens-Fox Bulletin — *"Nothing Compels Admiration More Than Victory and Success."* Undated.

American LaFrance Fire Engine Company Catalogue — *"Motor Fire Apparatus."* Undated.

Amoskeag Steam Fire Engines and Hose Carriages Catalogue 1895 (Reprint by Manchester, New Hampshire Historic Association. 1990).

Boyd, James, & Brother — *Fire Apparatus Catalogue.* Circa 1910.

Collins, D. — *Our Volunteer Firemen 1736-1882.* Science Press. Ephrata, Pennsylvania 1982.

Commercial Vehicles. 1921 (Reprint by R-Mac Publications. 1991).

Conway, W.F. — *Chemical Fire Engines.* Fire Buff House. New Albany, Indiana. 1987

Couple-Gear Freight Wheel Company Catalogue — *"Electric and Gas-Electric Trucks and Tractors for Fire and Commercial Uses."* Circa 1916.

Cowing and Company Pumps and Fire Engines. Seneca Falls, New York. 1840-1875. (Reprint by Seneca Falls Historical Society. 1973).

Daly, G. & Robrecht, J. — *An Illustrated Handbook of Fire Apparatus.* INA Corporation. Philadelphia. 1972.

Dunshee, K. — *Enjine!–Enjine!* Home Insurance Company. New York. 1952.

Early Unnumbered U.S. Patents 1790-1836. Research Publications. 1960.

Eckart, H. — *Mack Fire Apparatus. A Pictorial History.* The Engine House. Middletown, New York. 1990.

Erven, L. — *Fire Company Apparatus and Procedures.* Glencoe Press. Beverly Hills, California. 1974.

Hass, E. — *Ahrens-Fox: The Rolls-Royce of Fire Engines.* Sunnyvale, California. 1982.

Hass, E. — *The Dean of Steam Fire Engine Builders.* Sunnyvale, California. 1986.

Holloway Chemical Fire Apparatus Catalogue, 1881. (Reprint by Manor Publishing. 1975).

Kemp, F. — *Firefighting by the Seashore.* Seashore Fire Buffs. Atlantic City. 1972

King, W. — *History of the American Steam Fire Engine.* 1896. (Reprint by Owen Davies. 1960).

Lee, M. — *A Pictorial History of Seagrave Fire Apparatus.* Plymouth, Michigan. 1991.

Leggett, M. (editor). Subject-Matter Index of Patents for Inventions Issued by the United States Patent Office from 1790 to 1873, Inclusive. U.S. Government Printing Office. 1874.

Malecky, J.M. — *Mack Tilt Cab Fire Apparatus.* Fire Apparatus Journal Publications. New York. 1988.

McCall, W. — *American Fire Engines Since 1900.* Crestline Publishing. Sarasota, Florida. 1976.

Miller, F. — *Fire Department Supplies Catalogue.* 1884. (Reprint by Manor Publishing. 1978).

Miller, J. Fares Please. — *A Popular History of Trolleys, Horsecars, Streetcars, Buses, Elevators, and Subways.* Dover Publications. New York. 1960.

NFPA 1901: Pumper Fire Apparatus. 1991.

NFPA 1902: Initial Attack Fire Apparatus. 1991.

NFPA 1911: Service Tests of Pumps on Fire Department Apparatus. 1991.

Peckham, J. — *Fighting Fire With Fire.* Walter R. Haessner. Newfoundland, New Jersey. 1972.

Reading's Volunteer Fire Department. William Penn Association. Philadelphia. 1938.

Sorensen, W. & Wood, D. — *Motorized Fire Apparatus of the West 1900-1960.* Transportation Trails. Polo, Illinois. 1991.

Weaver, B. & Frederick, G. — *Hands, Horses, and Engines.* Baltimore County Fire Service Centennial Committee. Towson, Maryland. 1982.

Webb Motor Fire Apparatus Company Catalogue *"The Fireman and the Motor."* Circa 1910.

OTHER BOOKS

in the

Fire Service History Series

from Fire Buff House Publishers

- **CHEMICAL FIRE ENGINES** *by W. Fred Conway*
  The only book ever written about these amazing engines that for half a century put out 80% of all fires in spite of the fact that they never did perform as advertised! Hard cover. 128 pages, over one hundred photographs and drawings.

- **FIREBOATS** *by Paul Ditzel*
  The definitive work on the history of marine firefighting – a book so real you can almost hear and feel the engines and pumps throbbing! Relive the thrilling accounts of fireboats in action at spectacular fires. Hard cover. Over one hundred pictures.

- **FIRE ALARM!** *by Paul Ditzel*
  The story of fire alarm telegraphy - a nostalgic overview of fire alarm transmission in America. Includes the incredible story of an obscure telegraph operator, John N. Gamewell, who built a vast fire alarm empire with a 90% market share. Hard cover. Hundreds of pictures.

- **LOS ANGELES FIRE DEPARTMENT** *by Paul Ditzel*
  The fascinating complete history of the Los Angeles Fire Department. America's foremost fire historian details the major fires during the city's history, including the Bel-Air brush fire disaster, and the Central library blaze. Oversize book with 250 pages and hundreds of photographs of fires and apparatus. Soft cover.

- **A FIRE CHIEF REMEMBERS** *by Battalion Chief Edwin F. Schneider (Ret.)*
  The poignant, touching, moving account of his FDNY career. Now 82 years of age, Chief Schneider looks back at the series of incredible events in his 34 year career climb from buff to chief. Soft cover. Illustrated.

- **LAGUARDIA'S FIRE CHIEF** *by Kathleen Walsh Packard*
  The fascinating biography of New York City Fire Chief/Fire Commissioner Patrick Walsh, who served under the mercurial mayor, Fiorello LaGuardia. Chief Walsh and Mayor LaGuardia would race to see who could get to the fire first. This account of Chief Walsh's storybook career is written by his granddaughter. Soft cover. Illustrated.

- **DISCOVERING AMERICA'S FIRE MUSEUMS** *by W. Fred Conway*
  A comprehensive guidebook describing 150 museums in the United States and Canada displaying vintage fire apparatus, equipment, and memorabilia. Includes a history of firefighting apparatus and equipment from Colonial times to the present. 188 pages, 200 illustrations. Soft cover.

- **FIREFIGHTING LORE** *by W. Fred Conway*
  Probably the most unusual book of firefighting history ever written, it recounts dozens of amazing, strange, and incredible stories from firefighting's past and present – every one true. Soft cover. Illustrated.

These books are available wherever fire books are sold,
or you may order direct from the publisher.

Fire Buff House Publishers
P.O. Box 711, New Albany, IN 47151-0711